CAMBRIDGE LIBRARY COLLECTION

Books of enduring scholarly value

Life Sciences

Until the nineteenth century, the various subjects now known as the life sciences were regarded either as arcane studies which had little impact on ordinary daily life, or as a genteel hobby for the leisured classes. The increasing academic rigour and systematisation brought to the study of botany, zoology and other disciplines, and their adoption in university curricula, are reflected in the books reissued in this series.

The Fauna and Flora of Palestine

The Survey of Western Palestine was carried out under the auspices of the Palestine Exploration Fund between 1871 and 1877, and its results were published in a series of books between 1882 and 1888. This volume was written by H.B. Tristram (1822–1906), the naturalist, geologist and Anglican priest, whose *The Land of Israel* is also reissued in this series. Although he declined a bishopric in Jerusalem, he made four separate trips to Palestine in order to compile this catalogue of its flora and fauna. Including all native vertebrates, molluscs and plants known at the time, the work was first published in 1884 and laid the foundation for zoological study in Palestine. The catalogue offers a fascinating insight into the theories of late Victorian taxonomy as well the species it classifies here. Also included are detailed drawings and a scholarly preface summarising and tabulating Tristram's research.

The Fauna and Flora of Palestine

Henry Baker Tristram

CAMBRIDGE UNIVERSITY PRESS

Cambridge, New York, Melbourne, Madrid, Cape Town,
Singapore, São Paolo, Delhi, Mexico City

Published in the United States of America by Cambridge University Press, New York

www.cambridge.org
Information on this title: www.cambridge.org/9781108042048

This edition first published 1884
This digitally printed version 2013

ISBN 978-1-108-04204-8 Paperback

The original edition of this book contains a number of colour plates, which have been reproduced in black and white. Colour versions of these images can be found online at www.cambridge.org/9781108042048

THE SURVEY

OF

WESTERN PALESTINE.

THE FAUNA AND FLORA OF PALESTINE.

Pl.I.

J. Smit lith

Hanhart imp

HYRAX SYRIACUS

THE SURVEY

OF

WESTERN PALESTINE.

THE FAUNA AND FLORA OF PALESTINE.

BY

H. B. TRISTRAM, LL.D., D.D., F.R.S.,

CANON OF DURHAM.

PUBLISHED BY

THE COMMITTEE OF THE PALESTINE EXPLORATION FUND,

1, ADAM STREET, ADELPHI, LONDON, W.C.

1884.

PREFACE

ON THE

GEOGRAPHICAL AND GEOLOGICAL RELATIONS OF THE FAUNA AND FLORA OF PALESTINE.

THE present volume contains a catalogue of all the known Vertebrata, Terrestrial and Fluviatile Mollusca, and the Flora of Palestine. The Insect Fauna of the region has been so little studied, and the local collections hitherto formed are so meagre, that it has been thought better to pass over this branch of the subject, than to give lists which must necessarily be very imperfect, and therefore misleading.

The Committee of the Palestine Exploration Fund, which has already contributed so much to the general stock of information as to the geography, topography, history, and archæology of that country, felt that their work would be incomplete until supplemented by an account of its natural productions. Such an account it has been the endeavour of the compiler of this volume to supply, based chiefly upon his own observations and collections formed during four visits to the country; but he has not hesitated to avail himself also of the results of the researches of others, and must acknowledge with gratitude the aid he has received from Dr. Lortet, M. Locard, and M. Edmond Boissier.

In the geographical position and the geology of Palestine, we find special reasons why the consideration of its Fauna and Flora is exceptionally important; and a detailed examination discloses results even more interesting than could have been expected from the peculiar position of that region, as an outlying portion of the great Palæarctic region.

In 1858 Mr. Sclater laid before the Linnæan Society his views on the geographical distribution of the members of the class *Aves*, in which the learned author arranged the class, geographically, into six great regions, each area marked by its characteristic fauna, and by a general homogeneity of type. Dr. Günther shortly afterwards, in an elaborate paper before the Zoological Society, showed that an examination of the reptilian fauna gives us a division of the earth s surface into the same natural provinces as holds good with the class *Aves;* and since that time zoologists appear by common consent to have accepted Mr. Sclater's views as demonstrable in every province of life.

Accepting then Mr. Sclater's definition of boundary lines, Palestine forms an extreme southern province of the Palæarctic region, which includes Europe, Africa north of the Atlas, Western Asia (but not Arabia, which is Æthiopian), the rest of Asia north of the Himalayan range, Northern China, and Japan. An analysis of each class of its fauna and of its phanerogamic flora shows that while an overwhelming majority of its species in all cases belong to the Palæarctic region, there are in each class a group of exceptions and peculiar forms which cannot be referred to that region, and the presence of many of which cannot be explained merely by the fact of the Palæarctic infringing closely on the Æthiopian region, and not very distantly on the Indian ; but can only be satisfactorily accounted for by reference to the geological history of the country. These species are, almost all, strictly confined to the area of the Jordan valley and Dead Sea basin.

The Analysis of the Results of the various Collections made, including all Species hitherto recorded from Palestine, may be tabulated thus:

	Total Number.	Palæarctic.	Ethiopian.	Indian.	Peculiar.
MAMMALIA	113	55	34	16	13
AVES	348	271	40	7	30
REPTILIA, CHELONIA AND AMPHIBIA	91	49	27	4	11
PISCES:					
Fluviatile	43	8	2	7	26
MOLLUSCA	213	57	8	8	140
INSECTA:					
Rhopalocera	76	68	5	—	3
Noctuæ	166	—	—	—	25
Tineidæ	46	16	—	—	30
Coleoptera	380	—	—	—	—
Orthoptera	60	—	—	—	—
Hemiptera	59	—	—	—	16
FLORA:					
Phanerogamic and Filices	3002	—	161*	27†	251

* Exclusive of those common to Palæarctic region.

† Besides very many common to Indian and Ethiopian regions.

Of the Mammalia, the Indian 16 include 9 which are also Ethiopian. Five species are also Asiatic, but not Indian, and seven species included as Palæarctic occur only in Syria and Asia Minor. Those only elsewhere found in Arabia Petræa are included as peculiar.

Of these 113 species of Mammalia, a very much larger proportion belongs to the Ethiopian region than is found to be the case in the other classes, viz., 34 out of 113, or about $\frac{3}{10}$ of the whole. Of these, 9 species, including 6 *Felidæ*, a hyæna and a bat (*Rhinopoma microphyllum*), are equally Indian, to which region may also be assigned 4 others, two of which, however, *Camelus dromedarius* and *Bos bubalus*, introduced by man, can scarcely claim enumeration; and another, *Asinus hemippus*, equally pertains to the Palæarctic region, leaving only two doubtful species, *Gazella cora* and *Mus prætextus*. The traces of Indian immigration or introduction are thus reduced to an infinitesimal quantity.

Of the 34 mammalia which may be referred to the Ethiopian region, and which are certainly not Palæarctic, 4 antelopes, 2 hares, and 8 small rodents of the genera *Acomys*, *Gerbillus*, and *Psammomys*, are species strictly desert in their habitat, and therefore with powers of traversing the great African and Arabian sand-wastes, and settling themselves on their northern frontier.

The larger *Felidæ*, in which Palestine is so rich, possessing 6 species besides the Palæarctic *Felis pardina*, are specifically identical in the two regions, or if distinct, we have not yet ascertained to which races the Palestine specimens belong. They may have arrived by way of Egypt, or from the Euphrates valley. But there are more sedentary forms, as *Herpestes ichneumon*, and the cats, which seem to point to an earlier settlement than across the recent deserts. Besides, the transit from the east is not more physically difficult or distant than from Africa; yet we find no indication of such immigration in this class.

Of the 13 peculiar mammalian forms, three, *Ursus syriacus*, *Lepus syriacus*, and *Sciurus syriacus* are merely modifications of Palæarctic types; six, *Lepus sinaiticus*, *Gerbillus tæniurus*, *Psammomys myosurus*, *Acomys russatus*, *Mus prætextus*, *Gazella arabica*, are Ethiopian in their character, and probably extend further into Arabia and East Africa. *Eliomys melanurus* and *Dipus hirtipes* seem decidedly peculiar. One species, *Lepus* (*Eulagos*) *judeæ*, Gray—the hare of the Dead Sea basin—

is different in the form of its skull from either the European or the Syrian species, the back edge of the orbit in the former having a deep notch, while the edge of the orbit in *L. judeæ* is continuous, with an oval perforation, caused by the process of the notch being united to the skull at the end. The notch, if open, would be more than double the depth of that of the other species.

The last peculiar species is *Hyrax syriacus*, a member of a strictly Ethiopian genus, of which Dr. Gray enumerates 8 species, several of them merely local races. The other species extend from Abyssinia to the Cape. The Palestine coney, confined to the gorges of the Dead Sea and Arabia Petræa, must not be confounded with *Hyrax brucei* from Abyssinia = *H. syriacus.* Sch. Here we have one of the most peculiar and isolated forms of the mammalian class, exclusively confined to the Ethiopian region; but with this representative extending beyond that region, yet specifically differing from all its congeners. No theory of immigration or dispersion can account for its presence, especially when we bear in mind the sedentary character of the group.

The Avifauna of Palestine is, like the mammalian, unusually rich in number of species for so small an area, covering not more than 5,600 square miles. It consists of 348 known species, which may be thus classified. Palæarctic species, most of which occur elsewhere, 271; Ethiopian, 40, inclusive of 10 which are also Indian; Indian, but not Ethiopian, 7; and species so far as is yet known peculiar to Syria, 30. But the Avifauna is by no means equally diffused over the whole area. Of the Palæarctic species, almost every one (with the exception of the Accipitres, which are indifferently ubiquitous, and some Natatores, which are winter visitants) belongs to the coast area, and the highlands east and west of Jordan. The Ethiopian and Indian types are almost exclusively confined to the deep depression of the Dead Sea basin, which, with the exception of some winter migrants, affords us very few Palæarctic species.

Of the 30 birds pertaining to the Ethiopian Fauna, 18 species have not been found in Palestine out of the Dead Sea basin. The most remarkable of these are the sedentary *Cypselus affinis*, *Merops viridis*, *Cotyle obsoleta*, *Corvus affinis*, *Saxicola monacha.* Ten others are desert forms, probably common to Arabia, and reaching here their northern limits, as *Calandrella deserti*, *Certhilauda alaudipes*, *Pterocles exustus*, *Houbara undulata*, and formerly, though now extinct, *Struthio camelus.* As to the occurrence of these desert species no difficulty can arise, especially in the case of such of them as extend through the whole belt of sandy waste which girdles the whole Old World from Scinde to the Atlantic Coast of Africa.

The most interesting of the Indian non-Ethiopian species is *Ketupa ceylonensis*, and the occurrence of this great fish-eating owl is the more exceptional, as not only are there no Strigidæ in Africa bearing the least affinity to this well-marked genus, but because it has not yet been found in the Jordan valley, but sedentary by the streams of the coast. We have hitherto no record of its occurrence elsewhere west of India.

Of the Indian types, *Reguloides superciliosus* and *Sylvia nana* pertain equally as straggling migrants to the western Palæarctic region ; *Halcyon smyrnensis* and *Turtur risorius*, which are both sedentary in the Jordan valley, are the only other instances of so great a westward extension of purely Indian species. But both have appeared as stragglers in Asia Minor, where the former was known to Linnæus, but lost to science till rediscovered by Captain Graves, R.N.

Of the 30 species classed as either new or peculiar to Palestine, 13 are merely modifications or representative forms of familiar Palæarctic types, such as *Garrulus atricapillus*, *Picus syriacus*, *Saxicola melanoleuca*, etc., which take the place of the common western *G. glandarius*, *P. major*, *S. stapazina*, etc., and which are all found in the upper country or on the coast. Several of the other new species are closely allied to known desert or Oriental forms, and are found beyond the limits of the Dead Sea basin. Such are *Hypolais upcheri*, *Erithacus gutturalis*, *Saxicola finschii*, *Petronia brachydactyla.* These are also clearly African in their affinities. One, *Ruticilla semirufa*, inhabiting the hill country, is closely affined to the Indian group of *Ruticillinæ*, and not to the Palæarctic or Ethiopian members of the genus.

But there are 11 species belonging to as many different genera, peculiar to the Dead Sea basin, and not yet traced beyond its limits. Some of these belong to genera exclusively Ethiopian, most of them common to the Ethiopian and Indian regions; but of two at least the affinities are Indian rather than African. *Caprimulgus tamaricis* is perhaps most closely related to *C. asiaticus* of India, but with the characteristic plumage of *C. isabellinus* of Africa. *Passer moabiticus*, another very marked species, strictly confined to the lower end of the Dead Sea basin, though it belongs to a genus equally Ethiopian and Indian, yet must undoubtedly be classed among the Indian section of this group. By far the smallest species known of its genus; in its coloration and other peculiarities, it approaches in some respects to the Indian *P. cinnamoneus*, in others to *P. russatus* of China; but it is not affined to any known Ethiopian *Passer*. *Ammoperdix heyi*, a partridge limited in its range to the region round the Dead Sea and Arabia Petræa, belongs to a sub-genus of *Caccabis*, of which the only other member, *Ammoperdix bonhami*, is Indian. Four other species of the Dead Sea basin, *Drymœca inquieta*, *Sylvia bowmani*, *S. melanothorax*, and *Cercomela melanura*, belong to genera common to both regions, though all are more nearly affined to the African than the Indian members of their respective groups.

Of the others, *Ixus xanthopygius*, belonging to a genus widely extended through both regions, is yet by its sombre plumage and yellow vent close to five or six Ethiopian species, and more decidedly separated from any of its Indian congeners. *Cinnyris oseæ*, the only one of the sunbirds which reaches so far north, represents a family very numerous in both regions, and is confined to the Jordan valley, though occasionally in summer straggling a little beyond its limits. Though not far removed from *Nectarinia asiatica*, it approaches much more closely *C. affinis* of Abyssinia. *Argya squamiceps*, yet more circumscribed in its range to the lower part of the Jordan valley, beyond which it never ascends, is one of a peculiarly well-defined genus, comprising about 15 species, African and Indian; while *Amydrus tristrami*, the last to be named, limited in its range to the rocks that overhang the Dead Sea, is one of a restricted genus of starlings, of which the other species are Abyssinian and South African. None of the Indian *Sturnidæ* have any near affinities with this genus.

Thus the Avifauna of the Dead Sea is decidedly distinct and typical in its species, revealing sometimes Indian, more generally African affinities.

An examination of the Reptilian Fauna leads to the same conclusion, though we find here a less prominent intrusion of Ethiopian types. Dr. Günther has given it as his decided opinion, that herpetologically

Egypt must be embraced in the Palæarctic region, and many of the Egyptian snakes occur in Palestine. Of 91 known species of Reptilia, Chelonia and Amphibia, 49 are Palæarctic; among them *Eryx jaculus*, traced from Egypt to Siberia, four species of *Zamenis*, two of which, *Z. dahlii* and *Z. ventrimaculatus*, reach Kurdistan; 27 of the *Reptilia* are also African, among them *Monitor niloticus*, *Scincus officinalis*, *Uromastix spinipes*, and *Naja haje;* four are Asiatic exclusively, but do not extend into India beyond its frontier, as *Vipera euphratica.* One species of serpent, *Daboia xanthina*, belongs to a genus otherwise exclusively Indian, and eleven species are peculiar. One of these is an Ophidian, of the family *Oligodontidæ*, which stands as the type of a new genus, *Rhyncocalamus*, Günther, of which the affinities are rather obscure.

From this analysis it is evident that the herpetological fauna presents fewer anomalies than the other classes. But snakes, in particular, are more limited to the original locality of the individual. In these cases the agencies are wanting by which a species is rapidly spread over a larger portion of the globe in course of time, thus becoming mixed with foreign forms; and the groups, like individuals, are more stationary. Besides, this class of life is more susceptible of climatic changes than any other, and if any period of excessive cold, like the glacial epoch, had passed over the country, the reptiles would be the first to succumb, without any chance of their recovering their ground during subsequent modifications of the temperature. Yet even here we can clearly trace anomalies in the distribution, corresponding to the anomalies already mentioned, and to which we do not elsewhere find a parallel.

The fluviatile ichthyological Fauna, though limited in number of species, is beyond comparison by far the most distinct in its character. We find 43 species, of which only 8 belong to the ordinary ichthyological fauna of the Mediterranean rivers. But these belong to the rivers of the coast. In the Jordan system only one species out of 36 belongs to the ordinary Mediterranean Fauna, viz., *Blennius lupulus.* Two others, *Chromis niloticus* and *Clarias macracanthus*, are Nilotic. Seven other species occur in other rivers of South-western Asia, the Tigris, Euphrates, etc. Ten more are found in other parts of Syria, chiefly in the Damascus lakes, and the remaining 16 species of the families *Chromidæ*, *Cyprinodontidæ* and *Cyprinidæ*, are peculiar to the Jordan, its affluents, and its lakes. This analysis points at once to the close affinity of the Jordan with the rivers of Tropical Africa. The affinity is not only of species, but of genera, for *Chromis* and *Hemichromis* are peculiarly Ethiopian forms, while the other species are identical with, or very closely allied to, the fishes from other freshwaters of Syria. But the African forms are a very large proportion of the whole, and considering the difficulty of transportation in the case of freshwater fishes, the peculiarities of this portion of the Fauna are of great significance.

Turning to the Invertebrate Fauna, we find the Mollusca, terrestrial and fluviatile, to amount to 213 species, of which 57 are common European or Mediterranean forms, 8 Nilotic, and 8 other fluviatile species found also in the rivers of the Indian Ocean, the Tigris, or Euphrates. About 140 species occur which have not as yet been noticed out of Palestine. But many of these have no special significance in a geographical point of view. The land shells are for the most part merely modifications of wide-spread Palæarctic forms, such as the 28 peculiar species of *Clausilia*. It must be remembered that in no department of zoology do we meet with so many localized forms in limited areas as among the Pulmonifera, where variations appear in many cases rapidly to follow segregation, while other species extend with but slight modifications over an entire region. Thus most of the species of *Helix* in the Southern Desert show affinities to the Ethiopian type, while a peculiar group of *Helices* seems to have been developed in the highland districts, between the desert and Lebanon; of which *H. cariosa* may be regarded as the typical species. But in the Jordan valley are two groups of peculiar gasteropods. One group, found only by the Dead Sea, is a series of modifications of desert forms, exemplified in *Helix prophetarum* and *H. filia*. Another group, of which *Bulimus labrosus* may be taken as typical, is found through the whole length of the valley, and is a modification of a common Syrian and Asia Minor type.

But the fluviatile molluscs are far more distinct. Besides such species as *Melania tuberculosa*, extending from the west of Africa to China and Southern India; and *Melanopsis buccinoidea*, a common Mediterranean form, we find various peculiar *Unionidæ* and *Melaniadæ*, such as *Unio simonis* and *U. episcopalis*, which indicate very ancient separation from any adjacent district, and the affinities of which are certainly not Palæarctic.

The *Arachnidæ* have scarcely yet been sufficiently determined, to enable us to add any important facts to the above induction. But one extensive family, the *Drassidæ*, has been worked out by the Rev. O. P. Cambridge and Dr. Koch, who report that of the 13 known genera of this family, 8 are represented in Palestine, of which 7 are European, 5 being also found in Egypt, none being peculiar to Palestine. Of 46 species of Palestine *Drassidæ*, 24 are, so far as our present knowledge extends, peculiar, 13 are European, and 9 Egyptian. Admitting therefore that many of the new species will prove to have a wider range, we have here also a predominant Palæarctic character, with an infusion of African, and probably a few localized types.

Similar inferences may be drawn from an examination of the Insect Fauna, of which, however, our knowledge is merely fragmentary. The determination of 380 species of *Coleoptera* and of 60 species of *Orthoptera*, which are all that have been collected, no naturalist having as yet devoted himself to them, has not yet been accomplished.

But of the *Hemiptera*, 59 in number, 16 species appear to be new, the others with few exceptions being known from Northern Syria and Asia Minor. Of the *Lepidoptera* our knowledge is most imperfect. But the results of an examination of the existing collections are in harmony with those of the rest of the Fauna. Of 76 *Rhopalocera*, 68 belong to the Eastern Mediterranean, and are therefore Palæarctic, 5 are Ethiopian, being Nubian species, and 3 are new. These last 8 species are confined to the basin of the Dead Sea. Of 166 *Noctuæ*, and 46 *Tineidæ*, 25 *Noctuæ* and 30 *Tineidæ* are new. The 16 *Tineidæ* previously described, and all the *Noctuæ* hitherto known, belong to the Eastern Mediterranean, and do not indicate any Ethiopian affinities. But it must be mentioned that no entomological collector has worked as yet in the Jordan valley during the latter part of spring, when these insects would be most numerous; and even as it is, 14 of the 30 peculiar *Microlepidoptera* are from the plains of Jordan only.

It may here be stated that an examination of sand from the roots of *Anastatica hierochuntina* gathered at the north end of the Dead Sea, shows that the Rhizopod fauna was analogous to that of the Red Sea and Indian Ocean, being composed of *Gr. capreolus* and other Indian Ocean forms.

The Flora of Palestine is in its distribution parallel with its Fauna. About 3,000 species of phanerogamic plants are recorded from the district. Of these my own herbarium contains about 1,400, collected by Mr. B. T. Lownes, my companion in 1863-64. As might be expected, by far the larger proportion consists of the common Mediterranean forms. The Flora of the coast and southern highland region calls for no remark, as it is simply a reproduction of the Flora of Sicily, Greece, Asia Minor, and Northern Syria. Of about 250 species or varieties which have not yet been noticed elsewhere, there are none which call for special remark, as all are closely allied to other representative species, and the additions are distributed in fair proportion among the floras of each region, coast, plains, highlands; mountains; and Jordan valley, and deserts. Sir J. D. Hooker has remarked that though a vast number of plants are common to the whole country, and in no latitude is there a sharp demarcation between them, yet 'there is a great and decided difference between the floras of such localities as the (1) Lebanon at 5,000 feet, (2) Jerusalem, and (3) Jericho; or between (1) the tops of Lebanon, (2) of Carmel, and (3) of any of the hills bordering the Jordan; for in the first locality we are most strongly reminded of Northern Europe, in the second of Spain, and in the third of Western India and Persia.' For our present purpose we need only consider the Flora of the Dead Sea basin, and especially of its southern portion. In the little Wâdy Zuweirah, at the south-west corner of the Dead Sea, we have collected over 160 species of plants. Of these only 27 are common European forms, chiefly of very wide distribution, as *Tribulus terrestris*, *Emex spinosus*, etc. All of these 27, with one or two exceptions, also extend into North India. The remaining 135 species are African, scarcely any of them extending into Europe, and many of them extremely local. Thirty-seven of the Ethiopian are also Indian plants, chiefly belonging to the desert Flora of Scinde, and others, as *Cordia myxa*, though characteristic Indian plants, are equally Nubian and Abyssinian. Although the Dead Sea flora bears a very strong general similarity to the flora of Arabia Petræa, yet there can be no question of its distinctness from the adjacent floras of the same latitude, east and west of it.

A few of the plants claim especial notice. In the Jordan valley the *Cyperus papyrus* is locally abundant, covering many acres in the marshes of Huleh, though long since extinct in Egypt, and not now known in Africa further north than on the White Nile, lat. 7° N. *Calotropis procera* and *Salvadora persica* are never found except close to the Dead Sea, at Engedi, Safieh, and Seisaban, and are separated by many degrees of longitude and latitude from their other known habitats. It is also interesting to observe that of the 135 African species in Wâdy Zuweirah, 23 extend as far west as the Canaries, and 17 are decidedly Arabian, being included in the Flora of Aden. None of these African forms occur in other parts of Palestine.

While it is not probable that many other European plants have escaped observation on the shore of the Dead Sea, as they would naturally be early flowerers, many other decidedly tropical plants not included in those gathered in Wâdy Zuweirah in February have been noted in hasty visits to the district later in the year. Such are *Abutilon muticum* and *A. fruticosum*, *Zygophyllum coccineum*, *Indigofera argentea*, *Boerhavia plumbaginea*, *Conyza dioscoridis*, etc.

There are other very remarkable contrasts in the affinities of the Flora. Thus among the *Leguminosæ*, there are no less than 50 species of *Trifolium*, and 74 species of *Astragalus* in our list. Of the former only one, *T. stenophyllum*, is found in the south, and not one in the Jordan valley; and all of them are either European, or have European affinities. There is no indication of any species connected with the Himalayas or Central Asia. Of the 74 *Astragali*, on the contrary, only 3 have any Palæarctic affinities; all the others are either Indian, or belong exclusively to the Oriental or Ethiopian regions. Yet the Astragali are by no means confined to the Jordan valley. No less than 35 species are strictly limited to the mountain and alpine regions of Lebanon and Anti-Lebanon. The bulk of the remainder belong to the Jordan valley and Southern Desert, and if found elsewhere are either in Scinde or Africa.

The Flora of the Dead Sea area is remarkable for a small average number of species distributed through a large number of orders. In 250 species collected on the east side of the Dead Sea in February, 58 orders were represented, an average of 4½ species to each order. We may infer that in this borderland of Europe, Asia, and Africa, the more hardy and accommodating plants of each area hold their own, while those more readily affected by variation of soil or climate disappear.

To sum up our deductions, a review of the botany as well as the zoology of the Dead Sea basin reveals to us the interesting fact that we find in this isolated spot, comprising but a very few square miles, a series of forms of life differing decidedly from the species of the surrounding region, to which they never extend, and bearing a strong affinity to the Ethiopian region, with a trace of Indian admixture.

In order to form a just conclusion as to the mode by which this isolated region became peopled by animals and plants, it is necessary to take geological causes into consideration. Here is a patch of tropical character, containing southern forms so peculiar and unique, that we cannot connect their presence in it with any existing causes or other transporting influences. The basin, at the bottom of which they are found, is a depression in a mountainous country, sunk 1,300 feet below the level of the ocean, and occupied, with the exception of a few acres here and there, by the waters of a salt lake. It is hemmed in by two parallel mountain ranges, rising from 3,000 to 5,000 feet above its level, and these enclosing ranges are rarely more than 20 miles apart.

As it has been shown by Humboldt that zones of elevation on mountains correspond to parallels of latitude, the higher zones corresponding with the higher latitudes, so here we find a zone of depression, the only one known to us, producing similar phenomena, and exhibiting in generic correspondence, specific representation, and in some instances specific identities, the fauna and flora of much lower latitudes. As the flora and insect fauna of the Scottish mountain tops is Scandinavian, while the surrounding type is German, so we find this islet of Ethiopian flora in the midst of a Mediterranean district. If we had to deal only with a representative flora, it might have been assumed that it essentially depended on the law that climatal zones of animal and vegetable life are naturally repeated by elevation (in this case depression) and latitude. But the transmission of a transported flora requires another explanation, which can only be found by tracing geological history. That it became peopled by special creation within this area, or that its inhabitants can have had an independent origin on the spot, would not only be a most unreasonable assumption, but is negatived by the fact of the identity of many species of animal life, and of almost the entire flora, with species now living in the Ethiopian region.

That it was peopled by migration, or that wandering individuals in search of new homes, finding the conditions adapted for their existence, settled and colonized, and, in the case of birds, abandoned their migratory habits, is refuted by the fact of the co-existence of peculiar and unique forms, with others now found in regions widely separated from this colony. Besides which, there are many species which, after making all due allowance for all probable modes of migration at present in operation, could scarcely have been transported thither under present conditions, since either their physical characteristics, or the phenomena of their present distribution, forbid such a supposition.

It must be borne in mind that deserts such as those which isolate Palestine on the south and east are found to present far more insuperable barriers to the transport of species (excepting of course the case of desert forms) than either seas or mountain ranges. *E.g.*, it is the Sahara, and not the Mediterranean, which separates the Ethiopian from the Palæarctic fauna.

There remains, therefore, only the hypothesis of these species, and of all other peculiar inhabitants of the basin, having arrived there by migration or general dispersion before the character of the surrounding region presented the existing obstacles to their transport, and this at once invites consideration of the geological problem. If their position be mainly due to migration before the isolation of the area, it is necessary, if possible, to ascertain two fixed points in time between which this migration must have taken place. The migration must have been after the close of the Eocene period. The palæontological character of the most superficial deposits of all Southern Palestine is unquestionably Eocene. There are no beds of fossils synchronizing with the Meiocene deposits of Sicily, North Africa, and the Greek Islands. The whole of Syria and Arabia Petræa must have already emerged from the ocean, while the greater part of the Mediterranean and its adjacent coasts and islands was the bed of a Meiocene sea, and must have had a fauna and flora contemporaneous with the Meiocene flora of Germany. We have the clearest indications of this extension as far as Palestine. This Sir J. D. Hooker has shown in tracing the glacial moraines which stud the whole Lebanon range. There are also other indications of glacial action which we may notice presently.

But the general question is not materially affected by the precise epoch at which the lake of the Dead Sea became reduced to its present dimensions. It is certain that in its present general form it must have existed long before the glacial period.

We know from the Norfolk chalk that in our own country a much warmer climate existed previous to the glacial epoch; and we have every reason to infer that throughout the Northern Hemisphere a proportionate increase of temperature prevailed. The Ethiopian fauna and flora, admittedly more antique in type than the Palæarctic, must have had at that period a more northerly extension than at present. The peculiarities of the Dead Sea basin may be exactly paralleled with the traces of the Spanish flora yet lingering on the south-west of Ireland. They were probably synchronous in origin during the period of the great Meiocene sea which covered the Mediterranean and Western Europe. The great Meiocene land extended, as it would appear, south and west of this from Southern Asia as far as the Azores, or, as Professor Forbes has suggested, to the belt of Gulf Weed. There was then either continuous or closely contiguous land, which would enable South-west Ireland to be stocked by the flora of the Asturias, and Palestine by that of Abyssinia. The circumstances and chronology of these two isolated floras appear identical.

But during this epoch, the whole country was doubtless covered and peopled by the same forms, for which the warm climatal conditions were suitable; and the fauna and flora of Palestine were East African, either identical or representative. The actual present refuges of the remnants of this period, the nooks beside the Dead Sea, were still under water, for they do not now rise 200 feet above the lake. Towards the close of the Pleiocene period the area of the waters of the lake was diminished, as we may see by the marl deposits leaning against the inclosing ranges. Then supervened the glacial period. The climatal changes destroyed the mass of existing life; just as in South-west Ireland all the Spanish flora except the hardiest, such as the saxifrages and the heath, have perished.

But, as subsequently with the returning warmth in the British Isles, the Scandinavian remnant continued to survive on the tops of the Scottish mountains, so in the period of cold those species which were most tenacious of life, retiring to the depression of the Jordan valley, then, as now, proportionally warmer than the surrounding land, contrived to maintain the struggle for existence; and have survived to the present day, and form a tropical outlier, of which we have no other terrestrial instances, but which is exactly parallel with the northern outliers of arctic marine life which occur in our British Seas.

Here, when the bed of the glacial ocean was upheaved, that upheaval raised above water only such portions of it as had been formed at a moderate depth. The deeper tracts were still under water, and there the arctic forms would still live on, while climatal changes altered the zoological character of the shallows of our seas. So with our land flora. When the change of level connected the group of glacial islands, which stood out from the sea, crowned with the Scandinavian flora, now the mountain tops of Scotland, the original vegetation was preserved and survived in isolation; while on lower ground, the newly exposed land was covered with a new temperate or Germanic flora. 'Mutato nomine' for boreal marine outliers, read tropical terrestrial outliers, and the history of the biological isolation of the Dead Sea is solved at once.

The slight admixture of Indian types may be explained when we consider that at that time, with a continuous Meiocene continent north of the line, species would have a wider range than at present, and could migrate and spread without check. Since the change in the coast outline, some species may have become segregated in one part, and others in another. The glacial action must have affected the temperature of regions far beyond the reach of its actual glaciers and icebergs, and many species must have retired to the South, or perished utterly. The modifications of Ethiopian and Indian types, of which the avifauna presents several instances, may yet be found in Arabia, the interior of which is still to us a zoological blank. If absolutely peculiar, they must either be the descendants of species which inhabited the country with only a limited range prior to the glacial epoch, or are developments or variations of other species stereotyped through long isolation. Considering, however, their co-existence with other unchanged forms, and the comparatively recent date, geologically speaking, of the glacial epoch, the former would seem to be the more probable solution.

The fluviatile fishes claim special attention, dating, as they probably do, from the earliest time after the elevation of the country from the Eocene ocean. In the *Foramenifera* mentioned above as found in the

Dead Sea sand, such as *Gr. capreolus*, we have the relics of the inhabitants of that early sea. But of the living inhabitants, we must place the Jordanic fishes as the very earliest, and these, we have seen, form a group far more distinct and divergent from that of the surrounding region than in any other class of existing life. During the epochs subsequent to the Eocene, owing to the unbroken isolation of the basin, there have been no opportunities for the introduction of new forms, nor for the further dispersion of the old ones. These forms, as we have seen, bear a striking affinity to those of the freshwater lakes and rivers of Eastern Africa, even as far south as the Zambesi. But the affinity is in the identity of genera, *Chromis* and *Hemichromis* being exclusively African, while the species are rather representative than identical.

The solution appears to be that during the Meiocene and Pleiocene periods, the Jordan basin formed the northernmost of a large system of freshwater lakes, extending from north to south, of which, in the earlier part of the epoch, perhaps the Red Sea, and certainly the Nile basin, the Nyanza, the Nyassa, and the Tanganyika lakes, and the feeders of the Zambesi, were members. During that warm period, a fluviatile ichthyological fauna was developed suitable to its then conditions, consisting of representative and perhaps frequently identical species, throughout the area under consideration.

The advent of the glacial period was, like its close, gradual. Many species must have perished under the change of conditions. The hardiest survived, and some perhaps have been gradually modified to meet those new conditions. Under this strict isolation it could hardly be otherwise; and however severe the climate may have been, that of the Lebanon with its glaciers probably corresponding with the present temperature of the Alps at a proportional elevation (regard being had to the difference of latitude), the fissure of the Jordan being, as we certainly know, as much depressed below the level of the ocean as it is at present; there must have been an exceptionally warm temperature in its waters in which the existing ichthyological fauna could survive.

The glacial period has left its mark in the mountain range of Northern Palestine, not only in the moraines which stud the Lebanon; and the desolate heaps of which point out the position of the old glaciers as shown by Sir J. D. Hooker; but even in the existing forms of life.

Thus among mammals we find on the top of Hermon *Arvicola nivalis*, identical with the mountain vole of the Alps and the Pyrenees. Among the non-migrant birds occurs the exclusively alpine *Montifringilla nivalis*, traced on the mountain tops from the Pyrenees to Ararat; *Otocoris penicillata*, a slightly modified form of *Otocoris alpestris*, and which never leaves the snow-line; *Pyrrhocorax alpinus*, an equally sedentary form; and several others which move up and down the mountain sides according to the season are also found. These species on the Lebanon and Hermon form in fact a boreal outlier.

The flora, however, forms an exception here. The vegetation of the summits of Lebanon is not analogous to that of the Alps of Europe and India. More boreal plants may be gathered on the Himalayas at from 10,000 to 15,000 feet, than on the analogous heights of Lebanon, *i.e.*, from 8,000 to 10,000 feet. Three hundred flowers of the Arctic Circle inhabit the ranges of Northern India, while not half that number are found on Lebanon. Sir J. D. Hooker accounts for this partly by the heat and extreme dryness of the climate during a considerable part of the year; to the sudden desiccating influence of the desert winds; and to the sterile nature of the dry limestone soil; but still more perhaps to the warm period which succeeded the cold one, during which the glaciers were formed; and which may have obliterated the greater part of the traces of the glacial flora. Several of these causes do not apply with equal force to the fauna, with their powers of vertical migration which enabled them to remain. There are other traces of a glacial fauna now extinct, in the remains of *Cervus elaphus*, *C. tarandus*, and *Alces palmatus*, the Red-deer, Reindeer, and Elk, discovered in the breccia of cave floors in the Lebanon. We may take these traces of the glacial inhabitants as the representatives of the fauna which then overspread the whole country, synchronous with the introduction of the Scandinavian flora now lingering on the tops of the Scotch mountains, and with the deposition of the Pleistocene deposits of Sicily and Cyprus.

When afterwards the climatal conditions became less severe, the Mediterranean fauna and flora rapidly overspread the whole country, partly by way of Asia Minor and the Greek Islands, partly by way of Egypt, just as the Germanic flora overspread the British Isles, and has given its predominant character to the natural history of the country.

The conclusions at which we arrive are that while the fauna and flora of Palestine are decidedly Palæarctic in type, and belonging to the Mediterranean section of that type, there are traces of a boreal fauna in the north, and a large infusion of Ethiopian types in the Jordan valley: and that these exceptions can be satisfactorily explained only by a reference to the geological history of the country, which shows that the glacial period, though not extending in its intensity so far south, has left traces not yet wholly obliterated; while the preceding period of warmth has left yet larger proofs of its former northern extension in the unique tropical outlier of the Dead Sea basin, which is analogous both in its origin, and in the present isolation of its various assemblages of life, to the boreal outliers of our mountain tops and our deep sea bottoms; the concave depression in the one case being the complement of the convex elevation in the other.

COLLEGE, DURHAM,
December 17th, 1883.

THE

FAUNA AND FLORA OF PALESTINE.

MAMMALIA.

ORDER, PACHYDERMATA.

FAMILY, HYRACIDÆ.

1. *Hyrax Syriacus.* Hemp. and Ehr. Symb. Phys. Mamm. Pl. 2. Coney. Heb. שָׁפָן. Arab. طبسن, *Tubsun*, or الوبر, *el Wabr* (in Sinai).

PLATE I.

This singular little Mammal, neither ruminant nor rodent, but which is placed by systematists among the *Ungulata*, near the Rhinoceros, is one of the many peculiarly African forms which occur in Palestine. It is not uncommon round the shores of the Dead Sea, but is rare in the rest of the country, and not known in Lebanon. It is found throughout the Sinaitic Peninsula generally, but is not known to extend further into Arabia or Western Asia. It is represented by a very closely allied species in Abyssinia, and by another rather larger at the Cape. Several species, or varieties, occur in Eastern Africa, but this is the only one known beyond the limits of that continent. Its Hebrew name means 'the hider,' and its timid, cautious habits, and defenceless character are referred to in Scripture. The Syrian Coney is marked by a yellow dorsal spot on its otherwise uniformly tawny fur. It is scarcely so large as a full-grown Rabbit. Its teeth and toes resemble those of the Hippopotamus in miniature. It lives exclusively among the rocks in Wâdys, not generally burrowing, but utilizing fissures in the cliffs, where it has its inaccessible home, coming forth to feed only at sunset and at dawn.

Though not strictly gregarious, there are generally several in close neighbourhood. When feeding, a sentry is usually placed on some commanding outpost, who gives warning of approaching danger by a sharp bark. The Coney has four or five young at a birth.

FAMILY, SOLIDUNGULA.

Equidæ.

2. *Equus caballus.* L. Syst. Nat. i., p. 100.* The Horse. Heb. סוּס. Arab. حصان, *Hassān,* فرس, *Faras,* Mare.

The Horses west of Jordan are generally inferior to those on the east side, where may be found some of the finest bred specimens of the true Arab, with pedigrees going back several centuries. The Syrian Horse is generally a small animal of fifteen hands high, without much speed, but with great powers of endurance. Breeds of Horses were known to the Jews, the Hebrew using distinct words for saddle and for chariot Horses. Excepting in the plains, the Horse has never been much used in Palestine.

3. *Asinus asinus.* (L. Syst. Nat. i., p. 100.) The Ass. Heb. חֲמוֹר. Arab. حمار, *Homār.*

The most important beast of burden in the west and north. The Ass is taller, stronger, and fleeter in Palestine than in any other country I have visited. Much care is taken in the selecting both of sires and dams, and in the northern plains may be seen large herds of she-asses kept for breeding. The white Asses mentioned in Scripture are still highly prized, and command very high prices. A good Syrian Ass fetches about £40, the price of a good Horse. The origin of the Domestic Ass is from the African Wild Ass, the true Onager, a very much finer animal than the *Asinus hamār* (Smith) of Southern Asia.

4. *Asinus onager.* Pall. Act. Acad. S. Imp. Petrop. 1777, p. 258, t. 11. Wild Ass. Heb. עָרוֹד. Arab. حمار حشي, *Homār wahshi.*

This Wild Ass, the origin of the Domestic Ass, was formerly well

* The Edition of the *Systema Naturæ* quoted in this volume is the eleventh, Holmiæ, 1766.

known in Arabia, and is not extinct there, though very rare. I have seen this species in a state of nature frequently in the Sahara, and have handled captured though not tamed individuals. It no doubt, as the Arabs assure me, occasionally enters the Hauran. Their language, as well as the Hebrew, recognises two species of Wild Ass.

5. *Asinus hemippus.* St. Hilaire. Compt. Rend. xli., p. 1214. Syrian Wild Ass. Heb. פֶּרֶא. Arab. اخدر, *Akhdā.*

This, rather smaller than the true Onager, and confined to Syria, Mesopotamia and North Arabia, very rarely enters the north of Palestine from the Syrian desert, but is still common in Mesopotamia. It does not extend into India, but in summer herds of this animal frequently visit the Armenian mountains. It is the Wild Ass of Scripture and of the Ninevite sculptures.

FAMILY, SUINA.

6. *Sus scrofa.* L. Syst. Nat. i., p. 102. Wild Boar. Heb. חֲזִיר. Arab. حالوف, *Hallouf,* خنزير, *Khanzir.*

The Wild Boar is abundant in every part of the country, especially where there are marshes or thickets. It extends into the bare wilderness, even where there is no cover, nor other food than the roots of desert bulbs.

The Wild Boar is found throughout the whole of the Old World except South Africa. Some naturalists distinguish the Indian from the European. If they be distinct species, the Syrian must be classed with the European rather than the Eastern form, though Mr. Gray has held it to be a distinct species with the name of *Sus libycus.*

ORDER, RUMINANTIA.

FAMILY, TYLOPODA.

7. *Camelus dromedarius.* L. Syst. Nat. i., p. 90. Camel. Heb. גָּמָל. Arab. جمل, *Djimel.*

The One-Humped Camel is the only species used in Palestine. It is bred abundantly on the plains of Moab and in the south of Judæa ; but is not suited for employment in the hilly and central districts of the country. From the earliest records of man it was employed in Syria, and is still the great source of the wealth of all the Bedawin east of Jordan.

The Arabian Camel is the beast of burden of Egypt and all Southern Asia, as far as India. It was only introduced into Barbary by the Moors. In Central Africa it is universally employed for caravan traffic.

FAMILY, ELAPHII.

Cervidæ.

8. *Cervus capreolus.* L. Syst. Nat. i., p. 94. Roebuck. Heb. יַחְמוּר, 'Fallow-Deer,' A.V. Arab. يحمور, *Yachmur.*

I have seen the Roebuck on the southern edge of Lebanon, and found its teeth in bone caves. Captain Conder procured a specimen on Mount Carmel, which proves its identity with the English species, and it is also found further south, round Sheikh Iskander.

Palestine is the most southern and eastern region where the Roebuck still exists. It occurs through the whole of Europe. The North Asiatic species is distinct.

9. *Cervus dama.* L. Syst. Nat. i., p. 93. Fallow-Deer.

The Fallow-Deer, which is the Deer of Cilicia and Southern Armenia, still exists very sparingly in the north of Palestine. A few are to be found in the woods north-west of Tabor, and by the Litany river. Hasselquist mentions it on Mount Tabor, and I met with it not many miles north of the same place.

The Fallow-Deer is still found wild in Sardinia, Spain, and Tunis, as well as in Asia Minor.

10. *Cervus elaphus.* L. Syst. Nat. i., p. 93. Red-Deer.

11. *Cervus tarandus.* L. Syst. Nat. i., p. 93. Rein-Deer.

12. *Cervus alces.* L. Syst. Nat. i., p. 92. Elk.

The former existence of all these species contemporaneously with man is proved by the existence of their teeth and bones along with flints in the bone caverns of the Lebanon.

FAMILY, CAVICORNIA.

Antilopinæ.

13. *Antilope bubalis.* Pall. Spicil. Zool. xii. 16. The Bubale. Heb. תְּאוֹ (*generic*). Arab. بقر الوحش, *Bekk'r el wach.*

The Bubale I never saw in Palestine; but it certainly exists on the eastern borders of Gilead and Moab, and is well known to the Arabs, who assure me it sometimes comes down to drink at the headwaters of the streams flowing into the Dead Sea, where they not unfrequently capture it. It roams through Arabia and North Africa, where in the beginning of the last century, Dr. Shaw informs us, it was common, and where I have sometimes seen it. It is very like the *Hartebeests* of South Africa. It is the 'Wild Cow' (*Bekk'r el wach*) of the Arabs. In ancient times it must have been much more common.

14. *Antilope addax.* Licht. Act. Acad. Leopold. xii. Heb. דִּישׁוֹן. Arab. مها, *Meha.*

The beautiful milk-white Addax is a scarce and very large Antelope, but has a wide range through Abyssinia, Nubia, and Egypt, as well as Arabia. It is the *Strepsiceros* (Twist-horn) of Pliny, and is probably the 'Pygarg' of our Authorised Version. It is well known to the Arabs as 'Addas' or 'Akas,' and approaches the southern and eastern frontiers of Palestine. Its claim to be included here is rather historical than actual.

15. *Antilope leucoryx.* Pall. Spicil. Zool., fasc. xii. 17, 61. The Oryx, or White Antelope. Heb. תְּאוֹ.

Common in North Arabia, and found in the Belka and Hauran. Its horns may be purchased at Damascus. I have been near enough to identify it by its long horns. It is probably the *Teô*, or 'Wild Ox,' of the translators of our Bible. It is an inhabitant of Kordofan, Senaar, Upper Egypt, and Arabia, and, according to some authorities, extends into Persia.

16. *Gazella dorcas.* (L. Syst. Nat. i., p. 96.) Gazelle. Heb. צְבִי. Arab. غزال, *Ghazal.*

The Gazelle is extremely common in every part of the country south of Lebanon, and the only large game which is really abundant. I have even seen it on the Mount of Olives, close to Jerusalem. It is frequently mentioned in Scripture, and rendered 'Roe' in our translation; and is a favourite symbol both of fleetness and of beauty. There are many species or geographical races of Gazelle, some of them difficult to discriminate. The Dorcas Gazelle is found from Algeria through Egypt, and thence extends into Arabia and Syria.

17. *Gazella arabica.* (?) Ehrenb. Symb. Phys. Mamm. 1 r.

This species, larger than the Dorcas Gazelle, is found in the desert country east of Jordan. I had formerly identified it with Ehrenberg's species from South Arabia; but Sir Victor Brooke, while recognising its distinctness, is inclined to believe it another race, less widely separated from *G. dorcas.* The Persian *G. subgutturosa* and the Indian *G. benettii* are distinct.

Caprinæ.

18. *Capra hircus.* L. Syst. Nat. i., p. 94. The Goat. Heb. צָפִיר, He-goat; עֵז, She-goat. Arab. معز, *Ma'z.*

The Goat is more abundant in this hilly and scantily watered country than the Sheep, and constitutes its chief wealth. There are many different breeds or races. The ordinary Black Goat of Syria, universal throughout the country, with pendent ears a foot long, hanging down far below the recurved horns, has been distinguished as *Capra mambrica,* L. Syst. Nat. i., 95. The Mohair-Goat (*Capra angorensis,* L. Syst. Nat. i., 94) is occasionally bred in some parts of the north of Palestine.

19. *Capra beden.* Wagn. Schreb. Saug. V. a. 1303. (*C. sinaitica.* Ehrenb. Symb. Phys., t. 18.) Ibex. Heb. יְעֵלִים. Arab. بدن, *Beden.*

PLATE II.

The Syrian Ibex, or Beden, is still found, not only in the ravines of Moab, but in the wilderness of Judæa, near the Dead Sea. I have procured several specimens on both sides of Jordan. It is not now known in the north or in Lebanon, where I have found its teeth in cave-breccia, along with flint implements. The Beden is of a much lighter fawn colour than the European Ibex, with horns much more slender and recurved, wrinkled and knotted on the front face only. It is the 'Wild Goat' of Scripture.

J. Smit lith. Hanhart imp.

CAPRA BEDEN.

I obtained it twice at Engedi, where it is mentioned in connection with David's wanderings. Its range appears to be limited to Arabia Petrea and Egypt.

20. *Ovis aries.* L. Syst. Nat. i., p. 97. Sheep. Heb. צֹאן. Arab. غنم *Ghanam.*

Two varieties of Sheep are bred in Palestine; but by far the most common, and in most parts the only race, is the Broad-tailed Sheep (*var.: laticaudata*). The Palestine Sheep are generally piebald or skewbald; while the Goats are almost always black. The habits of the Sheep, the ways of tending them, and the life of the shepherds in Syria, remain unchanged even in the smallest particulars since the days of the Patriarchs.

Bovinæ.

21. *Bos taurus.* L. Syst. Nat. i., p. 98. The Ox. Heb. שׁוֹר, אַלּוּף, Ox; בָּקָר, Cow; פַּר, Bull. Arab. بقر, *Bakar,* Ox; بقره, *Bakara,* Cow; ثور, *Suwr,* Bull.

Neat cattle are not suited to the hilly central districts, and are not reared extensively, excepting in the sea-board and southern plains, and in the north. On the east side of Jordan they are much more general. The common cattle are an undersized race, not much larger than Scotch cattle. In the north there are much larger and better breeds. Everywhere Oxen and Cows are used almost exclusively for agricultural and draught purposes.

22. *Bos bubalus.* L. Syst. Nat. i., p. 99. The Buffalo. Arab. جاموس, *Djamus.*

The Buffalo is only used in the northern parts of the Jordan valley, especially about the marshes of Huleh, where both for the plough and for milk it supersedes the ordinary neat cattle.

In its wild state it is a native of India, but has been domesticated in all the warmer parts of Asia, from China to Syria, and along the whole of North Africa.

23. *Bos primigenius.* Bojan. Nov. Act. Leop. xiii. b. 422. The Aurochs. Heb. רְאֵם. Arab. ريم, *Reem.*

We have abundant evidence of the former existence of the great Wild Ox in Western Asia, and can with some accuracy fix the time of its final extinction. It is spoken of familiarly in the Bible, where the word *Re'êm* is unfortunately rendered 'Unicorn,' down to the time of David, B.C. 1000: and afterwards only once, in a prophetical passage. On the Assyrian monuments its chase is represented as the greatest feat of hunting in the time of the earliest dynasties of Nineveh; but does not appear in those of the later period of the Assyrian monarchy at Kuyonjik. It was seen and described by Cæsar in Germany, in the Hyrcinian forest; and did not become extinct in Central Europe till the Middle Ages. I obtained its teeth in bone-breccia in Lebanon, proving its co-existence there with man.

24. *Bison urus.* (L. Syst. Nat. i., p. 98.) Bison, or Lithuanian Aurochs.

The bones and teeth of this species have also been discovered in company with those of the former in Lebanon. It is known to have had an equally wide distribution in historic times, and even now is not quite extinct, a few being preserved in Lithuania and also in the Caucasus.

ORDER, RODENTIA.

FAMILY, LEPORIDÆ.

25. *Lepus syriacus.* Hemp. and Ehr. Symb. Phys. ii., t. 15. Heb. אַרְנֶבֶת. Arab. ارنب, *Arneb.*

The only Hare in the wooded and cultivated districts of Palestine. Down the coast it is found from Lebanon and Hermon to Philistia. I have also found it everywhere in the wooded and mountainous parts of Northern Syria. It is very little smaller than the English Hare, measuring about two inches less in total length and with rather shorter ears. It has four young at a birth. It has not been noticed beyond Syria.

26. *Lepus sinaiticus.* Hemp. and Ehr. Symb. Phys. ii., t. 15.

This species cannot be confounded with the preceding. It is much smaller, with a longer and narrower head, ears half an inch longer, and the fur a much lighter hue. It is very rare in Palestine, occurring only, so far as we yet know, in the Wâdys by the Dead Sea. It is the Hare of the Sinaitic Peninsula.

27. *Lepus ægyptius.* Geoffr. Descr. de l'Egypte. Mamm., t. 6. Egyptian Hare.

The Common Hare of the southern region of Judæa, of the wilderness of Beersheba and of the Jordan valley. It is smaller than our Hare, the body from nose to root of tail measuring only eighteen inches. The ears are long, fringed inside with white hairs. It is of a light sand colour above, and almost white beneath.

It is found through all the desert parts of Egypt, reaching Palestine from the south-west, as the preceding species does from the south-east.

28. *Lepus isabellinus.* Rüpp. Atlas, p. 52, tab. 20. Nubian Hare.

The Nubian Hare is very rare, only found in the sandy deserts of the south-east. I possess one specimen, which I found there, and never have seen another. It is of a rich fawn yellow colour, lighter than that of the Egyptian Hare, which it generally resembles, but is decidedly smaller. It is the Hare of Nubia and Senaar, but not of Abyssinia or Egypt. The distribution of these various Hares seems to be not so much geographical as dependent on the character of the soil, and their ranges overlap and cross each other most irregularly.

29. *Lepus judææ.* Gray. (?) = *L. craspedotis.* Blanf. Eastern Persia, vol. ii., p. 80.

Dr. Gray named as above a specimen collected by me, forming for it and *L. mediterraneus*, a new genus, *Eulagos* ('Ann. and Mag. Nat. Hist.,' Third Series, vol. xx., p. 222). It seems to me that it is so near the Persian species, *L. craspedotis*, described by Mr. Blanford, that I hardly like to separate them. It is found in the north-east of Palestine.

FAMILY, HYSTRICIDÆ.

Hystracinæ.

30. *Hystrix cristata.* L. Syst. Nat. i., p. 76. Porcupine. Heb. קִפּוֹד, *generic.* Arab. نيس, *Nis.*

The Porcupine is common in all the rocky districts and mountain glens, though from its nocturnal habits seldom seen. It is especially abundant in the gorges abutting on the Jordan valley, where its quills may be gathered in considerable quantity at any time. The Arabs pursue it for the sake of its flesh, which is considered a great delicacy. The Porcupine is found throughout Southern Europe, from Spain to Turkey, and on the south of the Mediterranean, from the Gambia and Morocco to Egypt. Syria seems to be its Eastern limit.

31. *Hystrix hirsutirostris.* Brandt. Mem. Acad. Petersb. 1835, i. 375. Asiatic Porcupine.

Wagner states that this species is a native of Syria, and that he obtained it near Jerusalem. I have not met with it, but cannot doubt the authority of so careful an observer. It is found in India and Persia.

FAMILY, MURIDÆ.

Mouse, *generic.* Heb. עַכְבָּר. Arab. فار, *Far.*

32. *Acomys cahirhinus.* (Geoffr. Descript. de l'Egypte, pl. v., fig. 2.) Porcupine Mouse.

PLATE III. 2.

Confined in Palestine to the Dead Sea basin and the ravines abutting on it. We trapped it as far up the country as Mar Saba. Common in Egypt.

33. *Acomys dimidiatus.* Rüpp. Atlas, p. 37, tab. 13.

PLATE III. 3.

More abundant than the former species, but, so far as our observation goes, not extending up the rocky ravines. It may be seen on the gravel and sand in the day-time. Its range extends through Nubia, Egypt, and Arabia Petræa.

J. Smit lith. Hanhart imp.

1. ACOMYS RUSSATUS. 2. ACOMYS CAHIRHINUS
3. ACOMYS DIMIDIATUS.

34. *Acomys russatus.* Wagn. Münch. Abhandl. iii. 195.

PLATE III. 1.

This diminutive species has only hitherto been found in Palestine near Masada, towards the south end of the Dead Sea. It is only known elsewhere from Sinai, where Wagner discovered it. These Porcupine Mice are amongst the most beautiful of diminutive quadrupeds, of a rich fawn colour, and their backs more or less covered with spines instead of fur, delicately marked black and white. The genus is a strictly desert one.

35. *Mus decumanus.* Pall. Zoog. Ross.-Asiat. i., p. 164. The Brown Rat. Arab., جردون, *Djardoon.*

This cosmopolitan pest has found its way to Palestine, and is as common there as elsewhere.

36. *Mus alexandrinus.* Geoffr. Descript. de l'Egypte.

This Egyptian species is the House-Mouse of the towns on the coast.

37. *Mus musculus.* L. Syst. Nat. i., p. 83. Mouse.

The European House-Mouse is common in all the towns.

38. *Mus sylvaticus.* L. Syst. Nat. i., p. 84. Field-Mouse.

Found in the plains. It is a native of all Europe and Western Asia.

39. *Mus prætextus.* Licht. Brants. Muiz. 125.

On the plain of Gennesaret, in the Jordan valley, and Dead Sea basin. Found through Arabia and Syria.

40. *Mus bactrianus.* Blyth. J. A. S. xv. 140. The Sandy Mouse.

Found about villages, and seems, indeed, to take there the place of *Mus musculus.* It was first described from the Punjaub and Affghanistan, and has since been noticed by Blanford throughout Persia.

41. *Mus variegatus.* Licht. Brants. Muiz. 102.

Said to inhabit the wilderness south of Judæa, but has not come under my notice. It is abundant in Egypt and Nubia, and found also in Sinai and Arabia.

42. *Cricetus phæus.* Pall. Zoog. Ross-Asiat. i., p. 163. The Hamster.

The Hamster is very common in the neighbourhood of cultivated ground, attacking especially beans and lentils. It is known from Russia east of the Volga, and throughout Northern Persia and Armenia.

43. *Cricetus nigricans.* Brandt. Bull. Acad. Peters. i. 42.

This Caucasian species was found by Dr. Roth near Lebanon.

44. *Cricetus auratus.* Waterh. Ann. and Mag. N.H. 1839, iv. 445.

I have frequently seen a large light-coloured Hamster about bushes, and making its escape from wild palms; but I never secured it. It cannot be mistaken for the much smaller *C. phœus*, and can be none other than this species, first described from Aleppo, and well known from Syria. I have seen specimens in the museum at Beyrout.

45. *Nesokia.* Sp. (?)

I obtained one specimen of a *Nesokia*, now in the British Museum. Mr. O. Thomas does not assign it positively to any of the Indian species, and in the absence of more examples hesitates to describe it as new. If distinct, it is not far removed from *Nesokia hardwickii*, Gray.

SUB-FAMILY, MERIONINI.

46. *Gerbillus tæniurus.* Wagn. Schreb. Säugth. iii. 471.

Found on Mount Carmel, and in the hill country generally. It is peculiar, so far as is yet known, to Syria.

47. *Gerbillus melanurus.* Rüpp. Mus. Senkenb. iii. 95.

In the Jordan valley and Dead Sea basin. An Abyssinian species, found also throughout Arabia Petræa.

48. *Gerbillus pygargus.* Wagn. Schreb. Säugth. iii. 475.

Occasionally met with in the southern wilderness. Found also in Egypt and Nubia.

49. *Psammomys obesus.* Rüpp. Atlas, p. 58, tab. 22, 23.

Extremely abundant in sandy places about the Dead Sea, and also in the plains and uplands of Southern Judea.

Pl. IV

J. Smit lith. Hanhart imp.

1. ELIOMYS MELANURUS. 2. ARVICOLA GUENTHERI.
3. ARVICOLA NIVALIS.

This Sand-Rat is found throughout the whole of the sandy regions of Barbary and North Africa.

50. *Psammomys myosurus.* Wagn. Wiegm. Archiv. 1848, 183.

This species, described first from Syria by Wagner, appears to take the place of the above in the higher ground. It has not been noticed beyond Syria.

51. *Psammomys tamaricinus.* Kuhl. Beitr. 69.

A small rodent met with beyond the south end of the Dead Sea appears to belong to this species, which is known from South-eastern Russia.

SUB-FAMILY, ARVICOLINI.

52. *Arvicola nivalis.* Blas. Wirbelth. Deutschl., p. 359. Alpine Vole.

PLATE IV. 3.

One of the most interesting discoveries we made was that of the Alpine Vole, which I took close to the snow-line on Mount Hermon in June. The specimen, identical with those from the Alps and Pyrenees, is now in the British Museum. The species was hitherto known only from the Alps, and rarely from the Pyrenees.

53. *Arvicola arvalis.* Gm. Syst. Nat. i., p. 134. Field Vole.

Very common everywhere on cultivated land.

54. *Arvicola socialis.* Desm. Mamm. Sp. 447.

Found in the desert of Sahara near Damascus, and probably also in Southern Judæa, where I have seen more than once a very light-coloured, short-tailed Field Vole.

55. *Arvicola amphibius.* Desm. Mammal., p. 180. Water Vole.

Only observed by us in the north. Found throughout Europe, Central and Western Asia.

56. *Arvicola guentheri.* Alston. P. Z. S., 1880, p. 62.

PLATE IV. 2.

This species, discovered recently by Mr. Danford at Marash in

Armenia, had long existed in the British Museum hidden and unknown.

In a specimen of the large Snake, *Cœlopeltis lacertina*, taken by me in 1863 on the plain of Gennesaret, Mr. O. Thomas lately found, on opening the stomach of the Snake, a perfect example of this Vole, clearly proving its existence in Palestine.

FAMILY, SPALACIDÆ.

57. *Spalax typhlus.* Pall. Zoog. Ross.-Asiat. i., p. 159. Mole-Rat. Heb. חֲפַר פֵּרוֹת. Arab. خلنت, *Khlŭnt.*

PLATE V.

The Mole-Rat is very common throughout the country, where our Mole (*Talpa europæa*) does not exist; nor has any species of the true Mole been found there. It lives chiefly about ruins, and is an exclusively vegetable feeder, attacking onions and other bulbs in gardens. It is much larger than our Mole, eight or nine inches long, and lives in societies in burrows, rarely, if ever, coming to the surface. It is mentioned in Scripture (Isaiah ii. 20).

The Spalax is found throughout the whole of South-eastern Europe, in Southern Asiatic Russia, and Syria.

FAMILY, DIPODIDÆ.

58. *Dipus ægyptius.* Licht. Springmäuse, 19. The Jerboa. Arab. جربوع, *Djerboa.*

The Jerboa is very common in all the desert districts. This species is found throughout North Africa and Arabia. It has not been observed further east than Palestine.

59. *Dipus sagitta.* Schreb. Mamm. iv., tab. 229.

Stated to be found in Syria. Has not come under my notice. It is a Mongolian and Central Asiatic species, with a very wide range.

60. *Dipus hirtipes.* Licht. Springmaüse, 20. Rough-footed Jerboa.

PLATE VI.

Found in the deserts east of Jordan.

J. Smit lith. Hanhart imp.

SPALAX TYPHLUS.

FAMILY, SCIURIDÆ.

61. *Spermophilus xanthoprymnus.* (Bennett. P. Z. S. 1835, p. 90.)

Exceedingly abundant on the sandy and stony plains of the uplands of Moab and Gilead, burrowing generally in the neighbourhood of ruins. It lives in large colonies, and when camped near their warrens, we heard their clear call-note, a sort of whistle, incessantly through the night. They are most difficult to catch, keeping very close to their holes, and dropping into them on the slightest alarm. I never met with this Marmot west of Jordan.

This species of Pouched Rat has generally been confounded with the European; but Alston (P. Z. S. 1880, p. 59) has pointed out its distinctness. It seems to occur through the Southern Steppes, Persia, and Asia Minor.

62. *Sciurus syriacus.* Ehrenb. Symb. Phys. Mam. i. cc. Syrian Squirrel. Arab. سنجاب, *Sinjāb.*

Extremely abundant in woods south of Hermon and throughout the Lebanon. I have never noticed it in the southern part of the country. In habits it is exactly like our Common Squirrel. Its range extends through Syria and Asia Minor.

63. *Sciurus russatus.* Wagn. Schreb. Säugth. iii. 155.

We collected in considerable numbers the variety discriminated as *S. russatus* by Wagner. But I cannot bring myself to believe in its specific value.

FAMILY, MYOXIDÆ.

64. *Myoxus glis.* Schreb. Mamm. iv. 825, tab. 225. Great Dormouse.

Very abundant in the oases of the Jordan valley, especially about Jericho, where it has its nest in every *dôm* tree. It is very lively in winter when disturbed.

This species is found throughout South Europe, from Spain to Greece, in Southern Russia, Asia Minor, the Caucasus, and the countries round the Caspian.

65. *Myoxus nitela.* Schreb. Mamm. iv., tab. 226.

This smaller species seem in Palestine to prefer cultivated ground, living chiefly in the olive groves.

It is a resident in Southern and Central Europe; found further north than the preceding species.

66. *Myoxus dryas.* (?) Schreb. Mamm. iv., tab. 225. B.

There is a third species of smaller Dormouse, differing in its habits from the former, and making its nest in very low thick bushes in the desert regions. There can be, I think, little doubt but that it must be assigned to this species, unless it be new, which is scarcely probable. It is an Arabian species.

67. *Eliomys melanurus.* Wagn. Abhandl. Akad. Wiss. iii. 176.

PLATE IV. I.

This beautiful little desert rodent, a link between the Dormouse and the Squirrel, I twice obtained among ruins on the upland plain of Moab. It inhabits holes in the rocks, and, until we found it, was only known by the two type specimens procured by Von Schubart on Sinai.

NOTE.—By far the richest part of the Fauna of the desert region adjacent to Palestine is its rodents, and it is certain that of these we know the least. Almost all the small Mammals of the stony region south of Judæa, and of the vast sandy and rocky expanses which stretch from Moab to Damascus, are crepuscular or nocturnal in their habits. Explorers are well content if they can by great good fortune shoot or trap a chance specimen; but it is impossible for them to note, still less to study, the habits of these most interesting creatures. The list I have given includes thirty-one species obtained by myself; but I am satisfied it could be very largely increased with time and opportunity, for I have observed many species I could not succeed in capturing, especially Dormice and Hamsters. One might easily be accused of exaggeration in describing the countless number of holes and burrows in regions which for a great part of the year present the features of utter desert. Sometimes for miles a district has the appearance of one vast warren of pigmy Rabbit burrows; yet for days, saving the bounding of a Jerboa here and there before one's horse, not another trace of rodent life is to be seen.

The vast number of these little rodents in apparent desert is explained by the nature of their food. This is chiefly supplied by bulbous roots. The greater part of the desert plants are tuberous or bulbous, and after nine months of utter barrenness, the first winter rains soon carpet the waste with a brilliant spangling of bulbous flowers—crocus, iris, squills, asphodels, cyclamens, and others. Their glory is soon over; but the large succulent roots remain, retaining their moisture through the summer, and affording abundant nutriment to the little burrowers.

ORDER, FERÆ.—CARNIVORA.

FAMILY, FELIDÆ.

68. *Felis leo.* L. Syst. Nat. i. p. 60. The Lion. Heb. אַרְיֵה; לָבִיא, *Old Lion;* כְּפִיר, *Young Lion;* שַׁחַל, *Dark Lion;* לַיִשׁ, *Strong Lion.* Arab. سبع, *Seba'.*

The Lion has long been extinct in Palestine, and among the inhabitants there is no tradition of its existence. Yet of its former abundance there can be no question. It is mentioned about 130 times in Scripture under five different Hebrew names. Within the historic period it was common in Syria, Asia Minor, and Greece. Not only by Homer, but by Herodotus, Xenophon, and Aristotle, it is spoken of as inhabiting Greece in the times of the respective writers. It seems to have disappeared altogether from Palestine about the time of the Crusades, the last mention of it being by writers of the twelfth century, when it still existed near Samaria. Though still found throughout Africa, from the south to the Atlas Mountains, it can scarcely be said now to exist in Asia west of the Euphrates, unless in Arabia, the latest trace being that a few years ago the carcase of one was brought into Damascus. It is still common in Mesopotamia, though becoming rare in India. The Arabs state it is found in Arabia. No specimens from thence have reached Europe, but there seems little reason to doubt its existence. The range of the Lion has become more circumscribed than that of other beasts of prey by the settlement of man, from its habit of resorting to open country. The sculptures of Nineveh abound with illustrations of Lion hunting as the most royal of sports.

The Asiatic Lion was formerly believed to be distinct from that of Africa, but though frequently smaller, its specific identity is now established.

69. *Felis pardus.* L. Syst. Nat. i., p. 61. The Leopard. Heb. נָמֵר. Arab. نمر ; *Nim'r.*

Unlike the Lion, the skulking Leopard still maintains itself in Palestine, though in very small numbers. It is found all round the Dead Sea, in Gilead and Bashan, and occasionally in the wooded districts of the west. I saw a fine pair which had been killed on Mount Carmel. It sometimes lurks near watering-places, to pounce at night upon the cattle. Its ancient abundance in the Holy Land is testified not only by the numerous allusions in Scripture, but also by the frequent occurrence of the word *Nim'r* in the names of places.

The Leopard extends over the whole of Africa, Southern Asia, Japan, China, and the islands of the Malayan Archipelago.

70. *Felis maniculata.* Rüpp. Zool. Atlas, i., p. 1, t. 1. Egyptian Wild Cat. Arab. قط الخلا, *Kot el khla.*

The Wild Cat is scarce west of Jordan, very common on the east side. Some of my specimens are the largest I have ever seen, the body being two feet in length, and the tail eleven inches, more bushy than in the Domestic Cat. This species is the probable original of the Domestic Cat, now as common in Palestine as elsewhere, though not mentioned in ancient times.

The Egyptian Cat is found throughout all Africa, Arabia, and Syria. The Syrian has sometimes been held to be a separate species, but I am unable to recognise its distinctness from Egyptian examples.

71. *Felis chaus.* Güld. Nov. Comm. Acad. Imp. Petrop., t. 20, p. 483, pl. 14. Jungle Cat. Arab. قط بري, *Kot buri.*

This large Cat, somewhat approaching the Lynx in its characteristics,

J. Smit lith.

Hanhart imp.

DIPUS HIRTIPES.

is not uncommon, especially in jungle and thickets, as by the Jordan. It can at once be recognised by its short tail and stout limbs.

The species is found throughout Northern and Central Africa, Western Asia south of the Caspian, and in India.

72. *Felis pardina.* Temm. Monog., vol. i., p. 186. Spotted Lynx.

The southern Spotted Lynx is confined to the wooded regions, where it is very rarely seen, but is well known to the natives, from whom I have obtained skins. These seem to identify it with the *Felis pardina* of Spain and Turkey, and which I have also obtained in the Taurid mountains of Asia Minor. It is a very beautiful animal.

73. *Felis caracal.* Güld. Nov. Comm. Ac. Petrop. xx. 500. The Caracal, or Red Lynx. Arab. عناق الارض, *Anag el ard.*

The Red Lynx is very rare in Palestine. I have seen skins obtained in Lebanon, and in Northern Syria it is not so uncommon. It has a wide range, extending through all Africa, Arabia, Persia, and India.

74. *Felis jubata.* Schreb. Säugth. iii. 392. The Cheetah, or Hunting Leopard. Arab. فهد, *Fahed.*

This graceful Leopard is scarce, but still haunts the wooded hills of Galilee and the neighbourhood of Tabor. East of Jordan it is far more common, and is much valued by the Arabs. It can be at once distinguished from the Leopard by its more slender build, its much longer limbs and tail, and by being dappled with black *spots* instead of rings.

Some writers distinguish between the Indian *Felis jubata* and the African *Felis guttata.* I am unable to recognise the distinction, or to say to which, if they be different, the Syrian Cheetah belongs. One or other species is found throughout Africa and the warmer parts of Asia.

FAMILY, HYÆNIDÆ.

75. *Hyæna striata.* (L. Syst. Nat. i., p. 58.) Striped Hyæna. Heb. צָבוּעַ. Arab. ضبع, *Debaâ.*

The Hyæna is common in every part of Palestine, and indifferent as to the character of the country. The old rock-hewn tombs and innumerable caves afford it convenient covert. Its wail may always be heard after nightfall. It attacks graves, and burrows into them, even in the close vicinity of towns. It is mentioned once in Scripture, but translated 'speckled bird' (Jer. xii. 9). The Hyæna is common through the whole of Africa from south to north, and through Southern Asia, being very abundant in India.

FAMILY, VIVERRIDÆ.

76. *Gennetta vulgaris.* G. R. Gray. P. Z. S. 1832, p. 63. (= *Viverra gennetta.* L.) The Genet. Arab. نسناس, *Nisnas.*

The Genet, well known from Spain and North Africa, is not unfrequent in Palestine. I saw it several times, and procured it on Mount Carmel.

Its occurrence is one of the many instances of the extension of the African Fauna into Syria; for though found in South Europe and along the whole Barbary coast from Tangiers to Egypt, this is the only recorded instance of its existence in Asia.

77. *Herpestes ichneumon.* (Fischer. Sym. Mamm., p. 163.) The Ichneumon. Arab. ذردى, *Zerdi.*

The Ichneumon is found among the scrub on the borders of all the cultivated plains, living in the rocks. It is very common. It was scarcely possible to take a walk at sunrise in any part of the country without meeting this little animal trotting away to its hole. Its range extends through North Africa. In Asia, excepting in Syria, its place is taken by representative species.

FAMILY, CANIDÆ.

78. *Canis lupus.* L. Syst. Nat. i., p. 58. Wolf. Heb. זְאֵב. Arab. ديب, *Deeb.*

The Wolf is still common in Palestine, but from the nature of the country and the scarcity of cover or woods, its habits are very different from those of the same species in Europe. It cannot here be considered gregarious, though two may often be found together. Probably from the abundance of food, it is rather larger and stronger than European specimens. It lurks during the day among the rocks, and prowls at night about the sheepcotes. But when disturbed during the day it is bold and fearless. Its habits are often referred to in Scripture. It is found in every part of the country. The range of the Wolf is through Europe and Northern Asia. Palestine is probably its South-eastern limit.

79. *Canis aureus.* L. Syst. Nat. i., p. 59. Jackal. Heb. שׁוּעָל· אִיִּים. Arab. واوي, *Wawi.*

The Jackal is most abundant in every part of the country. Round towns, in villages, fields, or wilderness, its voice is nightly heard, as it hunts in packs, wailing and howling. Ancient ruins are its special resort. It is often mentioned in Scripture as *Shu'al,* always in our version translated 'Fox.'

The Jackal is found in all the warmer regions of the Old World, from the Mediterranean countries, through the whole of Africa and all Southern and Central Asia.

80. *Canis familiaris.* L. Syst. Nat. i., p. 56. The Dog. Heb. כֶּלֶב. Arab. كلب, *Kelb.*

There is only one race of Dogs common in Palestine, the Pariah Dog, very similar in size and appearance to our Colley, or Scotch Sheep-Dog. Of these there are two classes—the Pariah Dogs of the town, which support themselves by offal as best they may, and live in small communities; and the same breed, belonging to private owners, generally to the shepherds, and used for guarding the flocks. No Dog surpasses the Pariah in instinct and intelligence, neglected and degraded though it be.

Dogs are rarely used in Palestine for hunting, excepting the large Persian Greyhound, with long silky hair on the ears, and a long fringe of the same on its tail. It is highly prized by the Bedawin, and is trained to chase the Gazelle.

The Dog is found in every country of the world, hot or cold, where man exists.

81. *Vulpes nilotica.* Rüpp. Atlas, p. 41, taf. 15. Egyptian Fox. Heb. שׁוּעָל. Arab. ثعالب, *Taalib.*

The Common Fox of the southern and central regions of Palestine, extremely abundant in Judæa and east of Jordan. In its habits it is very distinct from the Jackal, and in no way different from our Fox, which is somewhat larger. The Nilotic Fox ranges through Egypt, Arabia, and the Syrian desert.

82. *Vulpes flavescens.* Gray. Ann. and Mag. N.H. 1843, xi., p. 118. Tawny Fox.

This is the species of the wooded districts of Galilee and the north. It is considerably larger than the last species, and differs from the English Fox, of which perhaps it is only a local race, by its peculiarly bright light yellowish colour throughout, and finer and longer fur. It has black ears, and a splendid brush.

The Tawny Fox ranges from Syria to Central Asia, and the north side of the Himalayas.

FAMILY, MUSTELIDÆ.

83. *Putorius fœtidus.* Gray. P. Z. S. 1865, p. 108. Polecat. Heb. חוֹלֶד. Arab. فار الخايل, *Far el kheil.*

Occasionally found in the north, under Hermon and Lebanon. It extends through all Europe and Northern Asia.

84. *Mustela boccamela.* Bechst. Naturg. Deutschl. p. 819. Southern Weasel. Arab, سمعور, *Sammur.*

About Mount Tabor, and probably in other wooded districts. Extends through South Europe and the Mediterranean coasts of Africa and Asia.

85. *Martes foina.* (Gm. Syst. Nat. i., p. 95.) Marten.

I bought a skin at Beyrout, taken in the neighbourhood. It is found in the Taurid, as well as through the whole of the rest of the Palæarctic region.

86. *Lutra vulgaris.* Erxl. Syst. p. 488. The Otter. Arab. كلب الما, *Kelb el ma.*

The Common Otter was found by us only on the shores of the Lake of Galilee, where it has abundant food. We also heard its bark, but did not procure it, when camped by the seashore, close to the mouth of the Litany river.

The Otter extends from the British Isles and North Africa to the Himalayas.

FAMILY, MELINIDÆ.

87. *Meles taxus.* Schreb. Säugth., t. 142. Common Badger. Arab. عناق الرغر, *'Anak u'lard.*

The Badger is very common in all the hilly and wooded parts of the country, but has not been observed in the Jordan valley.

The Badger extends through the whole of Northern Europe and Northern Asia, but has not been found in North Africa. Palestine seems to be the southward limit of its range.

The words translated in the Old Testament 'Badgers' skins' (*oroth Techashim*), refer not to this animal, not procurable either in Egypt or in the wilderness of Sinai, but to the Dugong, *Halicore hemprichii*, common among the coral banks of the Red Sea, and pretty clearly identified by its Arabic name *Tucash*, the equivalent of the Hebrew *Tachash.*

FAMILY, URSIDÆ.

88. *Ursus syriacus.* Ehrenb. Symb. Phys. i. a, pl. I. Syrian Bear. Heb. דוב, Arab. دب, *Dub.*

The Bear has become very rare in Palestine, though still not uncommon on Hermon and the wooded parts of Lebanon. I only once saw it in Galilee, in a ravine near Gennesaret. It still exists on the east side of Jordan, in Gilead and Bashan. From the frequent references in Scripture, we see how familiar the Bear must have been to the ancient inhabitants, in the days when the Judean hills were still clothed with wood, and the primæval forests crowned the rugged heights of Galilee.

The Syrian Bear differs but slightly from the Brown Bear of Europe, *Ursus arctos*, and is still more closely allied to the Indian *Ursus isabellinus.* Horsf. It is found in Lebanon, the Taurid range, and Northern Persia.

ORDER, INSECTIVORA.

FAMILY, SORICIDÆ.

89. *Sorex araneus.* De Selys. Micromm. 18. Shrewmouse. Heb. אֲנָקָה. Arab. (*generic*) فار الغلا, *Far el kla.*

Frequently found dead in Northern Palestine, in the hilly districts.

Is a native of all parts of Europe except Scandinavia and Britain, and extends into the Caucasus and Armenia.

90. *Sorex tetragonurus.* Desm. Mamm. Sp. 234.

In the north, on Lebanon.

The Common Shrew of England, and inhabits the whole of Europe.

91. *Sorex pygmæus.* De Selys. Schinz. Eur. Faun. i., p. 27.

I obtained a specimen among the cliffs below Mar Saba, near the Dead Sea. It is known as a native of Central and South-eastern Europe, North Africa, and the countries bordering on the Caspian.

92. *Sorex crassicaudus.* Licht. Darstellg., t. 40, f. 1.

This pretty silver-grey species is found in the desert and southern ravines. It is also a native of Arabia, Egypt, and Abyssinia.

93. *Sorex fodiens.* Schreb. Säugth. p. 571. Water-Shrew.

I have several times seen, but did not procure, a Water-Shrew, both on the banks of the Wâdy Kurn and by the Litany river. It is most probably this species, known from Western Asia, as well as universal, though not plentiful, in all parts of Europe.

94. *Sorex.* Sp. (?)

I obtained a sixth species of Shrewmouse, which, I am unable to decide, the specimen being in a bad state.

FAMILY, ERINACIDÆ.

95. *Erinaceus brachydactylus.* Wagn. Schreb. Säugth. ii., 22. Hedgehog. Heb. קִפּוֹד. Arab. قنفود, *Kanfood.*

This species, rather smaller than our European, is common in the south of Palestine. It is the Common Hedgehog of Egypt, and very closely allied to *E. auritus*, Pall., from the Caucasus and Taurid.

96. *Erinaceus europæus.* L. Syst. Nat. i., p. 75. Common Hedgehog.

The Hedgehog of Lebanon and Hermon appears to agree with the European species. The Hedgehog in Scripture, under the name *Kippod*, translated erroneously in our version 'Bittern,' is mentioned several times. The Hebrews, like the Arabs, included the Porcupine and Hedgehog under the same name.

This species is found throughout the whole of Europe. That of North Africa scarcely differs from it.

ORDER, CHEIROPTERA.

FAMILY, PTEROPODIDÆ.

Bats.

Heb. עֲטַלֵּף (*generic*), Arab. واتواط, *Wat-wat* and طير الليل, *Their ellil.*

97. *Cynonycteris ægyptiaca.* Geoffr. Descript. de l'Egypte. ii., p. 135, pl. iii., fig. 3.

This fruit-eating Bat is found in large colonies in the caves of the

4

Wâdys in Northern Palestine, especially near the Plain of Acre, as in Wâdy Kurn. Those from Kurn are much smaller than the specimens from the hills behind Tyre, which are of the same size as those from Egypt and Cyprus.

The length of body varies from 4·2 inches, Wâdy Kurn, to 7·9 inches near Tyre, and expanse from 13 inches, Wâdy Kurn, to 22 inches Tyrian specimen.

The range of this Bat extends through Abyssinia, Egypt, Palestine, Northern Syria, and Cyprus.

Contrary to the usual arboreal habits of this genus, this species has in Palestine been found only in caves.

FAMILY, RHINOLOPHIDÆ.

98. *Rhinolophus euryale.* Blas. Archiv für Nat. i., p. 49.

Found in great numbers in the caves of the Wâdys opening on the Lake of Galilee, also in tombs behind Tyre.

It ranges through South Europe, North Africa, and Asia Minor.

99. *Rhinolophus blasii.* Peters. M. B. Akad. Berl. 1866, p. 17.

Abundant about Jerusalem, Bethlehem, and Hebron.

Found also in South Europe and North Africa.

100. *Rhinolophus ferrum-equinum.* Schreb. Säugth. i., p. 174.

The commonest Bat in Palestine, in all parts of the country.

Its range extends in Europe from England southwards through the whole of Africa, from Algiers to the Cape, and across Asia from Syria to the Himalayas and Japan.

101. *Phyllorhina tridens.* Geoffr. Descr. de l'Egypte. ii., p. 130.

Occurs in caves in the Dead Sea basin.

A common Egyptian species, and found as far south as Zanzibar.

FAMILY, VESPERTILIONIDÆ.

102. *Plecotus auritus.* (L. Syst. Nat. i., p. 47.) The Long-Eared Bat.

Very common in all the hill-country of Palestine, especially in caves and tombs about Bethlehem and Jerusalem, and by the Sea of Galilee.

This well-known Bat has a very wide distribution. It ranges from Ireland throughout Europe; and on the south of the Mediterranean, through Egypt and Syria as far as Nepal.

103. *Vesperugo serotinus.* Schreb. Säugth. i., p. 167, pl. 53. The Serotine.

Occurs in Lebanon.

Found throughout the whole of the northern hemisphere, the only Bat common to the Old and New Worlds. It ranges from England to Siberia, through Africa as far south as Gaboon; Arabia, Asia Minor, Persia, India; and in America from Lake Winnipeg to Guatemala and the West India Islands.

104. *Vesperugo kuhlii.* Natt. Deutsch. Fled. Ann. Wett. iv., p. 58.

Met with at Jerusalem, in caves near Bethlehem, and elsewhere.

Extends through Southern Europe, North Africa, Persia, and India, from north to south.

105. *Scotophilus temminkii.* (?). Horsf. Zool. Researches in Java (and plate). (Horsfield's work is neither paged nor numbered.)

I procured a Scotophilus under Mount Carmel, on the plain of Acre, which I believe belongs to this species, which occurs throughout India, South China, Java, and the Philippines. But it may possibly be the allied African species, *S. borbonicus.* Being much damaged, it was not easy to discriminate it.

106. *Vespertilio daubentonii.* Leisl. Deutsch. Fled. Ann Wett. Ges. Nat., p. 51.

In a tomb near Tibnin, Galilee. Found from the furthest north of Europe to Siberia, and south as far as Tenasserim.

107. *Vespertilio emarginatus.* Geoffr. Ann. du Muséum, vol. viii., p. 198. Notch-eared Bat.

In large colonies in tombs behind Tyre and on Mount Carmel. This species belongs to Southern and Central Europe, from France to Italy. It has not hitherto been noticed further east, except in Palestine.

108. *Vespertilio murinus.* Schreb. Säugth. i., p. 165, pl. 51. Mouse-coloured Bat.

Obtained at Tyre and Beyrout—not noticed inland. Its range is through Europe, from England and Denmark southwards; North Africa as far as Abyssinia; and also in India.

109. *Vespertilio mystacinus.* Leisl. Deutsch. Fled. Ann. Wett. Nat. iv., p. 55. Whiskered Bat.

In Southern Lebanon. Found throughout the whole of Europe, and has also been received from the Himalayas and from Pekin.

110. *Miniopterus schreibersii.* Natt. Deutsch. Fled. Ann. Wett. iv., p. 41.

In caves overhanging the Jordan valley. The specimens are of a very pale grey colour. Widely distributed through Southern Europe; through Africa from Algiers to the Cape and Madagascar; in all Southern Asia from Syria to Japan and the Philippines, the Malay countries; New Guinea, and Australia.

FAMILY, EMBALLONURIDÆ.

111. *Taphozöus nudiventris.* Cretzsch. Rüpp. Atlas. Reise. Nord. Afrik. Säug., p. 70, fig. 27 b.

We found this Bat in myriads in the caverns of the ravines opening on to the Sea of Galilee.

This and the following species belong to a tropical family of Bats, and are the only two species of the family found in so northern a region as Palestine. But even here they are confined to the subtropical ravines of the Jordan valley. The present species ranges from the Gambia to Egypt, and has also been brought from the Euphrates.

112. *Rhinopoma microphyllum.* Geoffr. Descript. de l'Egypte. ii., p. 123.

Swarms in the caves of the Jordan valley and the Dead Sea basin, and especially by the Callirrhöe (Zerka Maîn) on the east side. Like the preceding species, this Bat is tropical, belonging to the Indian Peninsula and Burma, out of which region it has only been found in Egypt, Kordofan, and the Jordan valley.

Both species are remarkable and peculiar, in having large deposits of fat at the base of the tail, laid on before the period of hybernation.

113. *Nyctinomus cestonii.* Savi. Nuov. Giorn. de' Lett., p. 230.

In caves over the Jordan valley. Found also in Southern Europe, in North-east Africa, and in China.

AVES.

ORDER, PASSERES.

FAMILY, TURDIDÆ.

1. *Turdus viscivorus.* Linn. Syst. Nat. i., p. 291. Missel-Thrush.

The Missel-Thrush barely claims a place in the Fauna of Palestine. I once saw it in winter, in a wooded district of Lebanon, but I have seen the young soon after leaving the nest on the southern spurs of the Taurid, near Marash.

It is found throughout Europe, the Barbary States, Asia Minor, the Caucasus, and the North-west Himalayas.

2. *Turdus musicus.* Linn. Syst. Nat. i., p. 292. Song-Thrush.

The Thrush is not uncommon in winter in the higher ground. I occasionally noticed it in the wooded districts of Galilee in spring, but never found the nest.

The Song-Thrush is found through all Europe, and in winter in North Africa. It also inhabits Western Asia and Siberia, and has been obtained in Northern China.

3. *Turdus pilaris.* Linn. Syst. Nat. i., p. 291. Fieldfare.

The Fieldfare is very rare in Palestine, but is occasionally met with in winter. I found one close to Jerusalem in February.

It is a bird of northern range, breeding in Northern Europe and Asia, and descending in winter, though rarely, as far as the Mediterranean countries, and to Cashmere in the east.

4. *Turdus merula.* Linn. Syst. Nat. i., p. 295. Blackbird.

The Blackbird is a permanent resident, scattered in every part of the country, and remaining to breed, even in the sultry Ghôr. It is nowhere abundant, perhaps one of the very rarest of all the resident species, and is one of the most retiring and shy of the inhabitants of the thickets.

The range of the blackbird is throughout all Europe, except its Arctic borders, and all North Africa, the Azores and Madeira. It does not appear to roam eastward of the Ural Mountains, but is found in Asia Minor, Syria, and sometimes, in winter, in Persia. The species of Affghanistan is distinct, and has been named *Merula maxima.*

5. *Monticola cyanus.* Linn. Syst. Nat. i., p. 296. Blue Thrush.

The Blue Thrush is, in the localities where it occurs, one of the most conspicuous birds of the country. In all its habits it is very different from the rock thrush. It resides throughout the year, singly or in pairs, among rocky wâdys, in ruins, and especially in cliffs by the sea-shore. It is in the habit of perching on conspicuous ledges, and does not avoid villages where there is a ruined keep, on the top of which it perches, uttering its somewhat monotonous song. Its breeding places are niches in caves or in the vomitoria of Roman amphitheatres. On the shore it feeds on shrimps. It is supposed to be 'the Sparrow that sitteth alone on the house-top' of Scripture.

The Blue Thrush ranges throughout Southern Europe, from the Pyrenees and Alps, through Northern and North-eastern Africa as far as Abyssinia, and eastward as far as Yarkand and North-western India. Beyond these Eastern limits its place is supplied by the Manilla Thrush, *Monticola solitaria.*

6. *Monticola saxatilis.* Linn. Syst. Nat. i., p. 294. Rock Thrush.

The Rock Thrush is in Palestine, south of Lebanon, only a passing traveller, tarrying but a night. It arrives in large flocks in the beginning of April, hopping rather than flying over the country, as it progresses northwards. I saw one of these flocks on the 8th April, passing over Mount Gerizim. In 1881 I saw another similar flock near Damascus. In Lebanon and Hermon it remains in pairs, and breeds on the bare rocky hills.

The Rock Thrush is a summer visitor to Southern Europe and Northern Africa, and spreads through Central Asia to Cashmere, Yarkand, Turkestan, as far as the Pekin Mountains.

7. *Saxicola œnanthe.* Linn. Syst. Nat. i., p. 332. Wheatear.

The Wheatear is only seen in Southern Palestine at the periods of the spring and autumn migrations. In the hill country of Galilee, and in the mountain ranges of Lebanon and Hermon, the Wheatear breeds and remains till autumn. The first specimens I noted were on 19th March, on Mount Carmel. On Mount Hermon I found it breeding in large numbers, close to the snow, in the beginning of June. Many of the Palestine specimens are exceptionally large and very bright in plumage, and were differentiated by Ehrenberg under the name of *Saxicola rostrata.* But, with a series of Palestine specimens before one, it is impossible to draw the line.

The Common Wheatear is the most widely distributed of its family, being found from Greenland and Iceland throughout all Europe and North Asia, and across Behring's Straits in Alaska. Southwards it extends from the Azores and Canaries to Kordofan and Abyssinia, and as far as the frontier of North India and Northern China.

8. *Saxicola isabellina.* Rüpp. Atlas, p. 52, pl. 34. Isabelline Chat or Menetries' Wheatear.

The Isabelline Chat resides throughout the year in Southern Palestine. In the north it visits the lower ranges of Hermon and Lebanon to breed, nesting about an hour or two's walk lower down than the Common Wheatear, and soon withdrawing with its young to the plains, where it is very numerous. In North Syria and Mesopotamia it is the most abundant of all the Passerine birds.

This bird only touches Europe in South-Eastern Russia, but extends through North-east Africa, Asia Minor, and Arabia to Persia, North India, Siberia, and North China.

9. *Saxicola aurita.* Temm. Man. d'Orn. i., p. 221. Eared Chat.

The Black-eared Chat returns to the Holy Land about the third week in March, always a few days later than its congener, *Saxicola melanoleuca,* and immediately spreads itself by twos and threes all over the plain country, affecting especially open tillage and cultivated fields. It is found in the same localities as its ally, but they never interbreed, though the habits, note, nest and eggs are precisely similar. Most of the immigrants arrive before they have assumed their bright black and white breeding dress, but in a few days they lose all traces of brown and grey.

I should be inclined myself to agree with Ehrenberg, and separate the Eastern form as *S. amphileuca*, never having met with the russet hue so characteristic of Western specimens ; but I am assured that such occur in Persia.

The Black-eared Chat is found in all the countries bordering on the Mediterranean, and in Persia, which appears to be its Eastern limit.

10. *Saxicola melanoleuca.* Güldenst. Nov. Com. Petr. xix., p. 468, pl. 15. Black-throated Chat.

This is the Eastern form of the Stapazine Chat of Western Europe, and returns to Palestine for nidification about 16th March. It is very numerous, and universally distributed in the lower and cultivated grounds, and less abundantly on the hills. On its first arrival it still wears the tawny hue of the Stapazine, but in a very few days the head and back become silvery grey, and then a pure silvery white, when the bird forms a conspicuous feature of the landscape, perched on the tops of the thistles and tall weeds of the plains. I never found a trace of buff on the breeding birds.

This species is found in North-East Africa and South-East Europe, Asia-Minor, and Syria. It occurs in Persia, and has been met with in Yarkand, its Eastern limit. It appears to winter south of Egypt.

11. *Saxicola deserti.* Temm. Pl. Col., pl. 359, fig. 2. Desert Chat.

The Desert Chat is only found in the desert portions of Palestine, especially among the sand wastes north and south of the Dead Sea, and south of Beersheba, in all which it is a permanent resident, nesting in the holes of Jerboas and other desert rodents.

As its name implies, it is a strictly desert form, ranging from the Sahara, through the desert regions of Egypt, Arabia, Persia, and Scinde, as far as Cawnpore. It is also found in Affghanistan, the Southern Kerghis Steppes, and Yarkand.

12. *Saxicola finschii.* Heugl. Orn. N.O. Afrik. i., p. 350. The Palestine Chat.

This bird, the characteristic Chat of Palestine, was discovered by me in 1863, but erroneously identified with *S. libanotica* (Hemp. and Ehrenb.), and, six years afterwards, was described by Heuglin from an Egyptian specimen. It is the only chat, besides *S. lugens*, which remains in the hill country in winter, scattered everywhere, but always solitary; very conspicuous with its white body and black wings.

Palestine is really its native country, but it is scattered, though very scantily, over Western Persia, Arabia, Egypt, and Algeria, in all which countries it is extremely rare.

13. *Saxicola mæsta.* Licht. Verz. Doubl., p. 33.
S. philothamna. Tristram. *Ibis*, 1859, p. 58. Tristram's Chat.

Tristram's Chat is one of the best known of its genus, occurring only in scrubby desert plains in regions far apart from each other. In Palestine it has been found only on the rolling plains south of Judæa: in Africa in the dayats of the Sahara. It has been found also in Egypt and Arabia. It breeds in the desert, making use of lizard or jerboa holes under the roots of bushes.

14. *Saxicola lugens.* Licht. Verz. Doubl., p. 33. Pied Chat.

The Pied Chat is very common throughout the year in the rocky regions overhanging the Jordan valley, in the Judæan wilderness, and on the highlands of Moab. I have never found it further north. Where it is found it affects the same districts as the Palestine Chat (*S. finschii*), but may be at once distinguished by its black back and buff under tail-coverts.

The headquarters of the Pied Chat appear to be Palestine, but it is also found in Arabia, Egypt, Algeria south of the Atlas, and occasionally in Nubia.

15. *Saxicola leucomela.* Pallas. Nov. Com. Petr., xiv., p. 584. Eastern Pied Chat. (=*S. morio.* Hemp. and Ehr. Symb. Phys., fol. aa.)

The Eastern Pied Chat is rare in Palestine, and as yet has only been found in the north, where the Pied Chat does not seem to occur. It breeds in the neighbourhood of Beyrout.

The range of the Eastern Pied Chat very slightly overlaps that of its congener. Eastward it extends through Cyprus, the Caucasus, Persia, North India, Mongolia, to North-West China. Westward it reaches in winter as far as Abyssinia.

16. *Saxicola monacha.* Temm. Pl. Col., pl. 359, fig. l. Hooded Chat.

The Hooded Chat is very limited in its range, and within that range is represented by very few individuals. It is to be found sparsely scattered over the salt hills and wastes at the south of the Dead Sea and the Arabah. Among the marl hills between Sebbeh and Jebel Usdum its presence is the only sign of life to be seen.

It has been taken in Egypt, Nubia, and the Sinaitic Peninsula, always in desolate districts; and eastward it is found in Baluchistan, but not in Scinde. This species is the most graceful and elegant of all the Chats. It is indeed the sylph of the family.

17. *Saxicola leucopyga.* Brehm. Vögelfang, p. 225. White-rumped Chat.

The White-rumped Chat, which has often been separated into two species, from the fact of its not acquiring the white head until its second moult, is found in Palestine only on the higher rocky ground of the ravines and wilderness on either side of the Dead Sea, where it is a constant resident.

The White-rumped Chat is found in the Algerian Sahara, Nubia, and Arabia. It is stated to have occurred at Sierra Leone, and probably ranges to the southern fringe of the Sahara.

18. *Cercomela melanura.* Temm. Pl. Col. No. 257, fig. 2. Black Start.

The Black Start is one of the most striking and characteristic birds of the bare ravines opening on to the Dead Sea and Jordan valley. Beyond these limits, in which it is sedentary, we never saw it. It has all the habits of a Stonechat, not of a Wheatear, perching on bare twigs, jerking and expanding its black tail, and not alarmed at the presence of intruders. Among the sparse desert shrubs and rocks, but not in the clumps or fertile oases, is its home. There is no distinction whatever in plumage between the sexes.

The Black Start is found in similar localities to his Palestine resorts throughout Arabia, and in Egypt, Nubia and Abyssinia, but not further east or south.

19. *Pratincola rubetra.* Linn. Syst. Nat. i., p. 332. Whinchat.

The Whinchat is not common in Palestine, and I believe only passes through the country at the season of migration.

This well-known bird is spread throughout the whole of Europe, even to its northernmost point in summer ; and in Africa is found as far south as the Gambia, Fantee, and Abyssinia, in winter. It does not enter far into Asia, Asia Minor being its ordinary limit, though occasionally procured in the Punjab.

20. *Pratincola rubicola.* Linn. Syst. Nat. i., p. 322. Stonechat.

The Stonechat is very common, scattered over the whole country in winter, but appears to leave for the north in April, returning in October.

This bird has a wide range throughout all Europe, except Scandinavia and Northern Russia, North Africa, and throughout Asia, to China and Japan, unless we separate the Indian form, as, in common with many naturalists, I am disposed to do, under the name of *P. maura* (Pallas. Reis. II. Anhang., p. 708). This race occupies India, Eastern Siberia, China and Japan. But if the species are distinct, the Palestine bird belongs to the Western form.

21. *Ruticilla phœnicurus.* Linn. Syst. Nat. i., p. 335. Redstart.

The Common Redstart is a very abundant summer migrant, arriving simultaneously all over the country the second week in March, but remaining to breed only in the more wooded localities and the neighbourhood of gardens. The Redstart is found in summer throughout the whole of

Europe. In North Africa and Egypt it is also only a summer migrant, wintering in Central Africa. Eastward it occurs as far as Persia.

22. *Ruticilla mesoleuca.* Ehr. Sym. Phys. Aves, fol. ee. Ehrenberg's Redstart.

I have found this species in the same places and at the same time as the closely allied Common Redstart. It is a summer migrant, and has been procured on Mount Carmel, among the oaks of Bashan, and on the plains of Moab.

The limits of this species are very restricted, viz., Asia Minor, the Caucasus, and Syria in summer; Arabia and Abyssinia in winter.

23. *Ruticilla semirufa.* Ehr. Sym. Phys. Aves, fol. bb. Palestine Redstart.

This bird is extremely close to the Indian *R. rufiventris,* but differs in its smaller size and black under-wing coverts. It resides in the Lebanon and Hermon, and has been found nowhere else. In Lebanon, though not uncommon, it has escaped the notice of all naturalists, excepting Ehrenberg. Whether it migrates, beyond descending and ascending the mountain ranges according to the season, is not yet ascertained. I found it sitting on eggs as late as the 26th of June, under the cedars of Lebanon.

So far as our present knowledge extends, this is one of the isolated forms peculiar to Palestine, and separated from its closely allied relative by the vast extent of continent between Syria and India.

24. *Ruticilla titys.* Linn. Syst. Nat. i, p. 335, 34 β. The Black Redstart.

The Black Redstart is, during winter, one of the most common and conspicuous birds on the rocky hills of Palestine, especially near the sea coast. It is partially migratory, ascending in spring to the spurs of Lebanon and Hermon, very few remaining further south to breed.

The Black Redstart inhabits Central and Southern Europe, North Africa, the Caucasus, and Persia.

25. *Cyanecula suecica.* Linn. Syst. Nat. i., p. 336. Red-spotted Blue-Throat.

This Blue-throat is widely distributed in winter throughout the marshy plain, but in very small numbers, generally consorting with the pipits.

I found it till the end of April, in 1881, and noticed it through the winter as more numerous than on my previous visits.

Though they meet in Palestine and in North Africa in winter, the summer range of the two Blue-throats is very different. This species passes annually over Heligoland, and breeds in Scandinavia and North-east Europe, wintering in North Africa and as far south as Abyssinia. In Asia it goes to India, Ceylon, and China.

26. *Cyanecula wolfi.* Brehm. Beit. zur Vögelk. ii., p. 173. White-spotted Blue-throat.

The White-spotted Blue-throat is by no means so common in Palestine as the last species, but is occasionally to be found in winter, generally on the maritime plains, with the habits characteristic of the pipits, with which it consorts. We never noticed it after April.

The White-spotted Blue-throat is an inhabitant chiefly of central Europe, breeding especially in Holland. It is common on passage in Spain, and occurs in winter along the North African coast, but has rarely strayed further east than Palestine.

27. *Erithacus rubecula.* Linn. Syst. Nat. i., p. 337. Robin.

Our familiar Redbreast is scattered everywhere by ones and twos throughout the country in winter, singing lustily when all else is silent, and as bold and familiar as at home. But not one remains after the end of February.

The Robin is spread throughout Europe, and is generally a migratory bird. In North Africa it spends only the winter. It breeds in the Caucasus, winters in Asia Minor and Syria, and has been taken in Persia. Westward it extends to the Madeiras and Canaries.

28. *Erithacus gutturalis.* Guérin. Rev. Zool., 1843, p. 162. White-throated Robin.

PLATE VII.

This very remarkable and beautiful bird is one of those which connect the ordinary Palæarctic fauna of Palestine with that of more tropical climes. There, as everywhere else where it occurs, it appears to be very scarce. We first discovered it on Hermon, among the vineyards near its northern base, and afterwards on Lebanon. Last year, in travelling through

Pl. VII.

J. Smit lith.

Hanhart imp.

ERITHACUS GUTTURALIS.

some wooded defiles in Southern Armenia, east of the Euphrates, I had several opportunities of noticing its habits, which are very Robin-like when in the bushes, but Chat-like on the open. The male has a magnificent bell-like note, not much inferior to the Bulbul, which it pours forth, perched among the thickest foliage. The White-throated Robin has a very limited range. It is found on the Abyssinian coast, in Asia Minor, Palestine, and Persia, always rare, among bushes in rocky valleys.

29. *Erithacus luscinia.* Linn. Syst. Nat. i., p. 328. Nightingale. Arab. بلبل, Bulbul (generic for fine songsters).

The Nightingale returns to Palestine in moderate numbers about the middle of April, and remains to breed, frequenting especially the fringe of trees by the banks of the Jordan. I have found it also on Tabor and in various wooded wadys. I never observed or heard it in the olive groves.

The Nightingale summers in Western and Central Europe, England being its Northern limit, and passes southward into Africa to winter. Asia Minor, Cyprus, and Palestine appear to be its eastward boundary.

30. *Erithacus philomela.* Bechst. Gem. Nat. Deutschl. iv., p. 536. Eastern Nightingale.

This bird was not obtained by me in Palestine; but as it is a native of Eastern Europe and Northern Asia, including the Caucasus and Asia Minor, and winters in Egypt, it is scarcely possible for it to escape passing through Palestine, where no doubt careful research will discover it.

FAMILY, SYLVIIDÆ.

31. *Sylvia cinerea.* Bechst. Orn. Taschenb. i., p. 170. White-throat.

Our familiar White-throat is very abundant everywhere, and remains through the year, though its numbers are considerably increased in spring. It is a very early breeder, its eggs being found from the beginning of March to the end of May. It seems equally at home in every part of the country.

The White-throat inhabits in summer the whole of Europe and Western Asia as far as Persia, wintering in the Mediterranean countries and in Africa, where it is found as far south as Kordofan and the Gold Coast.

32. *Sylvia curruca.* Linn. Syst. Nat. i., p. 329. Lesser White-throat.

The Lesser White-throat is only a summer visitant to Palestine, generally returning in great numbers in March, and breeding all over the country; but it remains throughout the winter in the warm nooks and glens surrounding the Dead Sea, especially on the eastern side. On Lebanon it was breeding as late as the middle of June.

The Lesser White-throat is a bird of most extensive range, visiting the whole of Europe and Northern Asia in summer, and Northern and Central Africa and India in winter. The Eastern race, which is slightly larger has, however, been separated as *Sylvia affinis*, Blyth. Palestine specimens are intermediate in size.

33. *Sylvia subalpina.* Bonelli. in Temm. Man. d'Orn. i., p. 214. Subalpine Warbler.

The Subalpine Warbler was obtained by me once near Mount Tabor towards the Jordan. It is very shy, and probably not so rare as has been supposed.

This bird is found in small numbers in all the countries bordering on the Mediterranean, frequenting thick low scrub. It has been procured in Kordofan, and on the Gambia.

34. *Sylvia conspicillata.* Marmora. fide Temm. Man. d'Orn. i., p. 211. Spectacled Warbler.

This bird is very generally spread, but in small numbers, over the bare highlands of Judea, and on the plains of Jordan, throughout the year, in dry stony places. Its nest, in a low bush, is a very neat open structure, and its habits are those of the Lesser White-throat.

The Spectacled Warbler inhabits all the countries bordering on the Mediterranean. Westward it extends to the Canaries, and eastward to Mesopotamia. It is abundant in the Sahara, but has not been noticed in Egypt.

35. *Sylvia melanothorax.* Tristram. *Ibis,* 1872, p. 296. Palestine Warbler.

One pair of this new Warbler, the type specimens, were obtained by me at Engedi, close to the shore of the Dead Sea, on 2nd February, 1872. I never observed any other individuals. Since then Lord Lilford has obtained several specimens in Cyprus in the month of May, and one has been procured on the coast of Phœnicia by Dr. Van Dyk, of Beyrout. No other specimens are known. It can be at once distinguished from the next species by its black throat and upper breast.

36. *Sylvia melanocephala.* (Gmel. Syst. Nat. i., p. 970.) Sardinian Warbler.

The Sardinian Warbler is very generally distributed in Palestine, and remains throughout the year in the scrub on the sides of the upland wadys, secreting itself after the manner of our Wood-Wren in the bushes.

The Sardinian Warbler is found in all the districts bordering on the Mediterranean, only migrating to a limited extent, and never wandering from the Mediterranean basin.

37. *Sylvia bowmani.* Tristram. *Ibis,* 1867, p. 85. Bowman's Warbler.

Very little is known of this Warbler, which I obtained in various parts of the country. It is not a migrant, and is very like the Sardinian Warbler in general appearance, but differs in various details, especially in the shortness of the tail, and in the iris being lemon yellow.

It was first obtained by Ehrenberg, who named it *Curruca momus,* but the description would be wholly unintelligible without the type. It has been supposed to be the same as the *Sylvia mystacea* of Menetries (Cat. Raisonn., p. 34), but I cannot agree with the identification for several reasons, and therefore retain the name of *Sylvia bowmani.* The bird has only certainly been found out of Palestine in Persia, and probably also in North-east Africa. It is the *Sylvia rubescens* of Blanford (*Ibis,* 1874, p. 77).

38. *Sylvia orpheus.* Temm. Man. d'Ornith. i., p. 198. Orphean Warbler.

A summer visitant to Palestine, returning in the beginning of April, and affecting chiefly the groves and olive-yards of the northern part of the country.

It lays its eggs about the beginning of June, and under Hermon is very plentiful. The Indian form is distinguishable by its larger size and bigger bill, and named *S. jerdoni* (Blyth, J. R. S. Bengal, xvi., p. 439). But both forms occur and interbreed in Palestine.

The Western form spends the summer in the Mediterranean countries, and the winter on the Gambia and in Central Africa. The Eastern form breeds in Persia and Turkestan, and winters in India and Arabia.

39. *Sylvia rueppellii.* Temm. Pl. Color. iii., 245, fig. 1. Rüppell's Warbler.

This beautiful Warbler is scarce, but generally distributed in Palestine. I have found it on Carmel, in Judæa and Gilead, and on Lebanon. It appears to reside permanently in the warmer parts of the country.

Rüppell's Warbler is confined within very narrow limits. It has seldom been noticed beyond the boundaries of Asia Minor, Syria, and Egypt, including the Sinaitic Peninsula. It is very rare in Greece and Algeria.

40. *Sylvia atricapilla.* (Linn. Syst. Nat. i., p. 332.) Black-Cap.

The Black-Cap is one of the commonest birds in Palestine. All through the winter it may be found in small flocks, the males having the brown cap, which in summer is characteristic of the female. In spring the numbers in the south are not much increased; but it breeds by hundreds about the Cedars of Lebanon and in the woods near Hermon. These mountain denizens do not arrive until after the birds in the south have begun the work of nidification.

The Black-Cap is found in every country of Europe, in the Caucasus, and Asia Minor. In winter it is spread over Africa as far as the Canaries, Senegal, and Abyssinia. Palestine may be considered its Eastern limit, though it has been once recorded from Persia.

41. *Sylvia hortensis.* (Gmel. Syst. Nat. i., p. 955.) Garden Warbler.

The Garden Warbler visits Palestine in spring, remaining to breed, but in very small numbers; this country appearing to be its extreme eastward limit.

The Garden Warbler ranges throughout Europe in summer, and through almost all Africa, even down to Damara-land and Caffraria, in winter.

42. *Sylvia nisoria.* (Bechst. Naturg. Deutschl. iv., p. 580.) Barred Warbler.

The Barred Warbler, nowhere a very common bird, is rare in Palestine, but visits the country in spring, I believe only on passage, as I never found it breeding.

The range of the Barred Warbler is limited from Central and South-eastern Europe as far as Turkestan and Persia, and southwards it has been noticed in Nubia.

43. *Sylvia undata.* (Bodd. Tabl. Pl. Ent., p. 40, No. 655.) Dartford Warbler.

The Dartford Warbler is very scarce in Palestine, being occasionally found in the scrub in the neighbourhood of the Spectacled Warbler, which it much resembles in habits. It remains, I believe, through the year.

Palestine is probably its extreme Eastern limit. It is found in the south of England, but not in Germany or Central Europe. Its head-quarters are the countries bordering on the Western Mediterranean. East of Italy it is very rare.

44. *Sylvia nana.* (Hemp. and Ehr. Symb. Phys. Aves, fol. cc.) Pygmy Warbler.

This curious little desert bird only asserts its claim to a place in our list by its occurrence in the desolate Sebkha, at the south end of the Dead Sea.

It is found in the deserts of the Sahara, the Sinaitic Peninsula, Southern Persia, Scinde and Turkestan.

45. *Regulus cristatus.* Hoch. Baiern. Zool., p. 199. The Gold-Crest.

The Gold-Crest is found in Lebanon, which is, perhaps, its southern limit in the East. It is spread through all Europe and North Asia, down to the Himalayas, and also inhabits North-West Africa, extending to the Canaries and Azores.

46. *Phylloscopus superciliosus.* (Gmel. Syst. Nat. i., p. 975.) The Yellow-browed Warbler.

The Yellow-browed Warbler rests its claim to its place here on a solitary specimen shot by myself at Jericho, 1st January, 1864. I have never seen or heard of another specimen. It was skulking like the Chiffchaff in the thickets by the waterside.

It is a strictly Asiatic species, breeding in Siberia and wintering in India and China; but stragglers have been obtained in England, Heligoland, and Germany.

47. *Phylloscopus rufus.* (Bechst. Naturg. Deutschl. iv., p. 682.) Chiffchaff.

The Chiffchaff swarms everywhere in Palestine during the winter, but is never seen after February.

This species is found in summer through all Northern Europe. In winter it is spread over Southern Europe, North and Central Africa, and Western Asia, as far as Persia.

48. *Phylloscopus trochilus.* (Linn. Syst. Nat. i., p. 338.) Willow-Wren.

The Willow-Wren positively swarms in winter, especially in the Jordan valley, but leaves for the North about the middle of March.

The Willow-Wren is a widely spread species, visiting the whole of Northern Europe and Asia as far as the Yenesei in summer; and in winter extending through Africa as far as the Cape of Good Hope, but in Asia not further south or east than Persia, where it is not common.

49. *Phylloscopus sibilatrix.* (Bechst. Naturg. Deutschl. iv., p. 688.) Wood-Wren.

The Wood-Wren appears in Palestine in great numbers about the last week in April, and has almost entirely disappeared by the second week in May. I had formerly believed that it was here only on passage, but in 1881 I found it breeding in the north near the Litany river.

The Wood-Wren is a summer visitor to nearly the whole of Europe, and is found in Barbary, Egypt, Nubia, and Abyssinia. It has been sent from the Gold Coast, but I have not been able to trace it further east than Palestine, excepting a statement by Ménétries that he procured it near the Caspian.

50. *Phylloscopus bonellii.* (Vieill. Nouv. Dict. xxviii., p. 91.) Bonelli's Warbler.

Bonelli's Warbler returns in considerable numbers at the beginning of April, and immediately disperses to its various haunts, especially frequenting olive-yards and wooded hills, where it builds its domed nest on the ground well concealed.

Bonelli's Warbler visits Central and Southern Europe from the Atlantic to Turkey in summer. It is not uncommon in Asia Minor, but has not been traced east of Palestine. It winters in North Africa as far as Senegal and Nubia.

51. *Hypolais olivetorum.* (Strickland in Gould's Birds of Europe, ii., pl. 107.) Olive-tree Warbler.

This, the largest of its genus, is a late summer visitant to the north of Palestine, where it breeds in small numbers in olive-groves.

The Olive-tree Warbler is a denizen of Greece and Asia Minor. In Algeria it is rare, though I have taken its nest and still possess the parent bird. It has been taken also once in Egypt, in Abyssinia, and is stated to be found in Fez. Nothing more is known of the distribution of this most local bird.

52. *Hypolais upcheri.* Tristram. P. Z. S., 1864, p. 438. Upcher's Warbler.

Though a specimen of this bird was taken in Palestine by Ehrenberg, who named it *Curruca languida*, yet as his description is unrecognisable, and has only been identified by the existence of the type at Berlin, the name is excluded by the Stricklandian Code.

Upcher's Warbler is confined, in Palestine, to the lower and cultivated slopes of Lebanon and Hermon. It is exactly intermediate in size between the Olive-tree and Pallid Tree Warblers. It is doubtful whether it migrates further than up and down the mountains. It breeds in Palestine, Egypt, Abyssinia, South-east Persia, Baluchistan, and Turkestan. In this last country it is only a summer visitor.

53. *Hypolais pallida.* (Hemp. and Ehr. Symb. Phys. Aves, fol. bb.) Pallid Tree Warbler.

This is the most common species of its family, especially in the lower and warmer parts of the country. It reaches the Jordan valley in March, and the slopes of Hermon in April, and commences to build its very neat nest immediately on its return, in low bushes well concealed. It prefers marshy situations.

The Pallid Tree Warbler appears to be confined to Greece, Asia Minor, Syria, and Persia, and in winter retires to Nubia and Abyssinia. In Western Europe and Central and Eastern Asia it is represented by closely allied species.

54. *Aëdon galactodes.* (Temm. Man. d'Orn. i., p. 182.) Rufous Warbler.

Of all the Warblers in Palestine, this is the most attractive and conspicuous, and perhaps the most abundant in summer. It returns about 14th April, and at once overspreads every part of the country. Its bright chestnut plumage, with its black and white tipped tail expanded like a fan, enlivens every thicket and thorn bush. Instead of skulking, it hops in the open, perches on the outmost bough of a bush or the stem of a tall reed, expanding and jerking its tail like a wren. The nest is placed very conspicuously in a tamarisk bush, and invariably has the cast skin of a serpent loosely twined in the interior, perhaps to intimidate the lizards, who prey on the eggs.

The Rufous Warbler breeds in Spain and Portugal and throughout North Africa as far as Abyssinia. In Greece and Asia Minor it is replaced by the following species:

55. *Aëdon familiaris.* (Ménétr. Cat. Rais., p. 32.) Grey-backed Warbler.

This species, the representative in North Syria, Asia Minor, and Greece, of the Rufous Warbler, to which it is closely allied, can only doubtfully claim to be of Palestine. Dr. Van Dyk, at Beyrout, showed me a specimen he had shot on the Phœnician plain, and it must pass through Palestine on its way to its summer quarters. It is most remarkable that north of Lebanon I never once detected *Aëdon galactodes*, so abundant everywhere to the south, while throughout North Syria, Mesopotamia, and Armenia, I never for an hour lost sight of *Aëdon familiaris*, most appropriately so named.

The Grey-backed Warbler is also found throughout Persia, the Caucasus, Turkestan, and Scinde.

56. *Acrocephalus streperus.* (Vieill. Nouv. Dict. xi., p. 182.) Reed Warbler.

The Reed Warbler returns about the beginning of March, and is common in all suitable localities.

This bird ranges through all Europe, excepting North Scandinavia and North Russia. It seems to be a permanent resident in North Africa; I have found it breeding in Algeria. In Egypt, Nubia, and Arabia it occurs on passage, and extends eastward to Persia, Baluchistan, and Turkestan.

57. *Acrocephalus palustris.* Bechst. Orn. Tasch. i., p. 186. Marsh Warbler.

The Marsh Warbler is scarce in Palestine, but may easily escape observation. It inhabits Continental Europe from Holland southward, and winters in Africa, where it goes as far south as Natal. Eastward it has been found in Persia.

58. *Acrocephalus arundinaceus* (Linn. Syst. Nat. i., p. 296) = *turdoides.* (Meyer. Vög. Liv. and Esthl., p. 116). Great Reed Warbler.

The Great Reed Warbler returns to Palestine about the beginning of March, and may generally be found where the reed-beds are of sufficient extent, there often being several pairs in the same swamp.

It extends throughout Central and Southern Europe, but not further east in Asia than the Caucasus, Syria, and Western Turkestan. Elsewhere it is replaced by *A. orientalis,* an allied and slightly smaller species.

59. *Acrocephalus stentoreus.* (Hemp. and Ehr. Symb. Phys. Aves, fol. bb.) Stentorian Reed Warbler.

This bird I have only observed in the marshes of the Huleh, where its unmistakable and discordant note, or rather scream, may often be heard from the wholly inaccessible papyrus swamps.

The Stentorian Warbler has a wide Eastern range from the Nile to Turkestan, India, Ceylon, and Java.

60. *Acrocephalus phragmitis.* (Bechst. Orn. Taschenb., p. 186.) Sedge Warbler.

The Sedge Warbler is very common wherever it can find suitable cover from the end of March. Probably some remain in the Jordan valley throughout the winter.

It inhabits Europe from the Arctic Circle southwards, and extends east as far as the Yenesei. It passes into Africa for the winter, and extends its flight to the furthest south, having been obtained in Damara land.

61. *Lusciniola melanopogon.* (Temm. Pl. Col., pl. 245, fig. 2.) Moustached Warbler.

Once only did I obtain the Moustached Warbler, on the plain of Gennesaret, 31st March, nor am I aware I ever observed it elsewhere.

It is only met with in South Europe, Asia Minor, North Africa, and thence eastward as far as the North-west Provinces of India.

62. *Locustella fluviatilis.* (Wolf. Tasch. Deutsch. i., p. 229.) River Warbler.

The River Warbler occurs in the upper valley of the Jordan, by the Lake of Galilee and at Lake Phiala, where I have procured it in April and May.

The River Warbler is everywhere a rare bird, and appears to be confined to South-eastern Europe (especially Galicia), Asia Minor, and Palestine. I can find no satisfactory evidence of its having been obtained in Egypt or North-east Africa.

63. *Locustella luscinioides.* (Savi. Nuov. Giorn. de' Lett. vii., p. 341.) Savi's Warbler.

Savi's Warbler appears to be a scarce summer visitor in Palestine.

I obtained it on the plain of Gennesaret on 1st May, and heard its note in the Huleh marshes later in the spring.

This rare Warbler visits England occasionally, and Holland ; but with this exception it would appear to be confined to the coasts of the Mediterranean, where it is very sparingly distributed, and only in some countries. Palestine is its Eastern limit. In Egypt it is comparatively not rare.

64. *Cettia sericea.* Temm. Man. d'Orn. i., p. 197. Cetti's Warbler.

Cetti's Warbler may be frequently heard, but rarely seen, among the willows by small watercourses. Its note is very fine and powerful, but suddenly broken, like the first part of a nightingale's abruptly cut short.

I formerly described the Palestine bird as distinct, *Cettia orientalis* (*Ibis*, 1867, p. 79); but though there is a slight difference in colour, and the bill is much longer and wider at the base than in any European specimens, the differences are scarcely sufficient on which to found a species.

Cetti's Warbler is found resident on both shores of the Mediterranean, in Egypt, Turkestan, and Scinde.

FAMILY, TIMELIIDÆ.

65. *Argya squamiceps.* (Rüpp. Atlas, p. 19.) Hopping Thrush.

Few birds have a more circumscribed limit than this Bush Babbler, one of the peculiar denizens of the Dead Sea basin. It does not even extend up the valley of the Jordan, but is strictly confined to the larger oases round the Dead Sea itself. Nowhere else did it come under our observation; and thus we find a distinct and most characteristic species belonging to a tropical family, limited to an area of forty miles by twenty, and not occupying more than ten square miles of that area. The Hopping Thrushes are sociable and noisy birds, always in small bands, though not in large flocks, hopping along the ground in a long line, with jerking tail, and then one after another running up a bush, where they maintain a noisy conversation till the stranger's approach, when they drop down in single file and run along the ground, to repeat the same proceedings in the next tree. The nest is a large structure of strips of bark loosely woven together and placed in the very centre of a zizyphus thorn-tree, containing four to six glossy dark-green eggs. Beyond the Dead Sea basin this bird is said to be found in Arabia, near Akabah, and in the Hedjaz in bushes and trees. Its food, so far as my own observation goes, is exclusively the berries of the zizyphus, which may be found all the year round.

SUB-FAMILY, ACCENTORINÆ.

66. *Accentor modularis.* (Linn. Syst. Nat. i., p. 329.) Hedge-Sparrow.

The Hedge-Sparrow resides in the Lebanon throughout the year, where, however, it is very scarce. I obtained it near Beyrout.

The Hedge-Sparrow is spread over all Europe, being a summer visitor to the north, but becomes very scarce in the south-east. It has been once taken near Smyrna, and was found by Mr. Blanford in Persia. With this exception Palestine is its Eastern and Southern limit.

SUB-FAMILY, DRYMŒCINÆ.

67. *Drymœca gracilis.* (Licht. Verz. Doubl., p. 34.) Streaked Wren-Warbler.

This bird is common all the year round in the plains, on the coast, and in the whole Jordan valley. It is a lively, active, and conspicuous bird. It builds a very neat domed nest near the ground, with four or five richly coloured pink eggs.

The Streaked Wren-Warbler is found in Asia Minor, its Northern limit, in North-east Africa as far as Abyssinia, in Arabia, Persia, and India.

68. *Drymœca inquieta.* (Rüpp. Atlas, p. 55.) The Hermit Fantail.

This bird, which I described (*Ibis,* 1867, p. 76) as *D. eremita,* not having recognised the very bad representation of it by Rüppell, is an Arabian and essentially desert form, which resorts throughout the year to the most desolate ravines round the Dead Sea, hopping among the retem bushes. It is very scarce wherever found, and so far has only been noticed in the Sinaitic Peninsula and South-east Palestine.

69. *Cisticola cursitans.* (Frankl. P. Z. S., 1832, p. 118.) Fantail Warbler.

The Fantail is to be found throughout the year in the moist maritime plains, where it will continually start up from the long grass, jerking up in the air for a few seconds as it rapidly repeats its single note, '*pink pink*,' and then drop perpendicularly again. Its nest is fixed among grass stems after the fashion of a Reed Warbler.

The Fantail has a wide range. It is common in all the countries bordering on the Mediterranean; through the whole of Africa to the Cape; and in Asia resides in India, Ceylon, China, and Japan. Many of the Fantails of the eastern islands seem scarcely separable from it specifically.

FAMILY, CINCLIDÆ.

70. *Cinclus rufiventris.* Hemp. and Ehrenb. Symb. Phys., fol. bb. Ruddy-bellied Dipper.

This Dipper is confined exclusively to the mountain streams in the deep gorges of the Litany river and of the Lebanon. It is particularly numerous about the magnificent springs of Afka, the ancient Adonis. It differs slightly from our Dipper. The white extends lower down the breast, the abdomen is of a lighter chestnut colour, and the back of the head and the shoulders have a rather rufous tint. I have therefore ventured to distinguish it specifically by the name which Ehrenberg gave it only as a variety. There is, perhaps, in reality only one true species of White-breasted Dipper, but as authors have made many species on no stronger grounds, I conceive that the Lebanon bird is equally entitled to specific rank, for it corresponds exactly with no other. The birds from Scandinavia, Switzerland, and Spain have been distinguished from peculiarities as trifling. Mr. Dresser groups the Lebanon form with that from Spain, which is decidedly a darker bird, and more like the Scandinavian form, while Mr. Sharpe unites it with *C. cashmeriensis*, but admits that it is a peculiar form, with more of a brownish red shade on the breast, which allies it somewhat to the *C. aquaticus* (the English bird) group. I cannot find that any Dipper has been found nearer Palestine than Greece, where *C. albicollis* is stated to occur: and Erzeroum, where the species is stated to be *C. melanogaster*.

FAMILY, PANURIDÆ.

71. *Panurus biarmicus.* (Linn. Syst. Nat. i., p. 342.) Bearded Tit.

The Bearded Tit has been taken in the reed-beds of *Arundo donax* in the maritime plain south of Beyrout. I have not myself seen or heard of it in any other part of the country.

The Bearded Tit has a considerable longitudinal but not latitudinal range. From the south-east of England it stretches through Holland to Southern Russia, Italy, and the Danube, and down to Greece and Smyrna. Palestine is thus its extreme Southern and Eastern limit.

FAMILY, PARIDÆ.

72. *Parus major.* Linn. Syst. Nat. i., p. 341. Great Titmouse.

This is the only Titmouse which is common in Palestine. It is found throughout the year in all the woods and olive-yards east and west, but never in the Jordan valley. Its coloration is brighter than in British specimens.

The Great Tit is an inhabitant of the whole of Europe, Barbary, Siberia to the Amoor, Asia Minor, Armenia, and Persia.

73. *Parus ater.* Linn. Syst. Nat. i., p. 341. Coal Titmouse.

The Coal Tit is very abundant in all the wooded parts of Lebanon, and especially about the cedar groves, where it breeds early in the season.

The range of this bird is not easily defined. The British Coal Tit differs slightly from the Continental, and in Eastern Asia *Parus pekinensis* also approaches it very closely. But the typical form ranges through Northern and Central Europe as far as the Caucasus, and extends into Western Siberia. It is very common in the Taurid range, but becomes very scarce in Southern Europe, and is not found south of the Mediterranean.

74. *Parus lugubris.* Temm. Man. d'Orn. i., p. 294. Sombre Titmouse.

The Sombre Titmouse, which is extremely plentiful in the wooded parts of Northern Syria and Southern Armenia, does not appear to extend further in Palestine than the Lebanon. It is to be seen and heard at the Cedars, and in all the few wooded districts.

The Sombre Titmouse is very different in its habits from our Marsh Tit, and affects the higher branches of trees. It is a bird of very limited range. Its centre is Greece, where it extends westward to Hungary, eastwards to Palestine and the Taurid range.

FAMILY, SITTIDÆ.

75. *Sitta neumayeri.* Michal. *Isis*, 1830, p. 814. Syrian Nut-hatch.

No one who has noticed and heard the Syrian Nut-hatch on the side of some rugged Lebanon ravine can ever forget it. The startling cheery note, and the movement of the bird among rocks and boulders, as easy in its motions over a moraine heap as a railway train on the rails, cannot be forgotten. This species is very common all through Lebanon, while north of these mountains I found both this and the true *Sitta syriaca*, the large species, identical with that of Central Asia, inhabiting the same localities.

Neumayer's Nut-hatch is only found in Greece, Asia Minor, and Syria, resorting exclusively to rocks and bare gorges, and never visiting trees.

76. *Sitta cæsia.* Wolf. Tasch. Deutsch. Vogelk. i., p. 128. Common Nut-hatch.

The Common Nut hatch is resident in the wooded parts of Galilee, round Hermon and throughout the Lebanon. It is especially abundant in the Leontes (Litany) glen, but never found further south. I have not been able to trace it east of Jordan.

The range of *Sitta cæsia* is peculiar. It is the only species of the genus found in England, but on the Continent it is not found north of Holland. It is sparsely scattered through all Southern Europe, and is not uncommon in Asia Minor and Armenia.

77. *Sitta krueperi.* Von Pelzeln. Sitz. Akad. Wissen. Wien. 48, Ab. i., p. 149. Krüper's Nut-hatch.

Krüper s Nut-hatch I have only noticed in the wooded walls of the stupendous gorge of the Leontes, where it may be seen among the trees which sparsely stud the chasm, while Neumayer's Nut-hatch gambols over the bare rocks.

This very peculiar species is only known from Asia Minor and Syria.

FAMILY, TROGLODYTIDÆ.

78. *Troglodytes parvulus.* Koch. Saüg. u. Vögel Baierns, p. 161. Common Wren.

The Wren occurs, but by no means plentifully, on the northern hills ot Palestine. It is found all over Europe; in the Barbary States from Morocco to Tunis; in North-Western and Central Asia, Asia Minor and Persia.

FAMILY, MOTACILLIDÆ.

79. *Motacilla alba.* Linn. Syst Nat.·i., p. 331. White Wagtail.

The White Wagtail is very abundant everywhere in winter, but becomes scarce as the spring advances, and is not seen at all in the south in summer. A few remain to breed in Galilee, where in the hills I have taken the nest.

The White Wagtail is universally spread over Continental Europe; North and Central Africa as far as Senegal and Abyssinia; Northern Asia as far as Lake Baikal, and southwards to Persia and Scinde.

80. *Motacilla vidua.* Sund. Oefv. vet. Förhdlg. 1850, p. 158. White-winged Pied Wagtail.

Has been obtained by Dr. Herschell in the Jordan valley, and I have seen a second specimen in Jerusalem. It is essentially an African species, found over the whole of that continent, except the Barbary States.

81. *Motacilla sulphurea.* Bechst. Gem. Naturg. Vög. Deutsch. ii. p. 459. Grey Wagtail.

This bird is not uncommon in winter by the banks of streams and swamps. Never observed after early spring. All I have examined are the long-tailed European, not the shorter-tailed Asiatic bird, *M. melanope*.

It is found throughout Europe, North Africa and Asia, as far as Persia.

82. *Motacilla flava.* Linn. Syst. Nat. i., p. 331. Blue-headed Yellow Wagtail.

Generally distributed in winter throughout the country, but not seen at any other season. This may be looked on as the typical form of Yellow Wagtail.

It breeds in Central Europe and Asia, and is found all over Central and Southern Europe; the whole continent of Africa; the whole of Asia, north and south; and Alaska.

83. *Motacilla cinereo-capilla.* Savi. Nuov. Gior. d. Lett., p. 190. Grey-headed Yellow Wagtail.

This species or variety also occurs commonly, but only in winter. I preserved several specimens in the winter of 1881.

This is a more northern form than the last, though trending eastward, and goes as far south as Abyssinia and as far east as China. In fact, it occupies the country enclosed by lines drawn from Finland to Abyssinia and China, thus crossing at an angle of 45° the region of the previous species.

84. *Motacilla melanocephala.* Licht. Verz. Doubl., p. 36. Black-headed Yellow Wagtail.

The Black-headed is by far the scarcest of the three species in Palestine, only occurring in winter in small numbers, but remaining later than any of the others. It was only on my fourth visit to Palestine that I ascertained this fact, which is accounted for by its breeding in Greece and Asia Minor.

The range of the Black-headed Yellow Wagtail is limited to Greece, Asia Minor, the Caucasus, and North-east Africa.

85. *Anthus pratensis.* (Linn. Syst. Nat. i., p. 287.) Meadow Pipit.

The Meadow Pipit occurs in small numbers throughout the winter, and a few pairs remain and breed in favourable localities. I found many young birds in the upper plains of the Hasbany in 1881.

The Meadow Pipit is a migrant in the North, and a resident in the South of Europe. On the southern shores of the Mediterranean it is scarce. Eastward it ranges to the Indian frontier, but only as a straggler.

Not uncommon in Asia Minor. Palestine appears to be its ordinary South-eastern limit.

86. *Anthus trivialis.* (Linn. Syst. Nat. i., p. 288, No. 5) Tree Pipit.

The Tree Pipit occurs throughout Palestine in winter. In spring I found it paired and very abundant in the north in 1881, in the very places where it had appeared scarce in 1864.

The Tree Pipit enjoys a wide range over the whole of Northern Europe and Asia in summer, and in winter as far south as Caffraria in Africa, and India and China in Asia, and has even been found at Batchian.

87. *Anthus cervinus.* (Pall. Zoog. R. A. i., p. 511.) Red-throated Pipit.

This is a scarce winter visitant to Palestine.

It is found throughout Europe, North Africa as far as Abyssinia, and the whole of Asia, but more plentiful in the eastern than western regions.

88. *Anthus spipoletta* ([*spinoletta* misprint]. Linn. Syst. Nat. i., p. 288.) Water Pipit.

The Water Pipit is found in winter in the lower and moist localities, especially in the Jordan valley.

It is found in moderate numbers in suitable situations throughout Central and Southern Europe and Asia, and North Africa.

89. *Anthus campestris.* (Linn. Syst. Nat. i., p. 288.) Tawny Pipit.

The Tawny Pipit is the Pipit of Palestine, residing throughout the year, and breeding in all the uplands and open parts of the country. In the semi-tropical Jordan valley I have not observed it.

The Tawny Pipit occurs scantily in Northern Europe, more plentifully in all the countries bordering on the Mediterranean; and extends very far south in Africa. In Asia it does not extend far north, but is found in Persia and the North-west Provinces of India, which appear to be its limit.

FAMILY, PYCNONOTIDÆ.

90. *Pycnonotus xanthopygus.* Hemp. and Ehr. Symb. Phys., fol. bb. Palestine Bulbul. Arab بلبل, Bulbul.

The Bulbul is one of the most characteristic birds of the warmer regions of Palestine. Never found in the hills or upper country, it is universally diffused through the Jordan valley and the lower parts of the adjacent wâdys, and throughout the whole of the maritime plain, from Gaza to Sidon, wherever there are olive-yards, groves, or even dwarf wood, and especially by the sides of streams. It is a permanent resident; never gregarious, but scattered throughout the year in pairs. It is the finest songster of the country, and the fame of the Bulbul as a musician is well deserved. It is inferior only to the Nightingale in power and variety of song. The nest is a small neat structure in the fork or on the branch of a tree, and the eggs are like those of the other species of the family: white, thickly covered with rich crimson, chocolate, and pink blotches and spots. It is by no means shy; is easily tamed, and is a favourite cage-bird with the natives.

The range of the Palestine Bulbul is very limited. Out of Palestine it is nowhere abundant, though I have found it occasionally in the warmer nooks of the lower valleys of Northern Syria. It inhabits the date and tamarisk groves of the Sinaitic Peninsula, and is stated to have occurred in Cyprus and Rhodes. Thus it may be looked on as one of the peculiar denizens of the Holy Land.

FAMILY, ORIOLIDÆ.

91. *Oriolus galbula.* Linn. Syst. Nat. i, p. 160. Golden Oriole.

The Golden Oriole passes through the country in May. Very few remain to breed.

The Golden Oriole passes the summer in the Mediterranean countries and Southern Russia, wintering in Central Africa, and occasionally crossing the line, migrating as far as Damara-land. It breeds commonly in Northern Persia; but its place is taken in India by the allied species, *O. kundoo.*

FAMILY, LANIIDÆ.

92. *Lanius aucheri.* Bp. Rev. et Mag. Zool., 1853, p. 294. = L. *fallax.* Finsch. Trans. Z. S., vii. 1872, p. 249, pl. xxv. The Pallid Shrike. Arab. ابو سروند, *Abou seround;* بورص, *Booras.*

The non-ornithological observer would probably pronounce this great Shrike to be the most common bird of the country. It certainly is everywhere except in the deserts and mountains, and at all times takes good care to be seen, perched motionless on a bare bough, or on the top of a tree, and placing its nest in the most conspicuous situations, but generally well protected by the masses of thorns which encircle it and defy the hawks. It is very confident and tame. In former writings I erroneously identified it with the European *Lanius excubitor.* The Palestine bird has usually been identified with the Indian *L. lahtora,* which it resembles so closely that only the most minute examination can detect the distinctions. These, however, appear to be permanent.

Mr. Dresser, in his 'Birds of Europe,' identified the Palestine and Indian birds, but Dr. Gadow, in the 'Catalogue of Birds, British Museum,' vol. viii., has by various subtle discriminations nearly doubled the number of species of this genus. I therefore, merely for convenience, distinguish my Palestine friend as separated from its congeners, only giving it Bonaparte's name, which claims priority over Finsch's. Thus restricted, *L. aucheri* has not a very wide range, being found in Abyssinia, Nubia, Palestine, Persia, and Baluchistan.

93. *Lanius minor.* Gm. Syst. Nat. i., p. 308. Lesser Grey Shrike.

This bird seems uncertain in its visits. In 1858 it was almost the first bird I shot on the plain of Sharon, where it was common. In 1863, 1864, and 1872 I never saw it, while in 1881 it was very common, and breeding as late as June, while the Pallid Shrike had hatched in March.

The Lesser Grey Shrike inhabits in summer South-eastern Europe and South-western Asia, not going further east than Persia, and in winter it retires to Eastern Africa, and reaches even to Damara-land.

94. *Lanius collurio.* Linn. Syst. Nat. i., p. 136. Red-backed Shrike.

The Red-backed Shrike returns to the northern mountain regions about the end of April in vast numbers, and immediately begins to build. I have not observed it south of Esdraelon.

This bird is found in summer throughout Europe, but is rare in the west, and not found in Portugal, Southern and Western Spain, or Ireland. Its North-eastern limit appears to be the Caucasus. It occurs all down the east coast of Africa, and breeds in South Africa.

95. *Lanius auriculatus.* Müll. Natursystem. Suppl., p. 71. Woodchat Shrike.

The Woodchat is a summer resident in Palestine, returning at the end of March, from which time it may be seen on every bush to an altitude of 4,000 feet. Above that line its place is taken by the Red-backed Shrike.

It is a bird of comparatively restricted latitudinal range, though it is found from Spain to Persia; but it does not reach Northern Europe, nor (so far as we know) proceed far south in Africa, Senegal in the west, and Upper Egypt in the east, appearing to be its limits.

96. *Lanius nubicus.* Licht. Verz. Doubl., p. 47. Masked Shrike.

This beautiful little Shrike returns about 20th March, confining itself to the bushes and scrub-covered hills and wâdys of the north, where it perches among the foliage, not on exposed twigs. The nest is very neat, built after the fashion of the Chaffinch.

The Masked Shrike is only found in Greece, very sparingly in Turkey and Asia Minor, and in winter in Egypt and Nubia, and sometimes as far as Abyssinia.

FAMILY, MUSCICAPIDÆ.

97. *Muscicapa grisola.* Linn. Sȳst. Nat. i., p. 328. Spotted Flycatcher.

Our well-known Flycatcher arrives the last week in April, and is spread at once over all the olive-yards, gardens, and stunted woods of the country.

The Spotted Flycatcher is a summer visitor to the whole of Europe and Western Asia as far as Persia and Ladakh. In winter it has been taken, though rarely, in North India. At that season it is spread over the whole of Africa.

98. *Muscicapa atricapilla.* Linn. Syst. Nat. i., p. 326. Pied Flycatcher.

The Pied Flycatcher visits Palestine for nidification in small numbers, returning about the last week of April.

It is rather a local bird everywhere, but is found in all the countries of Europe in summer, and in North Africa. Eastward it extends to Northern Persia. It does not appear to go far south in Africa.

99. *Muscicapa collaris.* Bechst. Gem. Naturg. Deutschl. iv., p. 495. Collared Flycatcher.

During a year devoted to ornithology in Palestine, in 1863-64, I only once had a glimpse of the Collared Flycatcher. In 1881 I came upon it overspreading all the country from Nazareth to Hermon, and for one individual of the Pied Flycatcher, which I remembered as so common, we had at least ten of this species. Its migrations, therefore, must be irregular.

The Collared Flycatcher visits Southern and Central Europe in summer, but does not appear to retire far into Africa in winter. With the exception of one recorded occurrence in Persia, Palestine is its Eastern limit. From its very limited range and its rarity in Egypt, it seems probable that its winter-quarters are in Arabia.

100. *Muscicapa parva.* Bechst. Gem. Naturg. iv., p. 505. Red-breasted Flycatcher.

I have not myself found this bird in Palestine, but have seen a specimen obtained near Beyrout.

The Red-breasted Flycatcher is found in Central and Southern Europe, South Russia, the Caucasus, Arabia, and Persia.

FAMILY, HIRUNDINIDÆ.

101. *Hirundo savignii.* Steph. in Shaw. Gen. Zool. x., p. 90. Oriental Swallow.

The Oriental Swallow, differing only from our Common Swallow in the rich chestnut red of the whole lower parts, is a constant resident in the Holy Land. Along the coast, in the maritime plains, and along the whole Jordan valley, it is numerous during winter, when not an individual of the other species is to be seen. In spring the numbers of this swallow rapidly increase, and from the middle of March they become distributed over the whole country, when along with them appears the common species. In the higher ground the latter predominates; in the lower, certainly this is the most numerous.

The only other country inhabited by the Oriental Swallow is Egypt, where also it is sedentary. It is not met with south of Egypt, nor in Syria north of the Lebanon. In its habits and architecture it closely resembles the Common Swallow, but never interbreeds with it.

102. *Hirundo rustica.* Linn. Syst. Nat. i., p. 343. The Swallow. Heb. דְּרוֹר (*generic*). Arab. سنونو, *Sununu.*

From the end of March the Swallow swarms all over the country. In winter not a solitary individual is to be seen.

The summer range of the Swallow is over nearly the whole northern Old World, though most naturalists would separate, I think rightly, the birds from Eastern Asia. In winter it seems to be scattered over the whole of Africa as far as the Cape.

103. *Hirundo rufula.* Temm. Man. d'Orn. iii., p. 298. Red-rumped Swallow.

This handsome Swallow returns to Palestine at the end of March, plentiful everywhere, but most numerous in the lower and warmer parts of the country. The nest is a remarkable structure, attached to the flat surface of the under side of the roof of a cave or vault, with a long neck a foot or more in length, like a retort, and large bulb, larger than a Thrush's nest. The eggs are pure white.

The Red-rumped Swallow is ordinarily only found in South-eastern Europe bordering on the Mediterranean, in Asia Minor, and Syria, and also rarely in North-east Africa.

East of the Caspian it is represented by *H. daurica,* and in India by *H. erythropygia* and other species.

104. *Chelidon urbica.* (Linn. Syst. Nat. i., p. 344.) Martin.

The Martin arrives in great numbers about the 5th April, and having no windows to be utilized, builds on the faces of cliffs in all the valleys and ravines.

The Martin inhabits Europe generally during summer, migrating in winter to Africa. It has not been traced east of Western Siberia and Persia.

105. *Cotile riparia.* (Linn. Syst. Nat. i., p. 344.) Sand Martin.

The Sand Martin returns about the end of March, and while numbers pass on, small colonies remain and breed in the localities adapted for the purpose.

No Passerine bird has such a world-wide range as the Sand Martin. It is found in summer in the whole of Europe, Asia, and North America, even as far north as Melville Island, and passes in winter through Africa as far as the Transvaal, into India, and in South America as far as Brazil.

106. *Cotile rupestris.* (Scopoli. Ann. I. Hist. Nat., p. 167.) Crag Martin.

The Crag Martin resides in all the glens of Palestine throughout the year, not generally in large numbers, and only in a few places appearing decidedly gregarious. It breeds early in March, laying, unlike most other Martins, spotted eggs.

Though confined entirely to the localities indicated by its name, the Crag Martin has a wide range from Spain to China, not extending, however, very far north or south of that line. So far as I have observed, in the Atlas, Greece, Asia Minor, and Palestine, it is sedentary.

107. *Cotile obsoleta.* Cabanis. Mus. Hein. i., p. 50. Pale Crag Martin.

This small species is in Palestine entirely confined to the Dead Sea basin, where it is sedentary. Round the sea itself it is the only species, but at the north end it mingles with *C. rupestris,* and they both breed in the same caves in Jebel Quarantania.

This is essentially a desert species, as the last is a mountain one. It

J. Smit lith. Hanhart imp.

CINNYRIS OSEÆ.

is found in Egypt, Nubia, Abyssinia, Arabia, and the Persian and Indian deserts. In habits it differs from its congener, sweeping the desert plains rather than soaring over the mountain cliffs.

FAMILY, CERTHIIDÆ.

108. *Tichodroma muraria.* (Linn. Syst. Nat. i., p. 184.) Wall Creeper.

The beautiful Creeper, the 'Butterfly Bird' of the French, is common throughout the year in all the rocky gorges of Central and Northern Palestine, from the glens opening on the plain of Gennesaret to the highest cliffs of Lebanon. No ornithological sight is more interesting than the movements of this richly coloured bird as it flits along the face of a line of cliffs, spreading its brilliant crimson wings at each sidling jerk.

The Wall Creeper is found in the mountain regions of all Central and Southern Europe and Asia, from Spain to the Himalayas and China.

FAMILY, NECTARINIIDÆ.

109. *Cinnyris oseæ.* Bonap. Comptes Rendus. xlii., Pt. 2, p. 765. Palestine Sun-Bird.

PLATE VIII.

To the naturalist this is perhaps the most interesting species of the whole Palestinian Avifauna. In the first place, it belongs to a truly tropical family. In the second place, it is absolutely peculiar, so far as we know, to the Holy Land, and within its limits is confined to a very narrow strip of territory; and lastly, we must travel very far from Palestine east or south to find another representative of the numerous Sun-bird family. We must go either to India or far up the Nile into Nubia. At least 135 species of Sun-bird are known, confined entirely to the warmer parts of the Old World, to Southern Asia and all its islands, as far as North Australia, to Africa, south of the Sahara, and to the Mascarene Islands and Madagascar. They are unknown in the New World and in Oceania. In habits they are more like our Titmouse than the Humming Birds, not hovering over the flowers, but clinging to the stems.

The nearest allies of *Cinnyris oseæ* are the West African species, *C. bouvieri*, and *C. venustus*, but it is very distinct in coloration from either of them. As has been said, its range is extremely limited, its head-quarters being the oases at the north-west and south-east extremities of the Dead Sea, while it spreads in smaller numbers up to the Lake Huleh. Beyond the gorge of the Jordan I never but once found it, and that was at the south of Mount Carmel, in the marshes of the Zerka river. It is everywhere very shy and restless, flitting in the foliage after the manner of a Tit, and with a note not unlike the call of the Blue Tit. Sometimes it perches on the top of a bush, and jerks for a moment into the air after an insect; but more generally it may be seen prying into flowers on the same quest.

The nest is pensile, always suspended from the extremity of a lower bough of some tree or shrub, generally but a few feet from the ground, and looking like a tuft of straw and weeds left entangled by some winter flood. The outside is studiously untidy, but the structure most compact and finished within; domed, with a small entrance at one side. Thus suspended, it is perfectly secure from the attacks of snakes or tree lizards, the great foes of small birds' eggs and young in this country.

FAMILY, FRINGILLIDÆ.

110. *Carduelis elegans.* Steph. in Shaw. Gen. Zool. xiv., p. 30. Goldfinch.

The Goldfinch is found in every part of the country at all times of the year; the great variety of composite plants, some of which are always to be found in seed, affording it an abundant supply of its favourite food. Olive-yards are its favourite breeding places.

The Goldfinch inhabits the whole of Europe, except the extreme north; North Africa and Asia, as far as Persia and Turkestan. At its Eastern limit it meets another species, *Carduelis caniceps*, Vig., and on the border line the two species appear gradually to run into each other, as may be seen by Mr. Seebohm's fine series from Central Asia.

J. Smit lith. Hanhart imp.

1. SERINUS CANONICUS. ♀.
2. „ „ ♂.
3. PASSER MOABITICUS. ♂.
4. „ „ ♀.

111. *Serinus hortulanus.* Koch. Saüg. u. Vög. Baierns., p. 229. The Serin.

The Serin, so far as I have been able to observe, is only a winter visitor to the wooded districts and little glens near the sea. It has not been noticed inland.

The Serin is the representative of the Siskin in Central and Southern Europe, the North African coast, and Asia Minor. Syria is its extreme Eastern limit.

112. *Serinus pusillus.* (Pall. Zoog. R. A. ii., p. 18.) Red-fronted Siskin.

The Red-fronted Finch occurs in Lebanon.

It is a bird of South-east Asia, a resident in the Caucasus and Taurid range, and along the mountain region as far as Ladak. It appears to be an uncertain visitor to the North-west Himalayas.

113. *Serinus canonicus.* Dresser. Birds of Europe, vol. iii., p. 555. Tristram's Serin.

PLATE IX. FIGS. 1, 2.

The name which I gave to this Siskin when I discovered it in 1864 was *Serinus aurifrons* (P. Z. S., 1864, p. 447). But this, having been once used, though afterwards abandoned by Blyth, for another bird, has been discarded by the purists of nomenclature.

This is one of the interesting peculiar forms of Palestine, though belonging not to the Dead Sea valley, but to the Lebanon and Anti-Lebanon exclusively. It is a true Siskin in its habits, note, and nidification. It never migrates, but descends to the villages on the edge of the snow-line in winter, re-ascending as high as there are bushes in spring. I cannot trace it on any of the spurs southwards, either from Hermon or Lebanon, and there it is very local. Its nest, very like a Goldfinch's, is rather conspicuously placed in the fork of a tall shrub. In winter it lives in little flocks, and is wild and shy. In spring the male bird always revealed the nest by his persistent return, after a minute or two, to recommence his song close to it.

114. *Coccothraustes chloris.* (Linn. Syst. Nat. i., p. 304.) The Greenfinch.

The Green Linnet is very common throughout the winter in the maritime plains, on Mount Carmel, and other wooded hills near the coast, but disappears in spring.

The Greenfinch is spread throughout Europe, except in the extreme north, but is not found further east than Asia Minor, the Caucasus, and the Syrian coast. It is also resident in Algeria, though the North African bird has been separated by many writers, and can always be distinguished.

115. *Coccothraustes chlorotica.* (Licht. Nom. Av., p. 46.) Syrian Greenfinch.

The Syrian Greenfinch has, as I believe, been rightly separated by Lichtenstein and Bonaparte. It is very much smaller and more brightly coloured than the Common Greenfinch, the length of the largest I can find being, culmen 0·4, wing 3·, tail 2·1, tarsus 0·6. The forehead, too, is of a rich gold, without the greenish tinge of the ordinary Greenfinch; and the brilliant yellow of other parts cannot be equalled by any other specimen I have seen.

But the distinction in habits is very noticeable. Very soon after the Common Greenfinch has disappeared, which is about the end of February, this bird about the middle of March makes its appearance in very great numbers among the olive-groves and gardens, where its habits and nidification in no way differ from those of its congener.

I can find no trace of any Greenfinch visiting Egypt or Persia, and must conclude, therefore, that the Syrian Greenfinch winters in Arabia.

116. *Coccothraustes vulgaris.* Pall. Zoog. R. A. ii., p. 12. The Hawfinch.

The Hawfinch may be more common than is generally supposed in Palestine, but it is very seldom seen. I have only twice detected it, once in Gilead, and once near Tabor.

The Hawfinch has a wide longitudinal range—from Spain to China and Japan; but it does not reach to the extreme north, and except in Algeria it is only a straggler to North Africa. It has not been found south of the Himalayas.

117. *Passer domesticus.* (Linn. Syst. Nat. i., p. 323.) House Sparrow. Heb. צִפּוֹר (*generic*). Arab. عصفور, *'Asfur* (*generic*).

The Sparrow of the Syrian cities is our own domestic species, and as abundant and bold there as here. It is found also in flocks in the southern wilderness of Beersheba in winter.

Assuming the Indian Sparrow to be identical with our own, though always a smaller and brighter bird, the Sparrow covers nearly the whole of Europe, North Africa, and Asia. But it is not found further east than Siam, being absent in China, East Siberia, and Japan. I have recently received it from the Albert Nyanza, Central Africa, agreeing exactly with Indian examples.

118. *Passer italiæ.* Vieill. Nouv. Dict. xii., p. 199. Italian Sparrow.

In some of the interior districts the Sparrows have the chestnut head of this species, and cannot possibly be separated from it. It has been generally stated that this bird is peculiar to the Peninsula, and that in all cases beyond its limits the *Passer hispaniolensis* has been mistaken for it. But as a matter of fact, both in North Africa and Syria, *Passer italiæ* does occur, without a vestige of the longitudinal *striæ* which mark the flanks of the other species. Moreover, the two differ widely in their habits, and no one familiar with them in life can mistake the two.

119. *Passer hispaniolensis.* (Temm. Man. d'Orn. i., p. 353.) Marsh Sparrow.

The Marsh Sparrow is chiefly confined to the Jordan valley, where it congregates at all times of the year in countless myriads, breeding in colonies so crowded that I have seen a jujube-tree broken down simply by the weight of their nests, while their noise is so deafening that it is impossible to carry on conversation in their rookeries. The Arab boys would bring in their eggs by the bushel. This bird feeds largely on the leaves of leguminous plants. In other parts of the country it is found, but not in such numbers, and never in the towns.

The Marsh Sparrow seems to be confined chiefly to countries bordering on the Mediterranean, but extends eastwards in small numbers to the western frontier of India.

120. *Passer moabiticus*. Tristram. P. Z. S., 1864, p. 169.

PLATE IX. FIGS. 3, 4.

So far as our present knowledge extends, this bird is the most limited in the world in its range, and the scarcest in number of individuals. And yet it is marked off from its allies more distinctly than any other member of the genus *Passer*. In three successive expeditions I have searched for it, but never obtained it except among the reeds in two spots on the west side of the Dead Sea, close to the shore, and again in the reed-beds of the Ghor es Safieh, at the south-east end of the Dead Sea; nor has it, so far as I know, been ever obtained by anyone except the members of our party in 1864. It is the most diminutive member of the Sparrow tribe, very shy and wary, and extremely restless, feeding on the seeds of the great feathery *Donax*. Its bright chestnut back and the bright yellow spot on each shoulder at once mark it as distinct. The female, in other respects clad in the same quiet hues as our hen Sparrow, has also the bright yellow spot on either side of the neck. The Yellow-necked Sparrow of India, *P. flavicollis*, has a yellow spot on the throat, none on the sides of the neck.

For the convenience of reference I append the original description and measurements of this rare bird:

Ex cinereo isabellinus, tectricibus alarum læte-castaneis: superciliis et dorso medio, cum remigum et rectricum maginibus rufescenti-isabellinis: dorso medio nigro-striato: gutture medio cum cervice nigris: maculâ suboculari et gutturis vittâ utrinque laterali albis: maculâ cervicali utrinque flavâ: ventre albo, crisso rufescenti: rostro superiore plumbeo, inferiore cum pedibus flavis.

Total length 3·8 inches, wing 2·3, tail 1·8.

121. *Petronia stulta*. (Gmel. Syst. Nat. i., p. 919.) Rock Sparrow.

The Rock Sparrow is not unfrequent in the open rocky country along the central ridge of Palestine up to the highest part of Lebanon. I had

Pl. X.

J. Smit lith. Hanhart imp.

PETRONIA BRACHYDACTYLA.

formerly stated it to be never found in winter, but was in error. In 1881 I met with it twice in February. It breeds down the wells.

The Rock Sparrow occurs over a wide range, from the Canaries across Barbary, Southern and Central Europe, Central Asia, North Persia, an Thibet as far as North China.

122. *Petronia brachydactyla.* Bp. Conspect. Gen. Av. i., p. 513. Desert Rock Sparrow.

Plate X.

This very plain and meanly coloured bird is very scarce and local. I first found it in a bare desert plain under Hermon, and took two nests in low bushes not two feet from the ground. The eggs are glossy white with a few marone spots, like a diminutive Golden Oriole's. It has only been found on bare desert ground in Arabia, North-east Africa, the Persian desert plateau, and Palestine.

123. *Montifringilla nivalis.* (Linn. Syst. Nat. i., p. 321.) Snow Finch.

Isolated and sedentary, a few pairs of the Snow Finch may be seen on the snowy tops of Hermon and Lebanon, descending in winter to the base of the mountains, a stranded relic, perhaps, of the glacial epoch, clinging, as it does, to these southern mountain tops, identical in species from the Pyrenees to the Caucasus.

124. *Fringilla cœlebs.* Linn. Syst. Nat. i., p. 318. Chaffinch.

The Chaffinch is very common in winter in flocks, the sexes apart, on the maritime plains and southern uplands, the male flocks appearing greatly to exceed the female in number. Early in spring they disappear, but numbers breed in the north among the mulberry groves under Hermon and Lebanon, and they are especially numerous at the Cedars. The species is identical with our own, while Algeria and the Canaries and Azores have their peculiar species.

The Chaffinch ranges from the North of Europe to the Mediterranean, and as far as the forest region of Persia, its Eastern limit.

125. *Linota cannabina.* (Linn. Syst. Nat. i., p. 322.) Linnet.

The Linnet roams through the lower country in flocks during the winter, and in summer ascends to the mountain regions, where it breeds, especially on the summits of Lebanon and Hermon, consorting with the Snow Finch and building in tufts of Alpine plants close to the snow. The plumage on the whole is more brilliant than in Western specimens.

It is widely spread throughout Europe, Barbary, and Western Asia, not passing into Siberia or beyond Persia.

126. *Carpodacus sinaiticus.* (Licht.) Bp. and Schl. Mon. Lox., p. 17. Sinaitic Rose Finch.

I have only seen this rarest of the Rose Finches in the desert south of the Dead Sea, and between that region and Beersheba, where it was in the company of small flocks of the Trumpeter Bullfinch. It is strictly a desert and ground bird, and has never been taken beyond the limits of the Sinaitic Peninsula.

127. *Erythrospiza githaginea.* (Licht. Verz. Doubl., p. 24.) Trumpeter Bullfinch.

The Trumpeter Bullfinch is confined to the southern wilderness of Judæa. I have never seen it in the Ghor. Its true home is the African Sahara, where it is widely spread, never coming north of the Atlas. It also inhabits the desert tracts of Egypt, Arabia, Persia and Scinde; and westward extends to the Canaries.

128. *Erythrospiza sanguinea.* (Gould. P. Z. S. 1837, p. 127.) Crimson-winged Finch.

This lovely Finch is extremely rare, even in its favourite districts. I only twice saw it in the Lebanon in 1864, and my fellow traveller, Mr. Cochrane, secured a nest of eggs with the parent cock-bird, which he shot off the nest, and which I still possess. In 1881 I again met with it in the very same place, among low fir trees.

The Crimson-winged Finch has been obtained in the Caucasus, once in the mountains of North Persia by Blanford, and in Turkestan by Severtzoff.

129. *Euspiza melanocephala.* (Scop. Ann. I. H. N., p. 142.) Black-headed Bunting.

The Black-headed Bunting returns to Palestine in the beginning of May, and from that time is very abundant in the upper country and on the coast; its bright plumage, powerful and cheery note, and habit of perching on the very top of the highest tree or bush in its neighbourhood, always attracting attention to it. Its nest is on or near the ground, and it lays blue eggs with fine russet spots. There is nothing of the Bunting in its habits and character.

Though taken in Heligoland, and it is said once in England, this is a strictly Eastern form, never having been found in Africa, and rarely west of Greece. It is abundant in Asia Minor, all through Syria and the Caucasus, and winters in North-west India.

130. *Emberiza miliaria.* Linn. Syst. Nat. i., p. 308. Common Bunting.

The Common Bunting is as common as the Skylark in England on all the lower plains throughout the year.

It inhabits all Europe, excepting Northern Scandinavia, but does not extend far east, though it has been taken in Turkestan. In Siberia it is not found, but southwards is common not only in Syria, but on the corn-plains of Mesopotamia and Persia.

131. *Emberiza hortulana.* Linn. Syst. Nat. i., p. 809. Ortolan Bunting.

The Ortolan is very abundant in spring, returning in the beginning of April, and resorting much to olive-yards and gardens.

The Ortolan, though only a straggler in England, is generally spread throughout Europe in summer, yet its distribution is perplexing. Generally speaking, its distribution is eastward, being rare in Holland and Denmark, common in Finland; yet found in Spain and Morocco, not in Algeria. It is plentiful in Southern Russia, but scarcely known in Egypt, though found on the Abyssinian Highlands; and eastward is found in Persia.

132. *Emberiza striolata.* (Licht. Verz. Doubl., p. 24.) Striolated Bunting.

The Striolated Bunting occurs on the bare desert hills and rocky ravines round the Dead Sea, remaining there throughout the year. Of course we should not expect to find a bird so strictly of the rocky desert in any other district. The Striolated Bunting has been found in restricted localities and in small numbers throughout the desert belt which girds the Old World from the Western Morocco coast to North-west India.

133. *Emberiza pusilla.* Pall. Reis. Russ. Reichs. iii., p. 697. The Little Bunting.

Rather a straggler than an inhabitant of the Lebanon, where I only know of one undoubted instance of its capture.

The Little Bunting is an inhabitant of North-east Europe and Siberia, migrating southwards in winter, principally to India, very few straggling westwards.

134. *Emberiza cia.* Linn. Syst. Nat. i., p. 310. Meadow Bunting.

Emberiza cia, certainly not a *Meadow* Bunting in Palestine, is found in the mountain regions in summer and winter alike, but in small numbers. We found it on Mount Carmel, about Galilee, on Hermon, and all through Lebanon. It is an inhabitant of the mountain districts of Southern Europe as far as the Caucasus, and in the Atlas range. The Taurid and Palestine appear to be its Eastern limits.

135. *Emberiza cæsia.* Cretzschm. in Rüpp. Atlas, p. 15. Cretzschmaer's Bunting.

Cretzschmaer's Bunting, which takes the place of our Yellow Hammer, returns simultaneously in great numbers about the third week in March, and peoples in pairs every part of the country, except the woods and olive-groves, preferring the scrubs or bare hill-sides or rocky wâdys. It builds a neat nest on the ground under a tuft, or low bush, and its eggs are easily distinguished from those of any other species. It is very tame, continually flitting in front of the traveller.

This Bunting appears to be restricted in summer to Greece, Asia Minor, and Syria, and in winter to North-east Africa.

FAMILY, STURNIDÆ.

136. *Sturnus vulgaris.* Linn. Syst. Nat. i., p. 290. Common Starling. Arab. زرزور, *Zerzour.*

The Starling is only a winter migrant, visiting the maritime plains in tens of thousands, with a few of the Sardinian Starling in their company. The latter does not, as in Algeria, remain to breed. The Starlings all depart at the end of February.

The Starling is found throughout Europe, the Atlantic Islands from the Azores eastwards, North Africa, all Northern Asia, and down to Persia and India.

137. *Sturnus unicolor.* De la Marm. Temm. Man. d'Orn. i., p. 133. Sardinian Starling.

This species, which is never spotted, is confined to the countries bordering on the Mediterranean, and is scarce in the eastern parts. It is much rarer than the other species, even where it does occur, and is much less migratory. I procured it three times in winter in Palestine.

138. *Pastor roseus.* (Linn. Syst. Nat. i., p. 294.) Rose-coloured Pastor.

The Rose-coloured Pastor is well known to the natives as the Locust Bird, from its habit of preying on that pest, whose flights it generally follows. It is very uncertain in its visits, being an erratic rather than a migratory bird. I found it in 1858, not in 1864 or 1872. In 1881 I came across marvellous flights of this bird in Northern Syria, which for three days (26-28 May) passed us on the Orontes, near the ancient Larissa, in countless myriads, all travelling to the westward. There must have been thousands upon thousands. The locusts were there, and on one occasion we rode over some acres alive with young locusts, which absolutely carpeted the whole surface. One of these flocks suddenly alighted, like a vast fan dropping on the earth and dappling it with black and pink. Soon they rose again. We returned, and not a trace of a locust could we find. See *Ibis*, 1882, pp. 410-414, for a full account of this marvellous migration. I may add that all these myriads were in fully adult plumage.

The Rose-coloured Pastor appears to range from India, east of which it is never found, through Persia, never going north of the Himalayas, to Syria, Asia Minor, Turkey, and Southern Russia. Westward of these regions and in North Africa it is only an occasional straggler.

139. *Amydrus tristramii Sclater.* Ann. and Mag. N. H., 1858, vol. ii., p. 465. Tristram's Grakle.

PLATE XI.

The discovery of this bird in the desolate ravines opening on the Dead Sea is one of especial interest, as it belongs to a group exclusively Ethiopian. This Grakle, known to the visitors to Mar Saba as the Orange-winged Blackbird, appears to be confined to the immediate neighbourhood of the Dead Sea, where it resides throughout the year in small bands, feeding at dawn and sunset. It has no varied notes, but a rich musical roll of two or three notes of amazing power and sweetness, which makes the cliffs ring again with its music. The Grakles are the wildest and shyest of the denizens of these desolate gorges, yet the monks of Mar Saba have succeeded in bringing them into a state of semi-domestication, while enjoying unrestrained liberty. I have never seen this bird elsewhere than round the Dead Sea. In the ravines of the Arnon and Callirrhoë it is more numerous than elsewhere.

Four other species of *Amydrus* are known from East Africa, one of which (*A. blythii*) has also recently been found by Professor Is. Balfour in the island of Socotra.

FAMILY, CORVIDÆ.

140. *Pyrrhocorax alpinus.* Koch. Saüg. u. Vög. Baierns. i., p. 90. Alpine Chough.

The Alpine Chough inhabits in small parties the higher grounds of Hermon and Lebanon, always keeping close to the snow. The Red-billed or Cornish Chough we never observed.

The range of the Alpine Chough is restricted to the highest mountains of Southern Europe and Asia, the Pyrenees, Alps, Apennines, Greek Mountains; rarely in the Caucasus, the Persian Demavend, and the Himalayas, beyond which it has not been traced.

Smit lith. Hanhart imp.

AMYDRUS TRISTRAMI.

141. *Garrulus atricapillus.* Isid. Geoffr. St. Hilaire. Et. Zool., fasc. i. Syrian Jay. Arab., حقاق, *'Akak* (the name elsewhere of the Magpie).

The Jay is very common in all the olive-groves from Lebanon to Hebron, and equally so in the true forests of Gilead and Bashan. It seemed to have increased greatly in numbers between 1864 and 1881. In its note and habits it in no way differs from the European Jay. It never descends into the Jordan valley.

The Syrian Jay is confined to Syria and the northern hills of Persia. In Asia Minor and the Taurid it is represented by a very closely allied species, *G. krynicki.*

142. *Corvus monedula.* Linn. Syst. Nat. i., p. 156. The Jackdaw. Arab., قاق, *Kak.*

The Jackdaw is very numerous in certain localities, and absent in others, as we note in England. There are populous colonies at Jerusalem and Nablus. Elsewhere it is scarce and local.

In the Jordan valley and in Gilead the place of our Common Daw is taken by the silvery white-necked variety, described by Drummond as *Corvus collaris.* While I quite agree with Mr. Dresser in declining to give specific value to this variety, it is worthy of note that there is here a distinct geographical line of demarcation between the two races or varieties.

The Jackdaw is found throughout Europe and Barbary, and reaching to Asia Minor, the Caucasus, and Palestine. It is only in Northern Persia, in what is in fact really Armenia, that the Jackdaw is found, and this is about its Eastern limit, though Jerdon states it is found in Cashmere. Other closely allied species take its place in Eastern Asia.

143. *Corvus agricola.* Tristram. P. Z. S., 1864, p. 444. Syrian Rook.

The Rook of Palestine is intermediate between our species, *C. frugilegus,* Linn., and the Chinese, *C. pastinator,* Swinh. The head of the Chinese bird is glossed with purple, of the English with blue-black, of the Palestine with green-black. The latter very rarely has the forehead, throat, and chin denuded, as in the English adult. In this it seems to agree with the Chinese.* Only two in twenty of adult birds we shot in spring had the slightest denudation. Both from this peculiarity and from the coloration, if specific rank be granted to the Chinese form it seems impossible to deny it to this one.

The rook in Palestine is very local, owing, no doubt, to the scarcity of well grown timber. Jerusalem and Nablus possess the chief rookeries, if groups on the tops of buildings and ruins may be so termed.

The Rook is the companion of cultivation throughout Europe and Asia, from Ireland to Japan, under one of its three forms above mentioned. It rarely is found, and then apparently accidental, south of the Mediterranean, nor does it extend beyond the western frontier of India. In Persia it is very rare. The Chinese form appears to run through Eastern Siberia, Japan, and North China.

144. *Corvus cornix.* Linn. Syst. Nat. i., p. 156. Hooded Crow. Arab., زاغ, *Zagh.*

The Hooded Crow is common in Southern and Central Palestine, a constant resident, but is never found in the Jordan valley, and I have rarely met with it in the north of the country. It is very plentiful on the east side of Jordan, both on the bare highlands of Moab, and in the undulating country of Gilead and Bashan. Its nest is the favourite foster-home of the eggs of the Great Spotted Cuckoo.

The Carrion Crow, *C. corone*, has never been noticed in Palestine.

The Hooded Crow is spread through the whole of Europe, North Africa, and Western Asia, as far as the Lena. The Carrion Crow thenceforwards supplants it as we go eastward. The two birds are generally now admitted to be specifically identical, but they have different ranges, though inter-breeding when they meet. But in Palestine, Asia Minor, and Egypt, the Hooded is the only form.

145. *Corvus affinis.* Rüpp. Neue. Wirb., p. 20. Fantail Raven.

This interesting and little-known miniature Raven is only found around the Dead Sea, in the most desolate and rugged cliffs, where it is very wild

* The 'British Museum Catalogue' in error marks my specimens shot in February as immature, an easy mode of settling the question.

and wary. There is a large colony in the ravines of the Zerka Maîn or Callirrhoë.

It has a rich musical note, and stately flight.

Palestine is its only known habitat out of Africa, but it will surely be found in the Sinaitic Peninsula. In Africa it appears to be confined to Abyssinia and Kordofan, thus giving another instance of the connection between the Jordan and Abyssinian Faunas.

146. *Corvus corax.* Linn. Syst. Nat. i., p. 155. The Raven. Hebr., עֹרֵב. Arab., غراب, *'Orab.*

The Raven is common in every part of the country, altitude or character of region being quite indifferent. In winter it is gregarious about the Mosque of Omar in Jerusalem, consorting by the hundred with the Brown-necked Raven, Hooded Crow, Jackdaw and Rook, all five species roosting together.

The Raven is found through the whole of Europe, Northern Asia, down to the Indian frontier, and across the whole of North America.

In Barbary it is replaced by a very closely allied species, *C. tingitanus.*

147. *Corvus umbrinus.* Hedenb. Sund. K. Vet. Ak. Handl. 1838, p. 198. Brown-necked Raven.

This is the common Raven of Jerusalem and the Jordan valley, but not of the coast or maritime plains, and only rarely seen north of Jerusalem. It is gregarious in winter, but breeds solitarily in cliffs. Its note is very different from that of any other species.

The Brown-necked Raven is a native of North-east Africa, but ranges as far as Baluchistan, being, however, very rare east of the Holy Land.

FAMILY, ALAUDIDÆ.

148. *Certhilauda alaudipes.* (Desf. Mem. Acad. Roy., 1787, p. 504.) The Desert Long-billed Lark.

This largest and most beautiful of the Lark tribe is not uncommon in the desert regions south of Judæa and east of Moab and Gilead, in both which districts I have frequently met with it.

It is a bird essentially of the sandy, not the rocky deserts of North Africa and Western Asia. But unlike most other desert species, it does not range higher than Scinde, not having been noticed in the steppes of Northern Asia.

It is a most aberrant Lark, both in its striking plumage, and flight, which is that of a Plover, and it seems in some respects a link between the Larks and the curious North Asiatic desert genus of *Podoces*.

Desfontaines first described this bird as *Upupa alaudipes*, and certainly its flight, and the white wings with their conspicuous black bars, would very naturally at first sight suggest a relationship to the Hoopoo.

149. *Alauda cristata*. Linn. Syst. Nat. i., p. 288. Crested Lark. Arab., قنبر, *Kenbar* (*generic*).

The Crested Lark is the commonest bird of the country in the open ground of the central, coast and northern regions, remaining all the year, but generally a late breeder. The pale form, *Galerida abyssinica*, Bp., is the form found in the south and in the deserts. But it differs only in colour.

The Crested Lark extends through Central and Southern Europe, North Africa, and eastward as far as India and China.

150. *Alauda isabellina*. (Bonap. Consp. Gen. Av., p. 245.) Isabelline Lark.

This small and short-billed species inhabits only the sandy desert at the south end of the Dead Sea. Elsewhere it is confined to the most arid parts of the Sahara and Egyptian deserts, and has not been noticed east of Palestine.

151. *Alauda arvensis*. Linn. Syst. Nat. i., p. 287. Sky Lark.

The true Sky Lark, of the European type, is found in large flocks on the coast-plains through the winter, but does not remain to breed, and never penetrates far inland. Notwithstanding the very able and exhaustive disquisition of my friend, Mr. Dresser ('Birds of Europe,' vol. iv., pp. 310-313), I cannot be persuaded to reject the claims of the next species, (*A. cantarella*) to specific rank, chiefly from my observation of the different habits of the two forms.

The Sky Lark is found all over the Palæarctic region, from the British Isles eastward to Siberia and Northern China.

152. *Alauda cantarella.* Bp. Consp. Gen. Av., p. 245. Southern Sky Lark.

This species (or form) congregates by thousands in the southern deserts, where there are none of the ordinary Sky Lark, during the winter hanging about the Bedawin camps and herds. We shot scores of them for food, and never detected a specimen of the other species. We did not discover them breeding.

This bird, the *Alauda intermedia* of Swinhoe, extends south of the line of *A. arvensis,* through North Africa (rarely in the South of Europe), Egypt, Southern Palestine, Southern Persia, India, and China.

153. *Alauda arborea.* Linn. Syst. Nat. i., p. 287. Wood Lark.

The Wood Lark remains all the year in the country, wintering in the hills of Benjamin and elsewhere in small flocks, and dispersing into the neighbourhood of the olive-yards and woods to nest in spring.

The Wood Lark is a summer visitor to Central and Southern Europe, and winters in the Barbary States. It is resident, but in very small numbers, in Turkey and Asia Minor, but does not reach further into Asia. Thus Palestine is its South-eastern limit.

154. *Ammomanes deserti.* Licht. Verz. Doubl., p. 28. Desert Lark.

The Desert Lark has been found in some plenty on the highlands on both sides of the Dead Sea and in the salt plains of the Ghor. It lives in small bands in winter, and pairs in spring, when it becomes more scattered. Palestine specimens are paler and less rufous than those from the Sahara.

The Desert Lark is confined to the south of the Atlas in Barbary, and thence spreads over the sands of Egypt and Nubia, and as far as Abyssinia. Eastward it inhabits the deserts as far as Scinde.

155. *Ammomanes fraterculus.* Tristram. P. Z. S., 1864, p. 434. Lesser Desert Lark.

The lesser species, which does not consort with its congener, is to be distinguished by its very short and conical bill, and by its throat, which is isabelline colour instead of white, as well as by its smaller size. It is far more widely spread over the barren and desert districts than *A. deserti.* I have not seen it from any other locality, though it is probably the Arabian form.

156. *Calandrella brachydactyla.* Leisl. Wett. Ann. iii., p. 357. Short-toed Lark.

The Short-toed Lark is a summer visitor to Palestine, re-visiting the central country and the north later in spring, and not occurring in the plains or desert in winter.

It inhabits all the countries bordering on the Mediterranean, and extends thence to India, but does not extend into Northern Europe or Asia, and seems to be an inhabitant of the plains, as *C. hermonensis* is of the mountains.

157. *Calandrella hermonensis.* Tristram. P. Z. S., 1864, p. 434. Mountain Short-toed Lark.

I regret that I must venture to differ from my friend, Mr. Dresser, and maintain the distinctness of this species, in which opinion I was more than ever confirmed when in the spring of 1881 I had opportunities, being on the snow-line of Lebanon in April, of observing the Short-toed Lark breeding lower down in the Buka'a, and this species on the highest parts of the mountains. The note and flight differ, especially the former, and this is a far more powerful and varied songster, pouring forth its melody, not on the wing, but perched on the top of a rock, a few yards from his nest. Its larger size, bright rufous coloration, and the distinctness of the black collar, are recognisable at a glance. This species extends over the Persian and Armenian Highlands, and is very probably to be found further west in mountain regions near the Mediterranean, according to the researches of Mr. Dresser.

158. *Calandrella minor.* Cab. Mus. Hein. i., p. 123. Lesser Short-toed Lark.

This, the smallest of the Larks of the country, is a strictly desert bird, less gregarious than most of its congeners, sedentary throughout the year in the few localities where it is found. It seems especially to affect salt plains and hard, not soft, soils.

The Lesser Short-toed Lark has been found only in the desert south of the Atlas, in those of Egypt, Nubia, and Arabia. Further east its place is taken by *C. pispoletta.*

159. *Melanocorypha calandra.* (Linn. Syst. Nat. i., p. 288.) Calandra Lark.

The Calandra Lark is extremely common in spring all over the cultivated open ground, whether on the plains or the hills, where it breeds abundantly. In winter it congregates in large flocks on the maritime plains.

The Calandra seems to be confined to the countries bordering on the Mediterranean; but in Egypt it is only accidental.

160. *Melanocorypha bimaculata.* (Ménétr. Cat. Rais., p. 37.) Eastern Calandra Lark.

The Eastern Calandra breeds abundantly on the higher slopes of Lebanon and Hermon. I did not at first, until Mr. Dresser directed my attention to the fact, detect the differences between this and the Common Calandra, which prove that we have here, in close proximity to the other species, the Oriental Mountain Calandra. It may be distinguished from the other by its shorter tail, and by having all the rectrices tipped with white, while the European species has the outer rectrix almost all white and this colour diminishing towards the centre of the tail.

The Eastern Calandra inhabits Abyssinia, the Caucasus, Central Asia, and North-west India.

161. *Otocorys penicillata.* Gould. P. Z. S., 1837, p. 126. Eastern Horned Lark.

The Eastern Horned Lark is confined to the heights of Lebanon and Hermon, where it is very numerous, descending in winter to the villages at the foot of the range, but not migrating further. I have always found it in winter, spring, and summer close to the snow-line. I found many pairs breeding on the top of Hermon on 2nd June, most having hatched out their young. The nests are very compact, neat, and deep, imbedded in a tuft of *Astragalus* or *Draba,* lined with grass roots.

The Indian *O. longirostris,* from the Himalayas, can always be discriminated, and the Chinese birds collected by Swinhoe are *O. albigula* of the Russian naturalists.

The true *Otocorys penicillata* appears to be confined to Palestine, the Taurid, Caucasus, and the mountains of Northern Persia.

ORDER, PICARIÆ.

FAMILY, CYPSELIDÆ.

162. *Cypselus apus.* (Linn. Syst. Nat. i., p. 344.) Common Swift. Hebr., סוּס and סִיס, A.V., erroneously, 'Crane.' Arabic, صيص, *Sîs.*

The Swift leaves Palestine in November and returns in countless myriads at the beginning of April. Clouds pass in long streams to the north, but still leave prodigious numbers behind. These swarm about all the towns, darting up and down the streets in pursuit of the gnats. It is less abundant in the more desolate parts of the country, though it may be found in flocks in the ravines, but it seems to prefer ruins, mosques, and houses for its nesting places.

I was enabled to detect the true rendering of the Hebrew word *soôs* or *sîs* in a curious way. I had noticed that the Swallow, or at least many individuals, remain through the winter, and had been therefore perplexed by the expression, 'the Crane and the *Swallow* (*Sîs*), observe the time of their coming' (Jer. viii. 7) : and by the soft note of the Swallow being used to symbolize the cry of pain, 'Like a Crane or a Swallow (*Soôs*) so did I chatter' (Is. xxxviii. 14 :) when in the beginning of April, being camped under Mount Carmel, the Swift suddenly appeared. We had shot several, which were spread out in front of my tent. I asked the Arab boys who crowded round, what the birds were, and they all called them *Sîs.* I asked them if they were not *Sununu* (Swallow). They took up a Swallow which was lying there and pronounced it to be the *Sununu.* Here, then, we have the local name handed down unchanged from the Hebrew, and my difficulties at once solved. The most unobservant Arab must notice the sudden return of the Swift, while its note admirably expresses the cry of pain.

The Swift is found in the whole of Africa, visits Europe in summer, and in Asia extends as far as Mongolia, but does not pass south of the Himalayas.

163. *Cypselus melba.* (Linn. Syst. Nat. i., p. 345.) White-bellied Swift.

The Alpine or White-bellied Swift, though very abundant in places, is rather a local bird in the Holy Land. It reappears in the middle of February, and soon the various flocks take to their respective quarters, generally some of the wildest and most inaccessible ravines in the Jordan valley, in the gorges of Moab, and those near the plain of Gennesaret. Their nests are in deep chinks, almost always in the most inaccessible cliffs. Their swiftness is amazing, far surpassing that of the Common Swift, our swiftest bird. They are known to feed often a hundred miles from their nearest resort.

This bird is a summer migrant to Central and Southern Europe, inhabits all Africa, and Asia as far east as India and Ceylon.

164. *Cypselus affinis.* J. E. Gray. Ill. Ind. Zool. i., pl. 35, fig. 2. White-rumped Swift.

This interesting little Swift was first described from Palestine by Antinori, under the name of *Cypselus galilæensis*, but it is proved to be identical with the Indian species, and seems to have had nine different names given to it by various writers. In Palestine, it is, unlike the two other species, a permanent resident, but strictly confined to Ghor or Jordan valley, which it inhabits from Lake Huleh to the south end of the Dead Sea. It flies at a great height; has, instead of the scream of its congener, a soft or melodious wail of three semitones, sharply repeated when alarmed. It breeds in colonies, and has laid its eggs when *C. melba* arrives, and hatched its young before the return of *C. apus.* The nest is most peculiar, attached to the roof of a cave or an overhanging ledge of rock, at a height of from 30 to 400 feet above any accessible standpoint. The nests are clustered side by side, or one under another, formed not like those of any other Swift, but of straw and quill-feathers, strongly agglutinated with the bird's saliva, and without any lining. Sometimes it appropriates the mud-nest of the different Swallows, especially the bottle-shaped structure of *Hirundo rufula*, to its own use, simply adding an agglutinated straw and feather entrance to the original edifice of clay. See *Ibis*, 1865, pp. 76-79.

The White-rumped Swift has a wide geographical range, though more circumscribed than that of the other species, but within that range is confined to comparatively few localities. It is found in tropical West Africa, and nearly to Cape Colony; and in North-east Africa, Arabia, Syria, Persia, India, Ceylon, China, Formosa, and Hainan. But in all these countries vast tracts may be traversed without one being seen. It is remarkable that the single sedentary member of the family should be the one which has the most limited range. In many genera of birds it may be observed that those species which have the most extended northerly have also the most extended southerly range, and that those which resort to the highest latitudes for nidification also pass further to the southward in winter than do the others. Thus the migrating Fieldfare and Redwing, visiting regions north of the Thrush and the Blackbird, on their southward migration leave their more sedentary relatives behind. The Brambling, which passes the Chaffinch in Norway, leaves it also in Europe, and crosses the Mediterranean every winter to the Barbary States. The Egyptian and Collared Turtle Doves remain throughout the year in North Africa and Syria; but the Common Turtle (*T. auritus*), so abundant in these countries in summer, never leaves a straggler behind in November, and yet in spring advances 1,000 miles nearer to the Pole than they do.

FAMILY, CAPRIMULGIDÆ.

165. *Caprimulgus europæus.* Linn. Syst. Nat. i., p. 346. Night-jar.

Visits Palestine in spring and summer. Not noticed in winter.

The Nightjar inhabits Europe and North-western Asia in summer, retiring into Northern and Central Africa in winter. Eastward it has been found in Persia and Turkestan.

3/4

J. Smit lith. Hanhart imp.

CAPRIMULGUS TAMARICIS.

166. *Caprimulgus ruficollis.* Temm. Man. d'Orn. i., p. 438. Red-necked Nightjar.

This Nightjar is only found ordinarily in South-western Europe and North-western Africa. It has straggled as far as England, and I have seen a specimen in Jerusalem, which I have every reason to believe was shot close to the city.

167. *Caprimulgus tamaricis.* Tristram. Proc. Zool. Soc., 1864, p. 170.

PLATE XII.

We obtained this bird, till then unknown, both at the northern and southern ends of the Dead Sea in the month of January, at Ain Feshkah and at Jebel Usdum. It must therefore be a permanent resident in this most desolate region. In form and size it somewhat resembles *C. asiaticus*, but is larger. *C. rufigena*, Smith, from South Africa, corresponds in size, but from both of them it differs decidedly in colouration and markings. I have seen a specimen in the collection of the late Rev. Dr. Herschel, which was obtained near Jericho. These are the only specimens known. It certainly differs from *Caprimulgus inornatus* from Abyssinia, with which it has erroneously been identified, and, so far, remains peculiar to the Dead Sea basin.

FAMILY, PICIDÆ.

168. *Picus syriacus.* Hemp. and Ehr. Symb. Phys. Aves, fol. v., note 5. Syrian Woodpecker. Arab., نقار الشجر, *Nakar esh shajar*, 'the Tree-Drummer,' and نقار الحشب, *Nakar el Hashab.*

This is the only species of Woodpecker found in Palestine, and of course only in the few wooded districts. To the Jordan valley it never descends. The general paucity of timber is quite sufficient to account for the scarcity of this beautiful group of birds, though it is very possible that further research may bring to light other species in Lebanon. It closely resembles our Greater Spotted Woodpecker in size and coloration, but may be at once recognised by the continuous white band from the bill through the eye and ear coverts to the nape of the neck.

The species seems to be strictly limited to Asia Minor, Syria, and Persia.

169. *Yunx torquilla.* Linn. Syst. Nat. i., p. 172. Wryneck.

The Wryneck is a summer migrant to Palestine. I have observed it occasionally wherever there is wood, and sometimes in mere scrub.

It extends from the British Isles to Japan, and down to Central Africa, India, and China.

FAMILY, ALCEDINIDÆ.

170. *Alcedo ispida.* Linn. Syst. Nat. i., p. 179. Common Kingfisher. Arabic, مخيط الما, *Mekhiet el ma.*

Our little English Kingfisher is scattered everywhere throughout the country where there are streams, and also on the shores of the Mediterranean. It is nowhere abundant, and is unaffected by climate, fishing indifferently in the little torrents of the Lebanon, among the ruined columns of Tyre, or in the seething swamps of the Jordan valley. In the Lebanon it is the only species. All the three indigenous species of Kingfisher resort to the shores of the Dead Sea, attracted by the shoal of fishes which are brought down by the fresh water streams, and stupefied by the brine of the lake.

The Common Kingfisher inhabits all temperate Europe, North Africa and Northern Asia, as far as Scinde. Eastward, in India, China, and Japan, it is represented by a closely allied species, *Alcedo bengalensis.*

171. *Ceryle rudis.* (Linn. Syst. Nat. i., p. 181.) Pied Kingfisher. Arabic, صيد السمك, *Saiad el semahk,* 'The Fish-hunter.'

This is the most conspicuous and common species in every part of the country where there is water, salt or fresh. It is particularly abundant about Tyre and Sidon, and round the lake of Galilee. On the plain of Gennesaret there is a great breeding place in the bank of a tiny streamlet, where I found thirty nests. The holes are burrowed a few inches above the water's edge, and unlike the burrow of the Bee-eater, which has a sharp turn about a yard from the entrance, the nest is in a hole scooped by the side of the little tunnel. It is a beautiful sight to watch a party of these birds hovering petrel-like over the water, and now and again making a sudden dive, and instantly resuming their places in the air, their silky plumage gleaming in the sunlight.

The Pied Kingfisher is only an occasional straggler to Europe. It is extremely common in Egypt, and throughout all Africa, south of the Sahara; is rare in Persia, and frequent throughout India, Burmah, and China.

172. *Halcyon smyrnensis.* (Linn. Syst. Nat. i., p. 186.) Smyrna Kingfisher.

The Smyrna Kingfisher was first noticed by Albin in 1760 as from that district, and so named by Linnæus. But it was never again detected in Western Asia till Captain Graves, R.N., re-discovered it near Smyrna, as noticed by Mr. Strickland in an interesting paper. (Ann. Nat. Hist., vol. ix., p. 441.)

We were the first to find it in Palestine, where it is strictly confined to the Jordan valley, though Russell in the last century mentions it near Aleppo. It is therefore not exclusively tropical in its habitat. In its habits it is very different from the lively Pied Kingfisher. It is shy and solitary, never hovers, and sits for hours on its perch over a swamp, its bright plumage well concealed by the foliage, and when alarmed, slinks away under the oleanders. Its food is not fish, but reptiles, frogs and locusts. Like all the other tropical birds of the Jordan valley, it remains throughout the year, from the Dead Sea to the Upper Jordan. It breeds in April in holes on the banks of streamlets on the plain of Gennesaret.

The Smyrna Kingfisher is strictly Asiatic, and as we have seen, most rare and local in Western Asia. Eastward it is more plentiful. I have seen it in Mesopotamia, and it inhabits Southern Persia, India, Ceylon, and China. It is one of the most interesting instances of the extension o the Indian Fauna to the Jordan valley.

FAMILY, CORACIIDÆ.

173. *Coracias garrula.* Linn. Syst. Nat. i., p. 159. Roller. Arab., شرقراق, *Schurkrak.*

The Roller appears in large flocks about the 1st April, and they very gradually disperse themselves over the whole country, breeding in burrows in sand or gravel banks, very often in small colonies, and more frequently a single pair by themselves, in a hollow tree or a rocky cleft. Brilliant and conspicuous, both in plumage, note, and manners, the Rollers attract attention everywhere, and are found in every kind of country alike, woodland, plain, desert, ravines, ruins, always perching where they can see and be seen.

The Roller visits all Europe in summer, though only accidental in the British Isles, and in winter goes as far as the Cape of Good Hope. Its Eastern range is more limited. I found it swarming in Mesopotamia in 1881, and it extends to Cashmere and the Altai mountains.

FAMILY, MEROPIDÆ.

174. *Merops apiaster.* Linn. Syst. Nat. i., p. 182. Bee-eater. Arab., وروور, *Warwar.*

A regular migrant, returning in great numbers in the beginning of April, and living in large societies, breeding in colonies in deep holes in low banks, in which, when the young are fledged, we may find handfuls of the *elytra* of beetles, on which they have fed. Since the introduction of the electric telegraph, its wires are the favourite perch of the Bee-eater, which returns to the same spot after short flights exactly like a Fly-catcher, for hours together.

The Bee-eater is a summer visitor to Southern Europe and Western Asia, as far as Persia; Scinde appearing to be its Eastern limit.

175. *Merops persicus.* Pallas. Reis. Russ. Reichs. ii., p. 708. Blue-cheeked Bee-eater.

The Bee-eater is rare in Palestine, and I have only seen it on passage. But it is extremely abundant in Mesopotamia, where I have found large colonies, both by themselves, and also, as at Jerabulus (Carchemish), in company with a great rookery of the Common Bee-eater, burrowing and nesting in the same bank; but the two species always hunted separately, the Persian bird generally skimming closer to the ground, of weaker flight, and often alighting on a thistle-tuft or flower-stalk.

It ranges down the West Coast of Africa and up to Natal; is very abundant in Egypt, rare in Algeria, extends eastward to India, but is only an accidental visitor to Europe.

176. *Merops viridis.* Linn. Syst. Nat. i., p. 180. Green Bee-Eater.

Not common, and apparently only on passage.

It is a native of Egypt and Abyssinia, is recorded from West Africa, and inhabits Persia, India, Ceylon, and Burmah.

FAMILY, UPUPIDÆ.

177. *Upupa epops.* Linn. Syst. Nat. i., p. 183. Hoopoe. Hebr. דוכיפת. A. V., 'Lapwing,' in error. Arab. هدهد, *Hudhud* (from its cry).

The Hoopoe, which leaves Palestine in winter, returns in the beginning of March; not seen in flocks, but suddenly spread over the whole country in pairs or in small parties. It resorts alike to the desert wâdys, the woods, gardens, and villages, where it is very tame, feeding on dunghills, indifferent to the presence of man. It does not appear to migrate far, as it remains all winter in Egypt and in the oases of the Sahara.

The Hoopoe is found in Southern and Central Europe, the whole of Asia, and Northern and Central Africa.

FAMILY, CUCULIDÆ.

178. *Cuculus canorus.* Linn. Syst. Nat. i., p. 168. Cuckoo. Hebr. (doubtfully) שַׁחַף. Arabic تكوك, *Tekook.*

The Cuckoo returns to Palestine at the end of March or beginning of April, when it is particularly obnoxious to the Bush Babbler (*Crateropus chalybeus*), which clamorously pursues it in the Jordan valley. It is spread generally over the whole country.

The Cuckoo has a very extensive range—through all Europe and

Asia up to the Arctic Circle in summer, and as far as South Africa and India in winter. It goes even as far as Celebes.

179. *Coccystes glandarius.* (Linn. Syst. Nat. i., p. 169.) Great Spotted Cuckoo.

This Cuckoo returns nearly a month earlier than its congener.

For a few days large flocks of them may constantly be seen on their passage northwards, but many remain scattered in the wooded parts of the country.

They have the same parasitic habit as most of the other members of the Cuckoo family. In Algeria they deposit their eggs in the nests of the Mauritanian Magpie, the eggs of which they very closely resemble. In the Holy Land I have found them only in the nest of the Hooded Crow (*Corvus cornix*), and that very frequently. No doubt they will also be found in the nest of the Syrian Jay, which is common in districts like Carmel, where there are no Crows, and where the Spotted Cuckoo abounds.

The Great Spotted Cuckoo has been twice taken in England. It migrates to South-western Europe in summer, and is found through all Africa, but not further east than Syria, excepting that it has been recorded once from Shiraz.

ORDER, STRIGES.

FAMILY, STRIGIDÆ.

180. *Strix flammea.* Linn. Syst. Nat. i., p. 133. Barn Owl. Heb. תַּחְמָס. A. V., Night-hawk.* Arab. معساما, *Masâsah*, and بومي ابيض, *Boomeh abiad*, 'White Owl.'

The Barn Owl may be more often heard than seen, but it is well known to the natives. We met with it occasionally, generally about ruins.

The range of this bird is almost world-wide. Excepting in the extreme northern regions of the Arctic Circle it has been noted everywhere in the Old and New World, and through the islands of the Pacific and Indian Ocean. It has only not been recorded from Japan, China, and New Zealand.

* Nat. Hist. Bible, p. 191.

FAMILY, ASIONIDÆ.

181. *Ketupa ceylonensis.* (Gmel. Syst. Nat., vol. i., p. 287.) Brown Fish Owl.

The appearance of this Great Indian Owl in Palestine is one of the most remarkable features of the singularly mixed character of the Fauna of the country. Prior to our discovery of its existence near the plain of Acre, it had not been noticed west of India. I obtained one specimen and saw three others in the Wâdy el Kurn, close to the great ruin of Kulat el Kurn, north of the plain of Acre, in December, 1863. The bird had been roosting, hidden among the dense foliage of a carob tree, under which we halted, and startled by our voices, scrambled out bewildered, and perched on a rock on the opposite side of the Wâdy.

The Wâdy possesses a perennial stream, well shaded by evergreen timber, and with its cliffs full of caves, while fish and crabs swarm in the water and supply abundant prey for the owl. I never saw the species elsewhere, but five years ago a traveller shot and preserved another specimen in the Wâdy Hamam, opening on to the plain of Gennesaret, thus rendering it probable that the bird will be found in all suitable localities throughout the country.

Ketupa is a peculiar Indian form. Only three species are known. They are Owls of the largest size, distinguished by their bare tarsi, adapted for diving into the water and seizing their prey. One species is confined to the sub-Himalayan region, a second to Java and Sumatra, Borneo, and the Malayan Peninsula, and the present, *K. ceylonensis*, to India generally, Ceylon, and the provinces on the east coast of the Bay of Bengal. It was also discovered by Mr. Swinhoe at Hongkong, not elsewhere in China, so that its range may possibly stretch across from Assam to China. But the present is the one isolated instance of its occurrence west of the Indian Peninsula.

182. *Asio otus.* (Linn. Syst. Nat. i., p. 132.) Long-eared Owl.

The Long-eared Owl is found in the wooded districts, especially in the north. It is rather plentiful in the forest district west of Safed, and there are several pairs which breed in the Cedars of Lebanon.

It inhabits the whole of Europe and Asia south of latitude 64° as far as North-west India, and also China and Japan. In North Africa it occurs in winter. The North American Long-eared Owl is barely separable.

183. *Asio brachyotus.* (J. R. Forster. Phil. Trans. lxii., p. 384.) Short-eared Owl.

The Short-eared Owl is only a winter visitant. I found it once in the hill country of the south, and sometimes in the north.

No Owl, except the Barn Owl, has so extensive a range. The whole world, except Australia and Oceania, is inhabited by it, and even in Oceania it is found in the Sandwich Islands.

184. *Syrnium aluco.* (Linn. Syst. Nat. i., p. 102.) Tawny Owl. Heb., לילית (probably). A. V. 'Screech Owl.'

Not uncommon in the forest districts of Gilead and Bashan, and also all over the wooded portions of Lebanon. It is seldom seen, but its unmistakable hoot can often be heard at night, when encamped in the woods. It descends into the Jordan valley, and I found a nest in a tree in Gilead. The Palestine specimens are, so far as I have noticed, always much greyer than those from England, and are not in the least tawny.

This Owl inhabits the whole of Europe, North Africa to the Atlas range, and Asia Minor and Syria, but has rarely if ever been found east of the Ural mountains.

185. *Scops giu.* Scop. Ann. I. Hist. Nat., p. 19. Scops Owl. Hebr., קפוז (probably). Arab., ماروف, *Maroof.*

Very common in spring about old ruins and olive-groves, returning about the middle of April. It breeds both in the walls of ruins and in hollow trees.

Its note is very peculiar, represented by its Arabic name, or more accurately like *kiu kiu*, repeated monotonously at regular intervals.

It is a summer visitor to Southern Europe, retiring to Africa in winter. It is found eastward as far as Turkestan. In India, China, and Japan it is represented by closely allied species.

186. *Bubo ascalaphus.* Savigny. Descr. Egypte, p. 295. Egyptian Eagle Owl. Hebr., יַנְשׁוּף. Arabic, بعفا, *Bafa.*

This takes the place of the Eagle Owl of Northern Europe, and occurs more frequently than most of the other species, except the Little Owl. In the rolling uplands of Beersheba it resorts to burrows in the ground. In Rabbah (Amman) it has its home among the ruins, and in the ravines of Galilee and in the Jordan valley it retires in security to the most inaccessible caverns.

The Egyptian Eagle Owl has a very limited range, extending from Algeria eastward to Palestine, and especially abundant in Egypt. It has not been observed in Persia or Asia Minor, nor does it seem to reach far into Africa.

187. *Athene glaux.* (Savigny. Syst. Ois. de l'Egypte, p. 45.) Southern Little Owl.

There are five Hebrew words employed to express the Owl, and though it is difficult to decide definitely which species is denoted by each word, it is reasonable to assign the names to the five species most common in Palestine. In the determination we are aided sometimes by the derivation and the context. The Hebrew כּוֹם, A. V., 'Little Owl,' lit. 'Cup,' and spoken of (Psalm cii. 6) as the Owl of ruined places, probably denotes this species; in Arabic, بومي, *Boomeh*, and often called 'the mother of ruins.'

It is one of the most universally distributed birds in every part of the Holy Land. In the olive-yards round the villages, in the rocks of the Wâdys, in the thickets by the water-side, in the tombs, in the wells, or on the ruins; among the desolate heaps which mark the sites of ancient Judah, on the sandy mounds of Beersheba, or on the spray-beaten fragments of Tyre, his low wailing note is sure to be heard at sunset, and himself seen bowing and keeping time to his own music. The Little Owl is a great favourite, and considered lucky by the Arabs. This bird was the symbol of ancient Athens, admirably represented on its coins, the badge of Minerva or of wisdom.

It is very like *Athene noctua*, the Little Owl of Europe, differing only in its smaller size and much paler coloration. It is not found north of the Mediterranean, and seems confined to North Africa, Syria, Persia, and Affghanistan.

ORDER, ACCIPITRES.

FAMILY, VULTURIDÆ.

188. *Gypaetus barbatus.* (Linn. Syst. Nat. i., p. 123, No. 6.) Lämmer-Geier, or Bearded Vulture. Hebr. פֶּרֶס, *i.e.*, 'the Breaker.' Arab. بلج, *Bidj;* and *generically* نسر, *Nissr*, along with the other large Vultures. A. V., 'Ossifrage' (*i.e.*, Bone-Breaker) (Lev. xi. 13).

This magnificent bird, the prince of Vultures, may still be seen in most of the mountainous regions of Palestine, but only singly, or a pair together. Its favourite resorts are the gorges opening on the Dead Sea and the Jordan Valley, especially the ravines of the Arnon and the Callirrhoe, where one of the grandest sights a naturalist can enjoy, as he stands on the brow of a gorge, is the spectacle of one of these majestic birds, with a stretch of wing ten feet across, sailing apparently motionless up and down the valley, close to the crest, and often within a hundred yards of the spectator. Though so conspicuous, the number of Lämmer-Geiers in Palestine might, probably, be counted on the fingers. They are carrion feeders, but marrow-bones, tortoises and snakes are also favourite delicacies. It carries these up to a great height in the air, and then drops them on a rock or stone, repeating the operation till the prey is thoroughly shattered. From this habit is derived its Hebrew name. The poet Æschylus met his death from a tortoise dropped on his bald head by one of these birds. The Lämmer-Geier's range extends from the Pyrenees, Alps, and Carpathians eastward to China and Tartary, south of Altai, being found in all the mountainous regions south of 45° N. Lat., and also in the Atlas range of North Africa. The species of Eastern and Southern Africa very slightly differs from the northern race. Everywhere, even in Tartary, its numbers are rapidly diminishing, and it will probably soon be added to the already long list of extinct species.

189. *Vultur monachus.* Linn. Syst. Nat. i., p. 122. Cinereous Vulture. Not distinguished specifically by the natives from the following species.

It occurs sparingly throughout the country, being chiefly seen in the wild uplands of the south and on the plains of Moab; seldom more than two together. Its black plumage distinguishes it at a glance from the other species, with which it frequently consorts. Its range extends from Portugal and Spain through Sicily, Greece, the whole of North Africa, the Caucasus, Himalayas, North-west India; and it has been found in China.

190. *Gyps fulvus.* (Gmel. Syst. Nat. i., p. 249.) The Griffon Vulture. Hebr. נֶשֶׁר. A. V. 'Eagle.' Arab. نسر, *Nissr.*

There can be no doubt of the identity of the bird called '*Nesher*' by the Hebrews with the Arabic '*Nissr*,' the Griffon Vulture, though rendered 'Eagle' by our translators. Not only is this evident philologically, but the expression in Micah (i. 16), 'Enlarge thy baldness as the eagle' (*Nesher*), can only apply to the Griffon. It is unfortunate that our language has only the one word 'Vulture' for the noble Griffon, and for the despicable, though very useful scavenger, 'Pharaoh's hen,' as Europeans in the East call the Egyptian Vulture.

The Griffon is employed by Orientals as the type of the lordly and noble. Nisroch, the Eagle-headed god of the Assyrian sculptures, was the deification of the *Nissr*, the standard of the Assyrian and Persian armies. 'Calling a ravenous bird from the East,' *i.e.*, Cyrus. From Assyria and Persia the Romans probably borrowed the ensign which has been adopted by so many modern nations, with more appropriateness of character than its bearers would be willing to acknowledge.

The Griffon is the most striking ornithological feature of Palestine. It is impossible in any part of the country to look up without seeing some of them majestically soaring at an immense height, and their eyries abound in great colonies in all the ravines of the country. The most notable colonies of eyries, some of them containing over one hundred pair of birds, are in the Wâdy Kelt, near Jericho, the ravines of the Jabbok, Callirrhoe, and Arnon, the gorge of the Litany river, some ravines near Carmel, and, the most numerous of all, the great 'Griffonries' in the Wâdy Hamam and the Wâdy Leimun, opening on the plain of Gennesaret. There toward evening every jagged rock in the cliffs is the perch of one or more of these noble birds. Many of the characteristics and habits of the Eagle (*i.e.*, Griffon) are alluded to in Scripture. Its soaring, 'They shall mount up with wings as eagles' (Isaiah xl. 31). Its swiftness,

'As swift as the eagle flieth' (Deut. xxviii. 49). 'As the eagle that hasteth to eat' (Hab. i. 8). Its power of sight, 'She dwelleth and abideth on the rock, upon the crag of the rock, and the strong place. From thence she seeketh the prey, and her eyes behold afar off' (Job xxxix. 28, 29). Its nesting in the cliff, 'O thou that dwellest in the clifts of the rock, that holdest the height of the hill, though thou shouldest make thy nest as high as the Eagle, I will bring thee down from thence, saith the Lord' (Jeremiah xlix. 16).

The range of the Griffon Vulture is most extensive. It is found in the whole of Europe from the Pyrenees and Alps southwards, in Southern Russia, through all Asia south of the Altai range, to India and Burmah, and throughout Africa down to the Cape of Good Hope.

191. *Neophron percnopterus.* (Linn. Syst. Nat. i., p. 123 (1766).) Egyptian Vulture, Pharaoh's Hen. Hebr. רָחָם. (A. V., Gier Eagle.) Arab. رخام, *Racham.*

This feeder on filth and offal is universally spread over the whole country in summer, but never seen in winter. It does not breed, like the Griffon, in colonies, but is scattered abundantly and almost equally over all parts of the country, returning from the south about the end of March. It is tame and fearless, and is found in pairs, hanging about the neighbourhood of man, whether in the Fellah village or the Bedawin camp.

The nests, huge clumsy structures of sticks, are generally in the lower parts of the cliffs, and easily accessible, in this respect differing very decidedly from the Griffon's. The birds in the brown plumage of the first year are rarely seen in Palestine, and probably do not migrate from the south.

The range of the Egyptian vulture extends from the south of France eastward to Western India, but it is only a straggler north of the Alps and the Caucasus. It is found throughout the whole of Africa and its islands down to the Cape. The Indian species is very closely allied to it, and has often been identified with it.

FAMILY, FALCONIDÆ.

192. *Circus æruginosus.* (Linn. Syst. Nat. i., p. 130.) Marsh Harrier. Arab. دريعة, *Deri'ah.*

The Marsh Harrier is very common throughout the year over marshes and in all the plains. As many as twenty may often be seen, not together, but quartering the plain independently. The males are generally in very rich plumage, with shoulders and tails silver grey, and the pinions black.

Its range is from the British Isles through Europe and Africa as far as the Transvaal, and eastward as far as Ceylon, Formosa, and Japan.

193. *Circus cineraceus.* Montagu. Ornith. Dict., vol. i., k. 3. Montagu's Harrier.

Probably not uncommon, though it has not often come under my observation. I obtained it twice by the Lake of Galilee in 1864, and on revisiting the same spot in 1881 I secured another, which rose, I believe, on the very same clump of rock as had the former bird. I also noticed it in Moab in 1872.

This bird is found sparsely through all the temperate regions of the Old World, as far as the Cape and China.

194. *Circus cyaneus.* (Linn. Syst. Nat. i., p. 126.) Hen Harrier.

Not at all uncommon on the plains at all times of the year.

It has a less extensive range than Montagu's Harrier, being found through Europe and Northern and Central Asia as far as China and Japan; but in North Africa it is a scarce winter visitant.

195. *Circus swainsoni.* Smith. S. Afr. Quart. Journ. i., p. 384. Pallid Harrier.

This Harrier is more abundant than either of the preceding species, though not a tenth in number of the Marsh Harrier. It is especially to be found along the sea-coast and the maritime plains. I noticed, last year, one of these birds, in almost white plumage, sail for a whole morning round the rock on which a Philistine village was perched, and in its measured beat approaching every ten minutes within a few yards of one of our party who was sketching.

It is a native of Southern Europe, Central and Southern Asia, and found throughout Africa.

196. *Buteo vulgaris.* Leach. System Cat. p. 10. Common Buzzard. Heb. רָאָה. A. V. 'Glede.'* (*Generic.*) Arab. عقاب, *'Ogab* (a term applied to all smaller Eagles and Buzzards).

The Common Buzzard is plentiful on the coast, in the plains, and on Lebanon in winter. It appears to migrate northwards in spring.

The Buzzard is found over the whole of Europe, but not generally beyond its limits. It occurs in winter on the north coast of Africa and in Asia Minor, while its Eastern limit appears to be Western Siberia.

197. *Buteo desertorum.* Daud. Traité d'Orn. ii., p. 162. African Buzzard.

This may probably be counted among the birds of Palestine (though I have never obtained a specimen), as it is found in all the surrounding countries.

It is found throughout Africa, and ranges eastwards as far as India.

198. *Buteo ferox.* Gmel. Syst. Nat. i., p. 260. Long-Legged Buzzard. Arab. شاهين, *Shahin.*

This magnificent and aquiline Buzzard is *the* Buzzard of the country, numerous in all parts and at all seasons of the year. It is gregarious in winter, segregated in pairs in spring, breeding in cliffs in the Wâdys.

The Long-legged Buzzard inhabits Southern Russia, Asia Minor, North-east Africa, Persia, India, and Turkestan.

199. *Aquila chrysaetus.* (Linn. Syst. Nat. i., p. 125.) Golden Eagle. Heb. עָזְנִיָּה (*generic* for all the Eagles). A. V. 'Osprey.' Arab. هقاب, *'Ogab* (*generic* for all the larger Eagles).

The Golden Eagle is not uncommon in winter over the whole country. In summer it is only found in the northern mountain ranges

* See Tristram, 'Natural History of the Bible,' p. 186.

of Lebanon and Hermon. It is scarcely necessary to state that the natives do not discriminate the various species of Eagles.

The home of the Golden Eagle embraces the whole of the northern hemisphere, from Lapland, Siberia, and Arctic America southwards to the Sahara, the Himalayas, China, Pennsylvania and California.

200. *Aquila heliaca.* Savigny. Descript. Egypte, p. 459. Imperial Eagle.

The Imperial Eagle is more numerous throughout the country than the Golden, nor does it, like its congener, disappear in summer, breeding probably in some of the isolated terebinths in retired districts. The Imperial, unlike the Golden Eagle, prefers trees to cliffs for its nidification. It is bold and comparatively indifferent to the near approach of man, and thus its rich dark plumage, white shoulders, and bronzy head may often be admired at leisure. I have had the young of this Eagle brought to me in the Lebanon, proving that it must breed in the neighbourhood.

The range of the Imperial is more limited than that of the Golden Eagle. It is found in South-eastern Europe, North-east Africa, and Southern Asia, from Asia Minor as far as China. In Spain it is replaced by the closely allied form, *Aquila adalberti.*

201. *Aquila clanga.* Pall. Zoogr. Rosso-As. i., p. 351. Greater Spotted Eagle.

This Eagle is not uncommon, especially in winter, when it may often be seen sailing over the plains. I have observed it two or three times in Lebanon in the spring, and found its nest once in a tree in the woods between Nazareth and Acre.

This bird is only a rare straggler in Western and Central Europe. It resides in Southern Russia, Turkey, North-east Africa, Asia Minor, Southern Siberia and India, where it is most numerous, but rarely wanders further east.

The Lesser Spotted Eagle, *Aquila pomarina*, Brehm., Vog. Deutschl., p. 27, is only doubtfully to be enumerated among the birds of Palestine. There is one specimen from Beyrout in the Norwich Museum.

202. *Aquila rapax.* (Temm. Pl. Col. livr. 76, pl. 455.) Tawny Eagle.

The Tawny Eagle is not uncommon in Palestine, chiefly in the wooded and inhabited districts. It breeds in cliffs, and is in the habit of plundering other birds of their booty. Palestine is the extreme Eastern limit of its range, it being an essentially African species, through the whole of which continent it is found, from Barbary to the Cape of Good Hope.

203. *Aquila pennata.* (Gmel. Syst. Nat. i., p. 272.) Booted Eagle.

This, the smallest of our Eagles, occurs, but not commonly, in Palestine. It appears to be confined to the wooded region of Galilee and Phœnicia, and to the Lebanon, where we frequently met with it.

It is found throughout Africa, in Southern Europe, and in Southern Asia as far as India and Ceylon.

204. *Aquila nipalensis.* Hodgs. Asiat. Res. xviii., pt. 2, p. 13. Steppe Eagle.

One specimen of this Eagle in Beyrout was procured on the Lebanon. But it is probably not uncommon, but confounded with its congeners.

Its range comprises South-eastern Europe, Southern Siberia and India.

205. *Aquila bonelli.* (Temm. Pl. Col. i., pl. 288.) Bonelli's Eagle.

This active and beautifully marked Eagle is not uncommon in the Wâdys and rocky terraces of Central Palestine, but appears to avoid the plains. It breeds in the ravines running up from the Plain of Gennesaret. Its general behaviour and habits are more like those of a Falcon than an Eagle.

Bonelli's Eagle is an inhabitant of warm and temperate climates, being found in Southern Europe, Northern and Central Africa, and ranging eastward to India. But it does not occur in China, nor in Africa south of the Equator.

206. *Circaetus gallicus.* (Gmel. Syst. Nat. i., p. 295, No. 52.) Short-toed Eagle.

This Eagle, which lives entirely on reptiles, is beyond all doubt the most abundant of the Eagle tribe from early spring to the commencement of winter in Palestine. A few remain throughout the year, but the greater number retire, probably to Arabia, during the period when the lizards and snakes hybernate. It is a very fearless bird, sitting on a tree or a rock till closely approached, and with its large flat head, its huge yellow eyes glaring round, and brightly spotted breast, is one of the most dignified of the Eagle tribe. The enormous number of lizards and serpents in Palestine accounts for its abundance. The tesselated scaling of its tarsi is a remarkable provision against the possibility of injury from its serpent quarry. It breeds in trees, not in rocks.

The Short-toed Eagle is found throughout Central Europe, but very sparingly. It is not unfrequent in all the countries bordering on the Mediterranean, in Arabia, Persia, Southern Turkestan, and India.

207. *Astur palumbarius.* (Linn. Syst. Nat. i., p. 126.) Goshawk.

No record of the capture of the Goshawk south of the Lebanon is known. But it has been obtained in the mountains near Beyrout, where I have seen it. I have observed it more than once in the mountains of Northern Syria.

Though only an occasional wanderer to the British Isles, the Goshawk is found throughout Europe and Asia, as far as North India and Northern China. In North Africa it is occasionally met with.

208. *Accipiter nisus.* (Linn. Syst. Nat. i., p. 130.) Sparrow Hawk. Hebr., נֵץ (*generic* for all small Hawks.) Arab., باشق, *Bashik.*

Very common about olive-groves and clumps of wood in winter in the south, and also in all the oases and shrubby places about the Dead Sea and Jordan valley. It feeds especially on the Marsh Sparrow (*Passer salicarius*), which swarms in these localities. It disappears in April from the lower country, but remains in the hilly parts of Galilee, in Lebanon, and Hermon.

The Sparrow Hawk is found in the Northern Old World, from Ireland to Japan, and as far south as Northern India, South China, Egypt, and Kordofan.

209. *Accipiter brevipes.* Severzov. Bull. Soc. Imp. Nat. Mosc. xxxiii., p. 234. Levant Sparrow Hawk.

The Levant Sparrow Hawk does not appear to be common in Palestine, but it may easily have been overlooked or mistaken for the other species. I did not notice it till the spring. It is more stealthy in its movements than the other, generally skimming the ground under bushes and up watercourses. When in the hand it is at once recognised by its short thick tarsi.

This bird has a very limited range, and is nowhere abundant. It appears to be confined to Southern Russia, Turkey, Asia Minor, and Syria.

210. *Milvus ictinus.* Savigny. Syst. Ois. d'Egypte, p. 28. Kite or Red Kite. Hebr., אַיָּה, A. V. 'Vulture.'* Arab., ساف, *Essaf.*

The Kite is very common in winter, and gregarious, tame, and fearless, often hovering over camps, or in rain sitting in rows motionless on rocks or trees. Only a few pairs remain to breed, and that only in the central and northern districts.

The Red Kite, once so common in England, is now found but sparingly throughout Europe and North-western Africa. Asia Minor and Palestine seem to be its Eastern limits.

It is perhaps the keenest-sighted of all the birds of prey, whence the allusion in Job xxviii. 7, where A. V. reads 'Vulture.'

211. *Milvus migrans.* Bodd. Tabl. Planch. Enl., p. 28. Black Kite. Hebr., דָאָה or דַיָּה.† Arab., حدايه, *Hadaiyeh.*

No sooner has the Red Kite begun to retire northwards than the Black Kite, never seen in winter, returns in immense numbers from the south, and about the beginning of March scatters itself over the whole country, preferring especially the neighbourhood of villages, where it is a

* See Tristram, 'Natural History of the Bible,' p. 188.

† Ibid., p. 181.

welcome and unmolested guest, feeding on garbage, and not, like its congener, making raids on the poultry. It breeds generally in trees, and is fond of decorating its nest with rags of various colours.

The range of the Black Kite is more limited than that of many of the larger birds of prey. It is found in Southern France, Spain, and Germany, scanty in Italy, common on the North-African coast, and occasional throughout that continent. It occurs in the Volga district, and is rare in Persia, beyond which it very rarely straggles.

212. *Milvus ægyptius.* Gmel. Syst. Nat. i., p. 261. Egyptian Kite.

This species, only distinguished from the former by its yellow bill and more deeply forked tail, replaces it in Africa, but is rarely found outside that continent, excepting in Arabia and Palestine. Here, however, it is by no means so abundant as the Black Kite, and is chiefly found in the hot Jordan valley and the adjacent ravines.

213. *Elanus cœruleus.* Desf. Mém. Acad. R. du Sci., 1787, p. 503. Black-winged Kite.

Rare in Palestine, but seen occasionally in different parts of the country, as in the Jordan valley, and in woods west of Nazareth.

This elegant little Hawk has its true home in Africa, throughout the whole of which continent it seems to occur, being most common in Upper Egypt. It has not been observed in Asia Minor, is only a straggler in Europe, but inhabits India in small numbers. Eastward its place is supplied by representative species.

214. *Pernis apivorus.* (Linn. Syst. Nat. i., p. 130.) Honey Buzzard.

The Honey Buzzard is a constant resident in Palestine, but rather scarce, though found in all parts of the country. I have noticed it near Jaffa, at Nazareth, and near Beyrout.

It is found throughout Europe, though nowhere very numerous, and has been procured in West Africa, Natal, and Madagascar. Palestine is probably its Easternmost limit.

GENUS, FALCO.

215. *Falco peregrinus.* Tunstall. Ornith. Britann. p. 1. Peregrine. Arab. طير الحر, *Tîr el hor*, *i.e.*, 'The noble bird.'

It is very interesting to observe the clearly defined geographical ranges of the different Falcons in Palestine, which they would appear never to transgress. The Peregrine, nowhere numerous, occurs at all times of the year, in all suitable localities near the coast, and on the western slopes of the watershed of Central Palestine. To the eastward of the crest we never observed it. It extends from the Lebanon to the olive-groves of Gaza. During winter I noticed it as far inland as Nazareth, and in March I obtained it in a garden at Jaffa, where it was evidently incubating. Inland it is replaced by the next species.

There seems to be no Hebrew word specially to denote the Falcon, while each species has its distinct Arabic name. This is easily explained by the fact that the Bedawin Arabs, who are keen falconers, and take great pains in training their Falcons, are well aware of the comparative merits and powers of the different species. But though regarded as valuable for the chase, the tame Falcons are far too sparingly scattered to claim a distinct notice in the catalogue of unclean birds in Leviticus.

The Peregrine is almost a cosmopolitan bird. It has been obtained in Greenland, and ranges over the whole Old World from the polar circle to the Cape of Good Hope, Java, and Sumatra, and through the New World from Hudson's Bay to the La Plata. Australia, the Pacific Islands, and the southern extremity of America, are inhabited by species barely separable from the Common Peregrine.

216. *Falco lanarius.* Schl. Rev. Crit., p. 2. *Falco fildeggi.* Schl. Abh. Geb. Zool. iii., p. 3. The Lanner. Arab. صعقر شاهين, *Sakkr Shaheen.*

This is by far the most common of the large Falcons in Palestine, universally distributed throughout the rocky Wâdys on both sides of the Jordan and the Dead Sea, and as far north as the foot of Hermon, being a permanent resident. No region is too desolate or dreary for it. It resorts to the most arid salt wastes south of the Dead Sea, and breeds in the ravines of Moab. It is highly esteemed by the Arab falconers, who train the young birds for the chase of the Hare and Bustard.

The range of the Lanner is somewhat restricted. Its true home is North Africa and Syria, but it occurs in Spain and Turkey. South of the Sahara and east of Palestine it is replaced by other species.

I have retained the name *lanarius* for this bird for convenience, though *feldeggii* has strictly the precedence.

217. *Falco sacer.* Gmel. Syst. Nat. i., p. 273. The Saker Falcon. Arab. صعقر الحر, *Sakkr el hor.*

The Saker Falcon appears to be confined in Palestine to the wild uplands of forests east of Jordan, where it replaces the Lanner. It is the most highly prized of all the species by the Arabs, and the chief tribe of the district, the Beni Sakk'r, take their name from it, and adopt it as their badge.

The Saker is only a straggler to Southern and Central Europe. Its home is east and south; from North Africa it ranges through Western Asia, Persia, Mongolia, to India and China. In North-western China it replaces the Peregrine.

218. *Falco subbuteo.* Linn. Syst. Nat. i., p. 127. The Hobby.

The Hobby is a summer visitant to Palestine, returning rather late in spring, and resorting only to the few wooded districts and the olive groves.

The range of the Hobby extends over the whole of Europe, Africa, and Asia, though rare in Northern Europe and Eastern Asia.

219. *Falco eleonoræ.* Gené. Rev. Zool., 1839, p. 105. Eleonora Falcon.

This fine large Hobby is another scarce summer visitant. I have only noticed it myself in the plain of the Buka'a (Cœle Syria).

It is a gregarious rock-frequenting bird, in contrast with all the habits of the Common Hobby.

Its centre seems to be the islands of the Greek Archipelago, and it seldom has been found beyond the Mediterranean coast, with the singular exception of Madagascar, which it inhabits.

220. *Falco æsalon.* Tunstall. Ornith. Brit., p. 1. The Merlin.

This northern bird visits Palestine in winter. I have found it occasionally in all parts of the country as late as the middle of March.

The Merlin breeds in Northern Europe and Asia, and in winter extends its visits as far as North Africa and China.

221. *Falco vespertinus.* Linn. Syst. Nat. i., p. 129. Red-legged Hobby.

A summer visitant, returning earlier than the Hobby, but scarce, there being few woods suited to its habits. Its food is beetles and locusts.

It is a native of South-eastern Europe, and is found throughout Africa, but not further in Asia than its westernmost countries, Asia Minor and Syria.

222. *Falco tinnunculus.* Linn. Syst. Nat. i., p. 127. Kestrel. Heb. נֵץ (*generic* for all small hawks). Arab. بعشيك, *Bashik.*

The familiar Kestrel is extremely common in every part of the country, east and west, to the confines of the southern desert, throughout the year. It abounds alike in the desolate gorges of the Dead Sea and in the sacred recesses of the Mosque of Omar. It is here more or less gregarious, and associates not only with its own kind, but shares caves with Griffons or Eagles, or utilizes the fringe of the huge nest of the Egyptian Vulture.

The Kestrel inhabits the whole of Europe and Asia, and North, East, and West Africa.

223. *Falco cenchris.* Cuv. Règne Anim. i., p. 322. Lesser Kestrel.

Unlike its congener, the Lesser Kestrel does not remain in Palestine for the winter, but returns in March, and at once consorts with the Common Kestrel. It is seen everywhere, especially towards evening, in the cultivated grounds about the villages, in pursuit of Cockchafers and Beetles. It breeds in communities, in very deep fissures in the rocks, choosing sometimes clefts in a town, as at Nazareth. The Arabs distinguish it from the other species, calling it very appropriately the 'White-nailed Bashîk.'

The Lesser Kestrel inhabits Southern Europe, North Africa, and

South-western Asia, migrating in winter into Central and Southern Africa. The species of India and China is very closely allied to it, and has only of recent years been discriminated.

224. *Pandion haliætus.* (Linn. Syst. Nat. i., p. 129 (1766)). Osprey. Heb. עָזְנִיָּה. Arab. كتاف, *Ketāf;* بو خاتن, *Bou-khatem.*

The Osprey affects especially the Syrian shores and the streams which flow into the Bay of Acre. It may always be seen by the little lagoons near the mouth of the Kishon. Though not numerous, it was likely, from its fish-eating habits, so different from those of other Eagles, to have a specific name among the Hebrews. It has not been observed on the Lower Jordan, but I have seen it on the Jabbok and in the Huleh marshes under Hermon.

The Osprey is almost literally cosmopolitan. It is found throughout the whole northern hemisphere, from the Arctic Seas southward; through Africa as far as the Cape; through Australia, New Zealand, and South America as far as Brazil.

ORDER, STEGANOPODES.

FAMILY, PELICANIDÆ.

225. *Phalacrocorax carbo.* Linn. Syst. Nat. i., p. 216. Cormorant. Hebr., שָׁלָךְ. Arab., عقق, *'Akak.*

The Cormorant is found on the coast and on all the inland waters very plentifully. It is always to be seen sitting on a snag at the mouth of the Jordan, watching for the fishes, stupefied by the brine as they enter the sea.

The Cormorant is found in suitable situations throughout Europe, Africa, and Asia. Some identify with it the species of Australia and New Zealand.

226. *Phalacrocorax pygmæus.* (Pall. Reise. ii., p. 712, Anhang.) Pygmy Cormorant.

The Pygmy Cormorant is to be found on the Leontes and other streams flowing into the Mediterranean. I did not observe it on the

14—2

Lake of Galilee, where our Cormorant is very common. I discovered in 1881 a great breeding colony of Pygmy Cormorant in the reedy islets of the Lake of Antioch, where this bird was nesting in hundreds in society with the Snake Bird of Africa (*Plotus levaillantii*), and the Common Tern (*Sterna hirundo*), hatching about the end of May.

The Pygmy Cormorant is found in South-east Europe from the Danube southwards, in North-east Africa, and Southern Asia, as far as India, Java, and Borneo; but it does not appear to extend towards China.

227. *Pelecanus onocrotalus.* Linn. Syst. Nat. i., p. 215. Roseate Pelican. Hebr., קָאַת. Arab., جمل البحر, *Djemel el bahr*, ابو جراب, *Abu djirab.*

The Roseate Pelican is frequently found on the Sea of Galilee, though I never myself was fortunate enough to find it there until my visit in 1881. I also observed a flock of this species mingled with the next off Tyre.

The Roseate Pelican has a limited range from the Danube to the east of the Mediterranean, North-east Africa, and Syria, to the Black Sea and the Caspian.

228. *Pelecanus crispus.* Bruch. *Isis*, 1832, p. 1109. Dalmatian Pelican.

The Dalmatian Pelican is more abundant than its congener. I have seen an immense flock pursuing their singularly gyrating flight near Mount Carmel. It is generally to be found at Lake Huleh.

The winter limits of the Dalmatian are much the same as those of the Roseate Pelican. But it extends further to the eastward, being frequent in Western India.

FAMILY, PLOTIDÆ.

229. *Plotus levaillantii.* Licht. Verz. Doubl., p. 87. African Darter.

PLATE XIII.

Though I have not actually obtained this bird within the confines of Palestine proper, yet, as I have discovered a great breeding colony on

Pl. XIII.

Smit lith.

Hanhart imp.

PLOTUS LEVAILLANTII.

the Lake of Antioch, which the bird could not possibly have reached without passing through Palestine, I think it fairly deserves a place here.

The vast shallow Lake of Antioch, extending over many miles, is very shallow, and swarms with eels. The Darters arrive from the south about the end of April, and have not hatched out till the beginning of June. As soon as the young can fly they at once suddenly depart, and are never seen again till the end of the next spring. The lake is full of islets, covered with coarse grass and a dwarf sort of marsh myrtle. For its nest the Darter merely seems to tread down a tuft of coarse grass or rushes, or press down the centre of a little bush. Wherever there is a stick it may be seen perched, hanging down its wings as if they were broken, or as if the bird were hanging itself out to dry.

What makes the discovery of this breeding colony most interesting is that Syria would appear to be entirely beyond the range of the African Darter, while the Indian species (*Plotus melanogaster*) might more reasonably have been looked for. *Plotus levaillantii* is known from South Africa, the Zambesi, the Niger, Senegambia, and the Suaheli country, but has not yet been noticed in North-east Africa, neither in Egypt, Nubia, nor the Red Sea; while in Madagascar the Indian species alone is found.

ORDER, HERODIONES.

FAMILY, ARDEIDÆ.

230. *Ardea cinerea.* Linn. Syst. Nat. i., p. 236. Common Heron. Hebr., אֲנָפָה. Arab., دنكالة, *Dunkaleh,* and غرنوق, *'Arnug.*

The Common Heron is scattered in suitable situations through every part of the country, especially about the marshes of Huleh, on the Jordan, Lake of Galilee, Kishon, and the coast.

The Heron inhabits Europe, all Africa, and all Asia.

231. *Ardea purpurea.* Linn. Syst. Nat. i., p. 236. Purple Heron.

The Purple Heron inhabits the same districts as the former species, but in smaller numbers.

It is found in Central and Southern Europe and Asia, and throughout Africa.

232. *Ardea alba.* Linn. Syst. Nat. i., p. 239. Great White Egret.

The White Egret is to be found in small numbers by the Sea of Galilee and the Lake Huleh throughout the spring and summer. I am not able to state whether it remains through the winter, never having observed it until March.

The Great White Egret inhabits South-eastern Europe, Western Asia, and North Africa.

233. *Ardea garzetta.* Linn. Syst. Nat. i., p. 237. Lesser Egret.

The Lesser Egret is to be found throughout the year scattered over the country, generally occurring singly by pools, ditches, or in stagnant swamps. It is a shy and solitary bird, but breeds in a large colony on the Huleh swamp.

The Lesser Egret is found in South Europe, the whole of Africa, all Southern Asia, the Malay Archipelago, and Australia.

234. *Ardea bubulcus.* Audouin. Expl. de l'Egypte, i., p. 298. Buff-backed Heron. Arab., ابو بقر, *Abou bekkr.*

Common, in large flocks, in the swamps of Huleh, and in smaller numbers on the marshy spots of the whole country.

The Buff-backed Heron is found in all the countries bordering on the Mediterranean, but is not recorded from Asia Minor. It extends through the whole of Egypt, but Palestine is the only country within the limits of Asia where it has yet been noticed.

235. *Ardea ralloïdes.* Scop. Ann. I. Hist. Nat., p. 88. Squacco Heron.

Found in the same localities as the preceding, but in much smaller numbers.

The Squacco Heron inhabits the whole of Africa and Madagascar, and all the countries bordering on the Mediterranean and Black Sea, and as far as the shores of the Caspian.

236. *Ardetta minuta.* (Linn. Syst. Nat. i., p. 240.) Little Bittern.

The Little Bittern is plentiful in the rushes and reeds round Lake Huleh, and I have occasionally flushed it in swamps in other parts of the country. It is probably far more numerous than it appears to be, from its habit of skulking among the rushes and refusing to take wing when disturbed. We found more nests than we saw birds.

The Little Bittern is found throughout temperate and Southern Europe, Northern Africa, and as far east as the Himalayas.

237. *Nycticorax griseus.* (Linn. Syst. Nat. i., p. 239.) The Night Heron.

The Night Heron is found in small numbers, never in flocks, about Lake Huleh and Gennesaret.

The Night Heron is found in Southern Europe, the whole of Africa, Asia, even to Japan and the Philippines, and North America.

238. *Botaurus stellaris.* (Linn. Syst. Nat. i., p. 239.) The Bittern. Heb., קִפֹּד.

Inhabits the marshes of Huleh, and probably other suitable localities.

The Bittern is found throughout Europe and Asia, except in the extreme North, and throughout the whole of Africa.

FAMILY, CICONIIDÆ.

239. *Ciconia alba.* Bechst. Nat. Deutschl. iii., p. 41. White Stork. Heb., חֲסִידָה. Arab., لقلق, *Laklak;* بالارج, *Balaredj.*

The White Stork is, in Palestine, a regular, though for the most part a passing, migrant. During the whole of April it covers the land, suddenly appearing in the south, and moving northwards a few miles a day. Thus we heard at Gennesaret that the country about Samaria was covered with Storks, when we had not seen one. Two days afterwards they overspread our neighbourhood; not close together, but scattered over hill and valley, plain and marsh alike, steadily quartering the ground about 100 yards apart, picking up snakes, lizards, frogs, or fish, according to the locality, and quite indifferent to the presence of man. A few pairs remain here and there to breed, notably about the ruins of deserted cities. They are never molested by the natives, and are looked on as a sacred bird.

The Stork, though now only a straggler in Britain, is a summer visitant to all the neighbouring countries of the Continent, wherever there are marshes. In North Africa also it is only a summer resident, wintering in Central and Southern Africa. Through Asia it is found as far as Japan, and in winter in India.

240. *Ciconia nigra.* (Linn. Syst. Nat. i., p. 235.) Black Stork. Arab, بالازان, *Balazan.*

The Black Stork is found all through the winter in small bands on the barren plains near the Dead Sea, never visiting the upper country. I have been told it breeds on oak trees in Bashan, but have not met with it there in my short visits to that region.

The Black Stork is found, though in scanty numbers, throughout Central and Southern Europe, from South Sweden and Denmark eastwards, especially near the Danube and the Caucasus. It is also an inhabitant of North and North-eastern Africa. It is rare in India, but common on the Amoor. I have frequently met with it on the Euphrates, but always solitary.

FAMILY, PLATALEIDÆ.

241. *Platalea leucorodia.* Linn. Syst. Nat. i., p. 231. Spoonbill.

The Spoonbill is only an occasional visitor to Palestine. I have seen it only in a local collection at Jerusalem.

It inhabits Central Europe, all the countries bordering on the Mediterranean, North-east Africa, South-western Asia, the Caspian, Persia, India, and China.

242. *Ibis falcinellus.* (Linn. Syst. Nat. i., p. 241.) Glossy Ibis.

I have only occasionally seen the Glossy Ibis in Palestine, and it is not there, as in Algeria, a certain companion of the Buff-backed Heron in the same proportion of the black sheep to the white in a flock.

The Glossy Ibis ranges over the greater part of the temperate and tropical regions, both of the Old and the New Worlds, including Australia.

The Hebrew תִּנְשֶׁמֶת, rendered A.V. 'Swan,' is probably the Purple 'Gallinule,' or the 'Ibis,' by which words it is rendered in the LXX. and other ancient versions.

FAMILY, PHŒNICOPTERIDÆ.

243. *Phœnicopterus roseus.* Pall. Zoog. Ross-As. ii., p. 207. Flamingo. Arab., نجاف, *Nihaf.*

The Flamingo is frequently seen in all suitable parts of the country, although its breeding place has not been discovered. I have obtained both adult and young near the mouth of the Kishon, where I have observed it at all seasons of the year, but never in large flocks.

The Flamingo inhabits Southern Europe, the whole of Africa, and Asia as far as India. It is especially abundant in salt lakes and lagoons.

ORDER, ANSERES.

FAMILY, ANATIDÆ.

244. *Anser cinereus.* Meyer. Tasch. Deutsch. Vög. ii., p. 552. Grey Lag Goose. Arab., وز, *Wuz* (*generic*).

Our Wild Goose is a winter visitant; I saw it at Jaffa in 1881.

The Wild Goose summers in all Northern Europe and Asia, wintering as far south as North Africa and India.

245. *Anser segetum.* Gmel. Syst. Nat. i., p. 512. Bean Goose.

Occasional in winter on the coast. It is brought into the markets of all the sea-board towns.

The Bean Goose breeds in Northern Europe and Asia, but does not go so far south in winter as the Grey Lag, seldom either crossing the Mediterranean or visiting India.

246. *Anser albifrons.* Scop. Ann. I. Hist. Nat., p. 69, No. 87. White-fronted Goose.

The White-fronted Goose has been to my knowledge procured off Beyrout, and I am told that it visits the coast every winter.

It inhabits the Northern Palæarctic region, is very common in winter in Egypt, and also reaches India.

247. *Anser brenta.* Pall. Zoog. Ross-As. ii., p. 229. Brent Goose.

Accidental in winter. I never heard of the Barnacle Goose being found.

The Brent Goose breeds in Arctic Europe, Asia, and America, migrating south in winter.

248. *Cygnus olor.* (Gmel. Syst. Nat. i., p. 501.) Mute Swan. Arabic, اردف, *Ardef.*

On passage in winter.

The Mute Swan inhabits Northern and Eastern Europe, North-east Africa, and Asia as far as Kashgar and the Punjaub.

249. *Cygnus musicus.* Bechst. Gem. Nat. Vög. Deutschl. iii., p. 830. Whooper Swan.

A fine specimen of *Cygnus ferus* was brought to me at Jerusalem on 26th December, which had been shot on the pools of Solomon a day or two before by Dr. Chaplin. The Swan, one or other species, is well known to the Arabs.

The Whooper Swan inhabits the whole of Northern Europe and Asia in summer, migrating in winter sometimes as far south as Egypt. It has not been observed south of the Himalayas.

250. *Chenalopex ægyptiaca.* (Linn. Syst. Nat. i., p. 197.) Egyptian Goose. Arab., وز, *Wuz.*

The Egyptian Goose is frequently to be seen by the Dead Sea throughout the year, and occasionally on the coast.

It is found through all Africa from North to South, and has been introduced and semi-domesticated in many parts of Europe. It has not been recorded from Asia, except in Palestine.

251. *Tadorna casarca.* (Linn. Syst. Nat. iii., App., p. 224.) Ruddy Sheldrake.

The Ruddy Sheldrake resides in the Sebkha Safieh, at the south end of the Dead Sea, and probably breeds in the cliffs of the Arabah. In the north it breeds among the Griffon Vultures, in the cliffs of ravines near the Lake of Gennesaret.

The Ruddy Sheldrake is an inhabitant of Southern and Eastern Europe, and of the greater part of Asia as far as India, China, and Japan. It is very different in its habits from most ducks, breeding in cliffs and holes in trees, and seems to affect especially glens in the neighbourhood of salt lagoons.

252. *Anas boschas.* Linn. Syst. Nat. i., p. 205. Wild Duck. Arab. بطا, *Batta ;* براق, *Brack* (*generic*).

Common throughout the country in winter.

The Wild Duck inhabits all Europe, North Africa, Asia and North America.

253. *Anas strepera.* Linn. Syst. Nat. i., p. 200. Gadwall.

The Gadwall occurs, mingling with other species, in winter.

This bird has an extensive range, being found throughout Europe, Asia, North Africa, and all North America.

254. *Anas angustirostris.* Ménétr. Cat. Rais., p. 58. Marbled Duck.

This rare Duck resides throughout the year in the swamps of the Huleh, in great numbers, very wary, and breeding in papyrus swamps wholly inaccessible. In summer it is almost the only Duck to be found there. In flight and manners it much resembles the Teal.

The Marbled Duck is found sparingly in Spain, Algeria, the Caucasus, and Scinde. I obtained a specimen at Alexandria. It appears to be very local in all the countries where it resides.

255. *Anas acuta.* Linn. Syst. Nat. i., p. 202. Pintail.

Not at all uncommon in winter on the coast, on the Jordan, and on all the little streams by the Dead Sea. My companion, Rev. Mowbray Trotter, shot a fine hybrid between Pintail and Mallard one morning, close to our tents, when encamped on the upland plains of Moab, in March, 1872.

The Pintail inhabits all Europe, Asia, North Africa, and North America almost to Panama.

256. *Anas crecca.* Linn. Syst. Nat. i., p. 204. Common Teal.

The Teal is scattered everywhere in small numbers, not only in swamps and by streams, but even by isolated little springs where there are a few patches of rushes. I think it must remain to breed, as I have found it in pairs in June.

The Teal extends over the whole of Europe, North and North-eastern Africa, and all Asia. The North American species is very closely allied.

257. *Anas circia.* Linn. Syst. Nat. i., p. 204. Garganey.

I have not taken the Garganey myself, but have seen it in local collections, both at Jerusalem and Beyrout.

The Garganey is an inhabitant of freshwater lakes, streams, and swamps throughout Europe, North and North-east Africa, and all Asia; sparingly towards the east, though it reaches the Philippines and even Celebes.

258. *Spatula clypeata.* (Linn. Syst. Nat. i., p. 200). Shoveller.

The Shoveller visits Palestine in winter, resorting to little pools and swampy spots by the streams, but never in large numbers together. I have seldom flushed more than a pair together.

The Shoveller inhabits all Europe, North Africa, the whole of Asia, and North America.

259. *Mareca penelope.* Linn. Syst. Nat. i., p. 202. Wigeon.

The Wigeon is common everywhere in winter, especially in company with the Pochard.

The Wigeon is an inhabitant of Europe, North Africa, Northern and Central Asia as far as China.

260. *Fuligula marila.* (Linn. Syst. Nat. i., p. 196.) The Scaup.

Occurs in winter on the coast.

The Scaup breeds in Northern Europe, Asia and America, retiring in winter as far as North Africa, India, China, and Mexico.

261. *Fuligula ferina.* (Linn. Syst. Nat. i., p. 203.) Pochard.

The Pochard is by far the most abundant of all the Duck tribe throughout the winter, and its numbers possibly exceed those of all the other species combined. Wherever there is a little bit of water screened from observation there is a flock of Pochards. They are found sparingly on the Dead Sea, where certainly they can find no food, but are themselves the chief prey of the Lanner Falcon. We once shot a Lanner with its captured Pochard, both of which fell into the Dead Sea.

The Pochard is found throughout Europe, North and North-east Africa, and Asia.

262. *Fuligula cristata.* (Leach. Syst. Cat. M. and B. Br. Mus., p. 39.) Tufted Duck.

The Tufted Duck occurs in winter in small numbers, mingled with flocks of other species, in all the reedy spots of the Jordan and elsewhere.

It is found in all Europe, Asia, and North Africa.

263. *Nyroca ferruginea.* (Gmel. Syst. Nat. i., p. 528.) White-eyed Duck.

The White-eyed Duck is very generally distributed in winter, consorting especially with the Pochards, but remaining when the latter have left. I believe it breeds in the upper parts of the Jordan valley, but have not found the nest.

The White-eyed Duck has a more restricted range than many of its congeners, extending through Central and Southern Europe, North Africa, and Central Asia as far as India.

264. *Œdemia nigra.* (Linn. Syst. Nat. i., p. 196) Common Scoter.

I have noticed a flock of Black Scoter in winter more than once, close under Carmel, on the sea. I felt certain they were of the common species, but it has been suggested to me that more probably they were the Black Scoter, yet were it so I could scarcely have failed to observe the white speculum.

Both Scoters are natives of Northern Europe and Asia, and rarely go so far south as Palestine in winter.

265. *Erismatura leucocephala.* (Scop. Ann. I. Hist. Nat., p. 65.) White-headed Duck.

The White-headed Duck may be seen at any time of the year on the Lake of Galilee and on Huleh, swimming and diving with its tail upright. It probably breeds in the recesses of Huleh, certainly nowhere on the Lake of Galilee.

It inhabits lagoons and lakes in Southern Europe and North Africa, as well as Asia Minor and Syria, and has been found as far eastward as Turkestan.

266. *Mergus serrator.* Linn. Syst. Nat. i., p. 208. Red-breasted Merganser.

The Merganser is common on the coast in winter.

It is a native of North Europe, Asia, and America, descending in winter as far as North Africa, North China, and the Southern United States.

267. *Mergus albellus.* Linn. Syst. Nat. i., p. 209. Smew.

I obtained one specimen on the coast near Carmel.

The Smew is a native of Northern Europe and Asia, ranging in winter to the Mediterranean, North India, and North China.

ORDER, COLUMBÆ.

FAMILY, COLUMBIDÆ.

268. *Columba palumbus.* Linn. Syst. Nat. i., p. 89. Ringdove or Wood Pigeon. Heb., יוֹנָה, *Pigeon;* גוזל, *Young Pigeon.* Arab., حمامة, *Hamamat* (*generic*).

The Ringdove is spread in countless myriads over the wooded parts of the country at the season of migration, both vernal and autumnal. In Gilead I have seen a migration which can only be equalled by the descriptions we read of the flights of Passenger Pigeons in America. The bulk of these prodigious flocks pass on, but a considerable number remain throughout the winter, and are taken in great quantity by the villagers by means of a decoy-bird with its eyelids sewn up, tied to a perch. Its struggles attract a crowd of its fellows. The Ringdove, like the other *Columbidæ*, feeds largely on the leaves of the leguminous plants, which abound in the country, coming into leaf in the Jordan valley throughout the winter.

The Ringdove is found throughout the greater part of Europe, and in the Barbary States, breeding in the Atlas, but generally there migratory. It does not appear to pass the Ural range, and Asia Minor and Palestine are its south-eastern limits.

269. *Columba œnas.* Linn. Syst. Nat. i., p. 279. Stock Dove.

The Stock Dove is rare in Palestine, but is to be met with any day in the regions of Gilead and Bashan, as well as near Jericho.

The Stock Dove inhabits the temperate parts of Europe, the Barbary States, Asia Minor, Syria, and Persia.

270. *Columba livia.* Bonnat. Encycl. Méth. i., p. 227. Rock Dove.

This species is extremely abundant on the coast and the highlands west of Jordan. The specimens from this district are identical with those fr m Northern Europe.

The Rock Dove is spread over the whole of Europe and North-Western Asia. Eastward its place is taken by closely allied species.

271. *Columba schimperi.* Bp. Consp. Gen. Av. ii., p. 48. Ash-rumped Rock Dove.

In the interior of Palestine and in the Jordan valley this species takes the place of the more northern form, *C. livia.* Its ashy rump, and the lighter hue of the lower parts, separate it at a glance. The myriads of these birds are beyond computation. The Wâdys, with precipitous cliffs of soft limestone, honeycombed by caves and fissures, are admirably adapted for them, and several are named from them 'Wady Hamam'—*i.e.*, Ravine of Pigeons. Their swift flight, and their roosting-places far in the fissures, secure them from the attacks of the many hawks which share the caverns with them. It is this bird which is alluded to by the Prophet Jeremiah: 'Ye that dwell in Moab, leave the cities and dwell in the rock, and be like the Dove that maketh her nest in the sides of the hole's mouth' (xlviii. 28).

This race of Rock Dove seems to be peculiar to Egypt, Abyssinia, and Palestine.

272. *Turtur communis.* Selby. Nat. Lib. Orn. v., p. 169. Turtle Dove. Heb., תּוֹר.

The return of the Common Turtle Dove is one of the most marked epochs in the ornithological calendar. 'The Turtle and the Crane and the Swallow observe the time of their coming' (Jer. viii. 7). 'The voice of the Turtle is heard in our land' (Cant. ii. 12). Search the glades and valleys in March, and not a Turtle Dove is to be seen. Return at the beginning of April, and clouds of Doves are feeding on the clovers of the plain. They stock every tree and thicket. At every step they flutter up from the herbage in front—they perch on every tree and bush—they overspread the whole face of the land, and from every garden, grove, and wooded hill pour forth their melancholy yet soothing ditty unceasingly from early dawn till sunset.

The native regions of the Turtle Dove are the countries surrounding the Mediterranean, not extending eastward of Persia.

273. *Turtur risorius.* (Linn. Syst. Nat. i., p. 285.) Collared Turtle Dove.

No birds better illustrate the geographical position of Palestine than the Turtle Doves. Here we find three species, one European, one Ethiopian, and one Indian (the present one) all meeting together. Of these the European, *T. communis*, is by far the most abundant, but only in spring and summer. *T. senegalensis*, on the contrary, the Ethiopian, is a permanent resident, not increasing its numbers by immigration, confined chiefly to the neighbourhood of the Dead Sea and the Lower Jordan, but residing throughout the year in the court-yards of houses in Jerusalem and the Temple Area, where, from its tame and confiding habits, it appears to be semi-domesticated. The present species, the Indian Collared Turtle, perhaps the handsomest as well as the largest of the group, is also a permanent resident round the Dead Sea, but in winter only in small numbers, very shy and wary. In spring its numbers are largely increased, and it spreads itself through the greater part of the country up to Galilee, and breeds everywhere, in trees and bushes, usually in small colonies of eight or ten together.

T. risorius has been obtained at Constantinople, but with this exception is a strictly Asiatic form, extending over the whole of Southern Asia.

274. *Turtur senegalensis.* (Linn. Syst. Nat. i., p. 283.) Palm Turtle Dove.

As stated above, this species is permanently resident in the warm nooks of the Jordan valley, round the Dead Sea, and about Jerusalem and Gaza. It never there migrates, and in summer is the least abundant of the three species.

The Palm Turtle Dove inhabits the whole of Africa, except the Barbary States, north of the Atlas. It especially affects palm groves. The Indian and Asiatic rorm, *Turtur cambayensis*, is, I consider, clearly separable.

FAMILY, PTEROCLIDÆ.

275. *Pterocles arenarius.* (Pall. Nov. Com. Petrop. xix., p. 418.) Black-bellied Sand Grouse. Arab. قطا, *Khata* (*generic*).

This large Sand Grouse inhabits the sandy wastes to the north-east and south of Palestine; more plentiful towards the north, in the Lejah, where we obtained it. I also met with it in Moab.

It inhabits Spain, the northern deserts of Africa, North Arabia, Persia, and the southern deserts of Central Asia as far as Scinde.

276. *Pterocles alchata.* (Linn. Syst. Nat. i., p. 276.) Pintail Sand Grouse.

In all the desert regions round Palestine; more common in the south, near Beersheba.

The Pintail Sand Grouse inhabits Spain, North Africa, Arabia, and thence to the Punjab and Scinde.

277. *Pterocles senegalus.* (Linn. Syst. Nat. i., p. 277 = Buff. Pl. Enl. 130.) Senegal Sand Grouse.

The Senegal Sand Grouse is the most universally distributed on all sides of Palestine, and the only one which actually breeds in the Jordan valley. It is scattered all over the highlands of Moab, where we obtained many specimens in spring.

This bird, of which *P. guttatus* (Licht.) is the female, is found in North-east Africa, Arabia, and the deserts west of the Tigris.

278. *Pterocles coronatus.* Licht. Verz. Doubl., p. 65. Crowned Sand Grouse.

Very rare, but found in the Syrian desert.

It inhabits the Sahara, North-east Africa, Arabia, and Syria.

279. *Pterocles exustus.* Temm. Pl. Col. 359, 360. Singed Sand Grouse.

This Sand Grouse is very common in the southern deserts and on the east of Jordan. It is amazingly abundant in the Syrian desert between Damascus and Palmyra

The Singed Sand Grouse is abundant in Egypt, and in the Sahara, to the southward of the Black-breasted and Pintail species. It is common in all the deserts of Western Asia and India, but not further eastward.

ORDER, GALLINÆ.

FAMILY, PHASIANIDÆ.

280. *Caccabis chukar.* (G. R. Gray, in Cuv. ed. Griff. iii., p. 54.) Chukor Partridge. Hebr., קֹרֵא. Arab., حجل, *Hadjel.*

The Chukor Partridge is the game bird, *par excellence*, of Palestine. It is wonderful to find it so plentiful everywhere in the hill country, among the rocks and bushes, where it has no protection and every conceivable foe. Yet from the Lebanon to the south of Judæa it is found, except in the low-lying plains and in the Jordan valley. The former it surrenders to the Francolin, the latter to the Sand Partridge (*A. heyi*). Its eggs are quite distinct from those of the closely allied Greek Partridge, and are smaller than those of any of its congeners, though it is almost the largest species of its genus. In the east of Moab we find only the very pale variety, which has been named *C. sinaica*, and is the characteristic form of the Sinaitic Peninsula, but the distinction is only in the paler coloration, doubtless from the arid and hot climate, and it can only be looked on as a climatic variety.

The only part of Europe where the true *Caccabis chukar* is found are the Greek Islands. It inhabits Asia Minor, Syria, North Persia, Scinde, and the region to the immediate north and south of the Himalayas, and extends into Northern China.

281. *Ammoperdix heyi.* (Temm. Pl. Col. 328, 329.) Hey's Sand Partridge.

This beautiful little Partridge takes the place of the Chukor in the Dead Sea basin and the ravines of the Jordan valley. In its restricted habitat it is very numerous. It lays its eggs in fissures of the rocks and holes in caves.

This is doubtless 'the Partridge in the wilderness' alluded to in the history of David, being plentiful about Engedi, where no other Partridge is found.

Hey's Partridge appears to be exclusively confined to the rocky regions of the Sinaitic Peninsula and the Dead Sea, very rarely in Nubia, and perhaps Abyssinia. In Northern Syria and Mesopotamia its place is taken by Bonham's Partridge, an allied species.

282. *Francolinus vulgaris.* Steph. Shaw. Gen. Zool. xi., p. 319. The Francolin. Arab., كحال, *Kohal.*

The Francolin is found in all the swampy parts of the Holy Land, in the plains of Sharon, Acre, and Esdraelon, Gennesaret, and Shittim, and also by the Huleh. It is not numerous, except in Gennesaret, but is widely spread.

The Francolin is now extinct in Europe, though it formerly inhabited Spain, Sardinia, Candia, and the Greek Islands. It still exists in Cyprus, Asia Minor, and Syria, and thence through Mesopotamia, where I shot it, Southern Persia, and North India by the Ganges as far as Assam.

283. *Coturnix communis.* Bonn. Enc. Méth. i., p. 217. Quail. Hebr., שְׂלָו. Arab., سلوا, *Salwa.*

A few pairs of Quail may be found here and there all through the winter; but in March they return by myriads in a single night, and remain to breed in all the open plains, marshes and cornfields, both in the Ghor and the upper country.

The geographical range of the Quail is immense, comprising the whole of Europe, Asia, and Africa, except the Arctic regions, being everywhere more or less migratory, and the only bird of its family which is so.

ORDER, FULICARIÆ.

FAMILY, RALLIDÆ.

284. *Rallus aquaticus.* Linn. Syst. Nat. i., p. 262. Water Rail.

The Water Rail is found in suitable situations in all parts of Palestine, at all seasons of the year. I obtained it even in a salt pool at the south end of the Dead Sea.

It is found through Europe, North Africa, Asia as far as China, but in Japan is replaced by a closely allied species.

285. *Porzana maruetta.* Leach. Syst. Cat. Br. Mus., p. 34. Spotted Crake.

I found the Spotted Crake in a little stream flowing into the sea, in a thin fringe of rushes, in 1881, not having previously met with it.

The Spotted Crake ranges through Europe, North Africa, and in Asia as far as India, but very scarce towards the east.

Baillon's Crake and the Little Crake (*P. bailloni* and *P. parva*) most probably occur, but I have not myself met with either of them.

286. *Crex pratensis.* Bechst. Orn. Taschenb., p. 336. Corn-Crake.

The Corn-Crake is universally diffused and met with at all seasons, being perhaps more common in winter than summer.

The Corn-Crake inhabits Europe, Western Asia as far as India, Africa down to the Cape.

287. *Porphyrio cæruleus.* Van delli. Mem. Acad. Real. Lisb. 1780, p. 37. Purple Gallinule.

Only observed in the marshes of the Huleh.

The Purple Gallinule inhabits the countries bordering both sides of the Mediterranean, and extends as far as the Caspian, but not south of that sea.

288. *Gallinula chloropus.* (Linn. Syst. Nat. i., p. 258.) Moorhen.

The Moorhen is very common in all suitable localities at all times of the year.

It is found throughout Europe; and the Moorhens of Africa down to the Cape, and of Asia to Japan, the Philippines, Celebes, Java, and Borneo scarcely differ, though they have been discriminated. The American Moorhen is also very nearly allied to the Old World form.

289. *Fulica atra.* Linn. Syst. Nat. i., p. 257. Coot. Arab., غرة, *Ghurrah.*

The Coot is common in all the waters of any extent in the country, and in the streams and fountains by the Dead Sea.

The Coot is found through Europe, Asia, the greater part of Africa, and Australia.

ORDER, ALECTORIDES.

FAMILY, GRUIDÆ.

290. *Grus communis.* Bechst. Vög. Deutsch. iii., p. 60. Crane. Heb., עָגוּר. Arab., كركي, *Kirkī.*

Large flocks of Cranes spend the winter in the open plains and downs of Southern Judæa, but I have never been able to find any evidence of their remaining to breed. The Cranes appear to have fixed roosting-places. There is one near Moladah, south-west of the Dead Sea, and another not many miles south-east of Gaza. Towards sunset the birds begin to return home, flying in order like Geese, with outstretched necks, keeping up a ceaseless trumpeting, which continues till morning, with only an occasional lull. During the whole night there is a succession of fresh arrivals. The howl of some wandering Jackal would rouse the whole camp; then, after a slight pause, the wail of an Hyæna awakened a deafening chorus, and before daylight began an angry discussion, perhaps on the next day's journey. Parties of some hundreds departed for the south with the dawn; others remained, probably to make up for their broken slumber, till the sun had risen for a couple of hours. The roosting-place at Moladah was a group of hillocks extending over several acres, and covered with the mutings of the birds as thickly as the resort of any sea-fowl.

The Crane inhabits Northern Europe and Asia in summer, migrating in winter to North Africa, India, and China.

FAMILY, OTIDÆ.

291. *Otis tarda.* Linn. Syst. Nat. i., p. 264. Great Bustard. Arab., رحاد, *Råad.*

The Great Bustard is not quite extinct on the Plain of Sharon, and I have several times seen it on the wide, grassy plains of Northern Syria, in the neighbourhood of the unfenced corn patches.

The Great Bustard, once a native of England, is rapidly yielding everywhere to the advance of human population. It is still found in Southern and Central Europe, Barbary, and Western and Central Asia.

292. *Otis tetrax.* Linn. Syst. Nat. i., p. 264. Little Bustard.

The Little Bustard is found on the plains, but not in great numbers. It breeds in the Plains of Sharon. I never met with it in the Ghor, where the Houbara is common. Eastward I met with it in 1881 at Oorfa, in Mesopotamia.

The Little Bustard inhabits all the countries which border on the Mediterranean and Black Sea, reaching eastward through Persia as far as the Punjab.

293. *Houbara undulata.* Jacq. Beitr. Gesch. Vög., p. 24. Houbara Bustard. Arab., حبارة, *Hubārah.*

The Houbara Bustard, though very shy and wild, is yet not at all uncommon in the plains of the Jordan valley, and at the south end of the Dead Sea. I have only once seen it on the uplands east of Jordan. It is a permanent resident.

The Houbara is a native of Northern and North-eastern Africa, but very rare in Egypt and Nubia. Its true home is the Sahara, and probably North Arabia. Syria appears to be its eastward limit, as the Houbara of the other side of the Euphrates is the Indian species, *H. macqueeni.*

ORDER, LIMICOLÆ.

FAMILY, ŒDICNEMIDÆ.

294. *Œdicnemus scolopax.* (S. G. Gmel. Reis. d. Russl. iii., p. 87.) Stone Curlew. Arab. كروان, *Kerouan.*

The Stone Curlew is plentiful in the Ghor, at the north end of the Dead Sea, remaining throughout the year and breeding there. When camped at Jericho we heard its cry continually through the night. In winter it is common in the southern wilderness, and we found it breeding in the north above Huleh.

The Stone Curlew inhabits temperate Europe and Asia as far as India, and North-eastern Africa.

FAMILY, GLAREOLIDÆ.

295. *Glareola pratincola.* (Linn. Syst. Nat. i., p. 345.) Pratincole.

The Pratincole is not found in winter, but is very plentiful in spring in colonies on all the marshy plains, especially Huleh, Gennesaret, and Acre, in all which it breeds, laying its eggs in the barest places.

The Pratincole inhabits Southern and Central Europe, Asia as far as India, and Africa almost to the Cape.

FAMILY, CHARADRIIDÆ.

296. *Cursorius gallicus.* (Gmel. Syst. Nat. i., p. 692.) Cream-coloured Courser.

Rare in Palestine proper. I twice obtained it near Acre, where the specimens are of a much deeper hue than any I have seen elsewhere. We also saw it in the southern wilderness and on the uplands of Eastern Moab.

The Cream-coloured Courser is really an inhabitant of the Sahara and of the deserts of Western Asia as far as Scinde. Elsewhere it is only a straggler, and is nowhere numerous.

297. *Charadrius pluvialis.* Linn. Syst. Nat. i., p. 254. Golden Plover.

The Golden Plover is very common in the plains and cultivated lands in winter, but leaves early. We never met with the Asiatic species.

The Golden Plover is found in Iceland and every part of Europe as far as the Caucasus, in Asia Minor and Syria. It has occurred in Persia. In winter it ranges through a great part of Africa. East of the Ural its place is taken by *C. fulvus.*

298. *Charadrius helveticus.* (Linn. Syst. Nat. i., p. 250.) Grey Plover.

The Grey Plover has been obtained on the coast south of Beyrout.

It is spread over the whole world except South America, breeding only in the far north.

299. *Ægialitis geoffroyi.* (Wagl. Syst. Av., fol. 4, p. 13.) Greater Sand Plover.

Very common in the southern wilderness in winter, mingling with the much more numerous common Dotterel. I have occasionally met with it in all parts of the country, in winter plumage only.

This is a truly Asiatic bird, only accidentally straggling westward. It reaches to Japan, and has been recorded as far south as Cape York, North Australia.

300. *Ægialitis asiatica.* (Pall. Reise Russ. ii., p. 715.) Caspian Plover.

Occurs occasionally in winter.

The Caspian Plover's habitat is the country round the Caspian, but in winter it passes through Syria into Nubia and Abyssinia, and has been taken in South Africa.

301. *Ægialitis mongolica.* (Pall. Reise Russ. iii., p. 700.) Red-throated Plover.

This, like the preceding, is a scarce winter visitor to Palestine.

It is a Central Asiatic species, but winters not only in India and China, but along the shores of the Red Sea and the East African coast.

302. *Ægialitis hiaticula.* (Linn. Syst. Nat. i., p. 253.) Ringed Plover.

A few Ringed Plover may always be seen in winter on the shores of the Lake of Galilee, and it breeds on the Upper Jordan. All I have procured are of the large race. It is not so common as either the Kentish or the Little Ringed Plover.

It is found throughout the whole of Europe, and sparingly throughout Africa, but does not appear to go further east than Asia Minor and Syria.

303. *Ægialitis cantiana.* (Lath. Suppl. ii., Gen. Syn., p. 66.) Kentish Plover.

The Kentish Plover is very common on the coast both in winter and spring, and breeds in the country. It is found, but not so plentifully, on the banks of the Jordan and its lakes.

The Kentish Plover is found in Central and Southern Europe, in Africa as far as the Cape, and in Asia as far as Japan.

304. *Ægialitis curonica.* (Gmel. Syst. Nat. i., p. 602.) Little Ringed Plover.

This pretty little wader is plentiful on all the gravelly banks of streams throughout the country. It must breed there, as we found it through the spring and summer, but I have not found its nest.

The Little Ringed Plover inhabits the edges of rivers throughout all temperate Europe and Asia to China and the Philippines. It goes as far south as Central Africa.

305. *Eudromias morinellus.* (Linn. Syst. Nat. i., p. 254.) Dotterel.

The Dotterel is found in vast flocks on the rolling pastoral plains in the south of Judæa, near Beersheba. Once during the course of a three days' ride they overspread the whole country continuously. The myriads of *Helices,* clustering on all the bushes and on every straw, till the whole looked like a sheet of white blossom, no doubt provided sustenance for all.

The Dotterel inhabits Northern Europe, Siberia, and Turkestan, migrating to North Africa and Palestine in winter.

306. *Pluvianus ægyptius.* (Linn. Syst. Nat. i., p. 254.) Egyptian Plover.

This Plover, sometimes called the Crocodile Bird, very rarely straggles far from the Nile, along the whole course of which it is found. It only accidentally visits Palestine. Mr. Herschell shot on the Jordan a specimen which I have seen.

307. *Hoplopterus spinosus.* (Linn. Syst. Nat. i., p. 256.) Spur-winged Plover. Arab., مقساق, *Zikzak.*

The Zikzak returns from the south just as the other Plovers are leaving, and vociferously proclaims its arrival in all the pools and marshy spots. It remains in pairs by streams and in little morasses, and seldom strays far from its selected home.

The Spur-winged Plover is a native of North-eastern Africa, occasionally reaching Greece and Asia Minor. Its range is thus very limited.

308. *Vanellus vulgaris.* Bechst. Orn. Tasch., p. 313. Lapwing.

The Lapwing is plentiful in flocks on the coast plains in winter, and I have shot a straggler by the Dead Sea.

It is a native of Northern Europe and Asia, descending in winter as far as North Africa and Northern China, but not crossing the Himalayas.

FAMILY, SCOLOPACIDÆ.

309. *Recurvirostra avocetta.* Linn. Syst. Nat. i., p. 256. Avocet.

The Avocet is scarce in Palestine, and resorts to the few shallow lakes. A few may often be seen near the north end of the Lake of Galilee.

It is found in Central and Southern Europe and Asia, and through the whole of Africa.

310. *Himantopus candidus.* Bonn. Encycl. Méth., p. 24. Black-winged Stilt.

The Stilt is not unfrequent in lagoons and shallow waters throughout the year. I found a colony breeding in a swamp near Jenin.

The Stilt is a native of Southern Europe, Southern Asia, and the whole of Africa.

311. *Scolopax rusticula.* Linn. Syst. Nat. i., p. 243. Woodcock.

The Woodcock is not rare in winter, and is found sometimes in unlikely places. Mr. Upcher on one occasion brought down *Bubo ascalaphus* and a Woodcock by a double shot, out of a cave high up in a bare ravine, near Gennesaret, whence they were startled together by my shooting a Wall-creeper.

The Woodcock is found throughout Europe, Asia, and North Africa.

312. *Gallinago major.* (Gmel. Syst Nat. i., p. 661.) Double Snipe.

I have only once met with the Double Snipe. A specimen was shot by Dr. Van Dyck near Tyre, in the winter of 1881.

The Double Snipe spends the summer in North-eastern Europe and North-western Asia, wintering as far south as Natal. In Asia it has not been noticed further than Northern Persia.

313. *Gallinago cælestis.* J. S. T. Frenzel. Gegend. um Wittenberg, p. 58. Common Snipe.

The Snipe is common in winter.

It is found throughout Europe, North Africa, and Asia.

314. *Gallinago gallinula.* (Linn. Syst. Nat. i., p. 244.) Jack Snipe.

The Jack Snipe occurs, though rarely, in winter.

It is a native of Northern Europe and Asia, visiting the Mediterranean coast of Africa in winter, and also India, which seems to be its Eastern limit.

315. *Tringa alpina.* Linn. Syst. Nat. i., p. 249. Dunlin.

The Dunlin is common on the coast in winter.

It inhabits Europe, Asia, North America and Northern Africa.

316. *Tringa minuta.* Leisler. Nachtr. zu Bechst. Nat. Deutsch. i., p. 74. Little Stint.

The Little Stint I obtained on the coast. A small flock was observed at the south end of the Dead Sea, on the salt flats, in February, and a pair secured.

The Little Stint breeds in Siberia, but in winter is scattered over Africa to the Cape, and over Asia as far as India, but not further east.

Temminck's Stint has not yet been noticed in Palestine.

317. *Tringa subarquata.* (Güld. Nov. Com. Ac. Petrop. xix., 471.) Curlew Sandpiper.

Occurs on the coast in company with the Dunlin in winter.

The Curlew Sandpiper is found throughout Europe, Asia, and Africa, but breeds only in the extreme north.

318. *Calidris arenaria.* (Linn. Syst. Nat. i., p. 251.) Sanderling.

Found among the other Sandpipers on the coast in winter.

The Sanderling breeds only in the far north of the Old and New Worlds, but is found at other seasons on the coasts of the whole world.

319. *Totanus hypoleucus.* (Linn. Syst. Nat. i., p. 250.) Common Sandpiper.

The Sandpiper is common in winter and spring. I have not found it breeding, though it remains late.

The Sandpiper inhabits Europe, Africa, Asia, and Australia throughout their whole extent.

320. *Totanus ochropus.* (Linn. Syst. Nat. i., p. 250.) Green Sandpiper.

The Green Sandpiper is the most generally distributed of all the genus during the winter, and remains until June, long after the other Waders have left.

It ranges through Europe, Africa, Northern and Central Asia.

321. *Totanus glareola.* (Gmel. Syst. Nat. i., p. 677.) Wood Sandpiper.

The Wood Sandpiper is not uncommon in winter in the inland parts of the country, though by no means so frequent as the Green Sandpiper.

It is found through the whole of Europe, Asia, and Africa.

322. *Totanus stagnatilis.* Bechst. Orn. Taschenb., p. 292. Marsh Sandpiper.

The Marsh Sandpiper is scarce, and appears to be confined to the southern part of the country.

The Marsh Sandpiper is a South Asiatic and African bird, only entering the south-east corner of Europe, and rarely straggling further.

323. *Totanus calidris.* (Linn. Syst. Nat. i., p. 252.) Common Redshank.

The Redshank is very common and vociferous in marshy places in winter, and extends to the few moist spots of Bashan and the Hauran.

It inhabits all Europe and Northern and Central Asia, migrating in winter to all parts of Africa.

324. *Totanus canescens.* (Gmel. Syst. Nat. i., p. 668.) Greenshank.

Obtained in winter on the coast and in the plains.

The Greenshank inhabits Europe, Africa, and Asia generally, breeding only in the northern latitudes.

325. *Numenius arquatus.* (Linn. Syst. Nat. i., p. 242.) Curlew.

The Curlew may often be seen and heard in winter. I found it at the south end of the Dead Sea in 1872.

It ranges over the whole Old World to Borneo.

326. *Numenius phæopus.* (Linn. Syst. Nat. i., p. 243.) Whimbrel.

The Whimbrel occurs in winter, but is scarce. I never obtained it until 1881, though I often saw it.

Like its congeners, the Whimbrel wanders over the whole of Europe, Africa, and Asia, as far as New Guinea.

ORDER, GAVIÆ.

FAMILY, LARIDÆ.

327. *Sterna fluviatilis.* Naum. *Isis,* 1819, p. 1848. Common Tern. Heb., שַׁחַף [A. V., 'Cuckoo' in error], (*generic* for all sea fowl).

The Common Tern is plentiful on the coast. It breeds in a great colony on the Lake of Antioch.

It is generally distributed on the coasts of Europe, Asia, and Africa.

328. *Sterna minuta.* Linn. Syst. Nat. i., p. 228. Little Tern.

The Little Tern may be found in winter on shallow lagoons near the coast.

It is inhabitant of temperate Europe, Western Asia, and Africa.

329. *Sterna media.* Horsf. Tr. Linn. Soc. xiii., p. 198. Allied Tern.

Once observed on the coast. It is found on all the African, Arabian, and Indian coasts.

330. *Sterna anglica.* Mont. Orm. Dict. Suppl. Gull-billed Tern.

Found on sand-spits and lagoons near Tyre and Beyrout, and believed to breed in the neighbourhood.

The Gull-billed Tern is an inhabitant of the Mediterranean coasts, the China and Australian seas, and the east coast of both Americas.

331. *Sterna caspia.* Pall. Nov. Comm. Petrop. xiv., p. 582. Caspian Tern.

Off the coast in winter, but not common.

The Caspian Tern has a wide range, occurring throughout the Old World and in the north of North America.

332. *Sterna bergii.* Lichst. Verz., p. 80. Swift Tern.

On the Sea of Galilee in winter.

This Tern extends from the Red Sea through the Indian and Eastern Pacific Oceans.

333. *Hydrochelidon hybrida.* (Pall. Zoog. Rosso-As. ii., p. 338.) Whiskered Tern.

The Whiskered Tern is seen in winter about the Sea of Galilee, and breeds in the marshes of Huleh.

It is spread in particular localities through South Europe, Africa, Asia, and Australia.

334. *Hydrochelidon leucoptera.* (Schinz. Mein. and Sch. Vög. der Schweitz.) White-winged Black Tern.

This beautiful little Tern may be constantly seen throughout the year, sailing up and down the little streams that run into the Mediterranean, generally near the sea.

The White-winged Black Tern is found in Central and Southern Europe and Asia, in North and North-east Africa, and has straggled as far as the Transvaal and New Zealand.

335. *Hydrochelidon nigra.* (Linn. Syst. Nat. i., p. 227.) Black Tern.

Not so common as *H. leucoptera*, but scattered on the coasts, generally about the marshes near the mouths of streams.

The Black Tern is found in Europe, North Africa, North-western Asia, and in America from the north to Chili.

336. *Larus ridibundus.* Linn. Syst. Nat. i., p. 225. Black-headed Gull. Arab., نورس, *Nurss* (*generic*).

The Black-headed Gull is scattered all over the country, not only where there is water, salt or fresh, but in the dreary waterless desert, where it feeds on the snails which cover the stems of the desert shrubs.

The Black-headed Gull ranges from Northern Europe and Asia to the North African coasts, and as far as the Punjab.

337. *Larus melanocephalus.* Natt. *Isis*, 1818, p. 816. Adriatic Gull.

The Adriatic Gull flits about the coast all winter, but appears to retire to Asia Minor for nidification.

This bird is almost confined to the Mediterranean and Black Seas.

338. *Larus ichthyaetus.* Pall. Reis. Russ. Reichs. ii., App., p. 713. Great Black-headed Gull.

This royal Sea Gull is to be found on the coast in winter plumage, and later on is frequent on the Sea of Galilee in its full nuptial dress. It is indeed the Eagle of the Gull tribe. But it leaves the lake in May, and where it breeds I know not.

The Caspian appears to be the headquarters of this Gull, which ranges through Egypt and Nubia, and is found on most of the rivers of India.

339. *Larus canus.* Linn. Syst. Nat. i., p. 224. Common Gull.

Frequent on the coast in winter.

It ranges through Europe and Asia, as far south as Northern China.

340. *Larus gelastes.* Licht. in Thein. F. Vög. Eur., pl. v., p. 22. Slender-billed Gull.

Occurs on the coast in winter.

It inhabits South Europe, North Africa, and South-western Asia.

341. *Larus leucophæus.* ((Licht.) Bruch., J. F. O., 1853, p. 101, No. 18.) Yellow-legged Herring Gull.

Common on the coasts.

This bird is almost confined to the Eastern Mediterranean.

342. *Larus argentatus.* Gmel. Syst. Nat. i., p. 600. Herring Gull.

I obtained this bird off Tyre. It appears to be not rare in winter, but not so common as the yellow-legged species.

343. *Larus fuscus.* Linn. Syst. Nat. i., p. 225. Lesser Black-backed Gull.

This Gull is very common on the coast, on the Lake of Galilee, in the Southern Desert, and on the uplands of Moab. In the latter localities it feeds on the desert snails.

Its range, according to the season, is through Europe, North Africa, and Asia down to the South China coast.

ORDER, TUBINARES.

FAMILY, PROCELLARIIDÆ.

344. *Puffinus anglorum.* Temm. Man. d'Orn. ii., p. 806. Manx Shearwater.

I picked up a specimen in a fresh state, after a gale in winter, under Mount Carmel.

It is an inhabitant of the North Atlantic and the Mediterranean.

ORDER, PYGOPODES.

FAMILY, PODICIPIDÆ.

345. *Podiceps cristatus.* (Linn. Syst. Nat. i., p. 222.) Great Crested Grebe. Arab., خطيس, *Ghutîs* (*generic*).

All through the winter and spring, to the middle of April, the placid surface of the Lake of Galilee is dotted over from end to end by Grebes in amazing numbers. They begin to assume the breeding plumage by the end of February, but I could not ascertain that there were any nests in the neighbourhood of the lake; in fact, there is no cover for them. As I found them swarming on Huleh in May, they probably retire thither for safe quarters. In winter the Grebe goes as far as the south end of the Dead Sea.

The Great Crested Grebe inhabits Europe, Asia, Africa, Australia, and New Zealand; but in the tropics it is very scarce.

346. *Podiceps nigricollis.* (L. Brehm. Vög. Deutschl., p. 963.) Eared Grebe.

This is the most abundant species on the lakes of Galilee, remaining in hundreds on the Lake of Gennesaret long after it has assumed the nuptial dress, but retiring in May to Huleh.

The Eared Grebe inhabits South Europe and North and South Africa, and the greater part of Central and Southern Asia.

347. *Podiceps fluviatilis.* (Tunstall. Orn. Brit., p. 3.) Little Grebe.

The Little Grebe, though not so abundant on the Lake of Galilee, is found on almost every little piece of fresh water in the country, and breeds everywhere.

It has a wide range through Europe, Asia, and Africa. The Australian bird can hardly be separated.

ORDER, STRUTHIONES.

FAMILY, STRUTHIONIDÆ.

348. *Struthio camelus.* Linn. Syst. Nat. i., p. 265. The Ostrich. Heb., נוֹצָה. Arab., نعامة, *Nâameh.*

The Ostrich only just claims a place in the Fauna of Palestine, by its occurrence in the further parts of the Belka, the eastern plains of Moab. It is no doubt now but a straggler from Central Arabia, though formerly far more abundant. Xenophon speaks of its abundance in his time in Assyria (Anab. i. 5), and we have traditional accounts of its former existence as far as Scinde.

All the other wild animals named by Xenophon are still found in the region of the Euphrates and Tigris, a fact which makes the rapid retrocession of the Ostrich the more remarkable.

I possess a portion of an Ostrich's skin, the back, neck, and wings, captured in the Belka by the late Sheikh Aghyle Agha, and given by him to my friend, T. B. Sandwith, C.M.G., then Consul at Haiffa.

REPTILIA.

ORDER, OPHIDIA.

FAMILY, TYPHLOPIDÆ.

1. *Typhlops syriacus.* Jan. Icon. Oph., p. 15, livr. 3. Figured, Lortet. Arch. Mus. H. N. Lyon, t. iii., pl. xix.

This blind burrowing Snake is very common all over the country. Dr. Lortet mentions that it is also found in Mesopotamia. The species east of the Tigris (*T. persicus*) is distinct. It is nocturnal or crepuscular, and feeds on worms and very small insects.

2. *Onychocephalus simoni.* Böttg. Ber. Senck. Nat. Gesells. Frankf., 1878-79, p. 58.

This Blindworm, about six inches long, was found by Hans Simon both at Jaffa and at Caiffa. These are the only specimens known.

FAMILY, ELAPOMORPHIDÆ.

3. *Micrelaps muëlleri.* Böttg. Ber. Senck. Ges., 1879-80, p. 137. Lortet. Rept. de Syrie., pl. xix., fig. 2.

This graceful little Snake, about sixteen inches long, black, with white rings from head to tail, is found in the hill country of Judea and Galilee. The type is from Jerusalem. Dr. Lortet also found it near Lattakieh.

FAMILY, OLIGODONTIDÆ.

4. *Rhyncocalamus melanocephalus.* Günther. P. Z. S., 1864, p. 491.

PLATE XVI., FIG. 1.

The new genus, *Rhyncocalamus*, was established by Dr. Günther (P. Z. S., 1864, p. 491), for the reception of this species, brought by me from the neighbourhood of Lake Huleh. It has since been found by Dr. Lortet, both near Jericho and near the Lake of Gennesaret. It thus inhabits the whole Jordan valley. The Family in the Old World is represented by over fifty species, inhabiting South-eastern Asia and its islands. Only one species is found as far west as Persia.

FAMILY, COLUBRIDÆ.

5. *Ablabes coronella.* (Schlegel. Essai. s. l. phys. de Serp., Bnd. 2, s. 48.) Lortet. Rept. de Syrie, pl. xix., fig. 3.

This prettily marked Snake is common in every part of the country; found in Lebanon, Hermon, Huleh, Gennesaret, Tyre, and Nablus. It lives under stones, and is most rapid in its movements. It is widely distributed on the eastern Mediterranean shores.

6. *Ablabes modestus.* (Martin. P. Z. S., 1838, p. 82) = *Eirenis rothii.* Jan.

Found throughout Galilee, and in Lebanon and Hermon. It is of a brilliant yellow colour, with bluish metallic sheen on the back, which disappears after death. *Eirenis rothii*, Jan., Dr. Günther does not consider specifically distinct.

This species has been found in Syria, the Caucasus, Mesopotamia and Persia.

7. *Ablabes fasciatus.* (Jan. Arch. Zool. Genov. ii., p. 260.)

This Snake, originally described from Palestine, is also found in Persia. It inhabits dry stony places on the hills, and is not common. I am not aware of specimens having been procured in other countries, though it doubtless will be found in the intervening regions of Syria and Mesopotamia.

8. *Ablabes decemlineatus.* Dum. and Bibr. Erp. Gen. vii., p. 327.

Collected in Galilee, on the Plain of Phœnicia, under Lebanon, and at Huleh.

This species differs constantly from *Ab. modestus* in having the posterior and anterior chin-shields of equal length, while in *Ab. modestus* the anterior are much longer than the posterior. Moreover, it never has the black markings on the head and neck.

9. *Ablabes collaris.* (Ménétr. Cat. Rais., No. 228, p. 67.)

This species, very closely allied to the last, is found on the coast near Beyrout.

It was originally described from the Caucasus, and inhabits also Mesopotamia and Persia.

10. *Coronella austriaca.* Dum. et Bibr. vii., p. 610.

This Snake has been found near Beyrout, and elsewhere in Palestine.

It has an extensive range through Central and Southern Europe, especially the eastern part, the Caucasus, Persia, Egypt and Algeria.

11. *Coluber æsculapii.* Sturm's Fauna, iii., Heft. ii., f. a.

This European Snake has been found near Beyrout.

It is local in many parts of Central and Southern Europe, and in Transcaucasia.

12. *Coluber quadrilineatus.* Pall. Zoog. Ross.-As. iii., p. 40.

This Snake occurs in Northern Palestine. It inhabits the Caucasus and Southern Russia, Greece and Turkey, and is said to be found also in Sicily.

13. *Zamenis diadema.* (Schl. Ess. Phys. Serp. ii., p. 148) = *Periops parallelus.* Geoffr.

Found on the Phœnician and Philistian plains. This species has a wide range, from North-west India through Persia and Mesopotamia. It has been noticed also on the eastern shores of the Caspian.

It is rather a handsome Snake, and very gentle, never attempting to bite. Its food consists of small insects.

14. *Zamenis ventrimaculatus.* Gray. Ind. Zool. ii., pl. lxxx., fig. 1.

Found round the Dead Sea.

This Snake extends from Baluchistan through Persia into Egypt, but in each country the varieties are more or less distinctly marked. The Palestine form is the true typical *Z. ventrimaculatus.*

15. *Zamenis caudælineatus.* Günther. Cat. Col. Sn. Br. Mus., p. 104.

This Snake has a wide range through the country in stony places. I have taken it near Jerusalem and Nazareth, and drew a gigantic specimen out of a chink in the masonry of Hiram's Tomb, above Tyre, hybernating in December.

It is found in Transcaucasia, Persia, and in the region east of the Caspian.

16. *Zamenis viridiflavus.* Dum. and Bibr. vii., p. 686 = *Z. atrovirens.* Shaw.

This very large species is abundant among brushwood throughout the country, both in the Jordan valley and on the hills. It climbs the trees, and is most destructive, devouring the nestlings of the arboreal-breeding birds.

It is common throughout Southern Europe.

17. *Zamenis viridiflavus,* var. *carbonarius.* Bonap. Amph. Europ., p. 435.

This, which is really only a black race of the preceding species, is very common in the warmer parts of the country. In the Jordan valley it is the commonest, but it also occurs on the maritime plains, where the other is much more abundant.

18. *Zamenis dahlii.* Dum. and Bibr. Erp. Gén. vii., p. 692. Lortet. Rept. de Syrie, pl. xix., fig. 4.

This, one of the most graceful of Snakes, attains a length of over three feet.

It is not found in the hills, but abounds among grass and bushes, and in moist, but not marshy places. I never noticed it among stones or rocks. When alarmed it glides away among the grass, and may be traced by the gentle motion of the blades over its sinuous track.

This Snake has rather a wide range, from Dalmatia and Greece to Persia, and northward to the Talisch mountains near the Caspian.

19. *Zamenis ravergieri.* (Ménétr. Cat. Rais., p. 69.)

Common among the rushes and rank herbage round the extinct crater which forms the little Lake Phiala, now Birket er Ram. I do not presume to say whether this is a species or a mere variety of *Z. caudæ-lineatus.* Both forms, however, are found.

20. *Zamenis algira.* (Jan.) (Lortet. Poiss. et Rept. du lac de Tiberiade, p. 88.)

Found by the banks of the rivers, the Litany, Nahr el Khebir, and others.

This Snake did not occur to myself.

21. *Tropidonotus hydrus.* (Pall. Itin. i., p. 459, No. 18.)

This water Snake swarms not only in the lakes, but in little ponds and ditches throughout the country. It is especially abundant in Lake Phiala.

It is found round the Caspian and Black Seas, as well as in Northern Syria.

Böttger unites this and the following species.

22. *Tropidonotus tessellatus.* Laur. Schinz. Faun. Europ. ii., p. 39.

This Snake, which attains a considerable size, is of a greyish green colour, with black spots, living among the thistles and herbage, generally in marshy places. Dr. Lortet found it near the Nahr el Khebir, as well as on the plain of Gennesaret, and in gardens at Sidon. It feeds chiefly on small Mammals.

This Snake inhabits all Southern Europe, from Spain to Greece. Syria appears to be its Eastern limit.

R. Mintern del. et lith.

Mintern Bros. imp.

CŒLEPELTIS LACERTINA.

23. *Tropidonotus natrix.* (Linn. Syst. Nat. i., p. 380.) Ringed Snake.

Our well known common Snake, found throughout the whole of Europe and Asia Minor, did not come under my own observation, but is reported from Palestine by Böttger, Fr. Müller, and Bedriaga.

This is its furthest extension southwards and eastwards.

FAMILY, PSAMMOPHIDÆ.

24. *Cœlopeltis lacertina.* Wagl. Syst. Amphib., p. 189.

PLATE XIV.

This rock and desert Snake, a native of North Africa, Arabia, and Persia, is not uncommon. I found it near Jerusalem, and in Galilee. But it seems to be equally common round the Lake of Gennesaret and Lake Huleh, where Dr. Lortet collected it.

25. *Psammophis moniliger.* Daud. Rept. vii., p. 69.

Found among the scrub at Tiberias, where it hunts for its prey. Dr. Lortet also obtained it at Solomon's Pools, near Jerusalem, on an olive tree. It hides among the leaves of trees and bushes, and darts upon birds or mice from its retreat. In these habits it resembles its congener, *P. leithi,* from Persia.

FAMILY, DIPSADIDÆ.

26. *Tachymenis vivax.* (Fitz. Neue Classif. Rept., p. 57.)

This pretty species, of a bluish-grey colour, with black spots, is common under stones. Noticed near Jerusalem, Tabor, Tiberias, and other places.

It appears to be crepuscular or nocturnal in its habits, feeding chiefly on lizards.

This Snake inhabits South-eastern Europe from Illyria eastwards, the countries bordering on the Black Sea and the Caspian, Syria and Egypt.

FAMILY, ERYCIDÆ.

27. *Eryx jaculus.* (Linn. Syst. Nat. i., pp. 390, 391.)

This Snake was first noticed by Hasselquist in the Holy Land. It is common, but so rapid in its movements that it is not easily captured. It feeds on myriopodes and beetles.

Eryx jaculus is an inhabitant of Greece, Turkey, and the islands of Asia Minor and the Caspian region, Syria and Egypt. Persia appears to be its Eastern limit.

FAMILY, ELAPIDÆ.

28. *Naja haje.* (Linn. Syst. Nat. i., p. 387.) The Cobra.

Happily the Hooded Cobra is rare in Palestine. I am not aware of its occurrence in the cultivated districts, but in the plains and downs beyond Beersheba it is well known. I met with it near Gaza, on the sandy plain.

The Hooded Cobra is a native of Egypt and of the deserts of the Sahara. In India it is represented by an allied species, *Naja tripudians.* The Cobras are the most deadly of venomous serpents.

FAMILY, VIPERIDÆ.

29. *Vipera euphratica.* Martin. P. Z. S., 1838, p. 82.

This large Viper, one of the most poisonous of its family, was first described from the Euphrates. I found it in Galilee. Dr. Lortet also procured it near Jericho, so that it is probably generally distributed through the country.

It inhabits also Transcaucasia and Persia. It basks in the sand, and conceals itself in little tufts of herbage, preying on small quadrupeds.

30. *Vipera ammodytes.* (Linn. Syst. Nat. i., p. 376.)

This Viper appears to inhabit the higher ground. I obtained it on the lower slopes of Lebanon.

Pl. XV.

R. Mintern del. et lith. Mintern Bros. imp

DABOIA XANTHINA.

It is a native of Eastern and South-eastern Europe from the north of Italy, Austria, and through Greece. I do not find it recorded from Persia, though it has been met with in the Transcaucasian Provinces of Russia.

31. *Daboia xanthina.* Gray. Cat. Sn. Br. Mus., p. 24.

PLATE XV.

I twice obtained this poisonous Serpent, once on the Plain of Acre, and once near Tiberias. On one occasion it had swallowed a full-grown hare whole, and was unable to move. On the other it had just struck a quail, which dropped down dead as I came up with it, with no other mark of injury than a slight scratch close to the tip of its wing.

It is very interesting to find this peculiarly Indian, and not African, genus in Palestine. It has also been found near Lake Urumiah, in Armenia.

32. *Cerastes hasselquistii.* Strauch. Syn. d. Viperid., p. 112.

The Horned Cerastes, well known as an inhabitant of Egypt and the Libyan desert, is also found in the desert country of Southern Judæa. I have known my horse rear and shake with terror on descrying this little but deadly Serpent, coiled up in the depression of a camel's footmark, on the path before us.

The Persian species has been discriminated as *C. persicus.* The Cerastes is not known further east.

33. *Echis arenicola.* Boie. *Isis,* 1827, p. 558.

This poisonous little Serpent I have frequently found on the dry sands both north and west of the Dead Sea, but not in the upper country.

It is an African species, well known in Egypt, and differs very slightly from the Indian *Echis carinata,* which occurs as far west as Persia.

ORDER, LACERTILIA.

FAMILY, AMPHISBÆNIDÆ.

1. *Amphisbæna cinerea.* (Vaud.) Dum. and Bibr., vol. v., p. 500.

Not uncommon under stones on the Plain of Gennesaret and on the maritime plains. The same species occurs in Spain and Barbary, as well as in Asia Minor, Rhodes, and Cyprus.

FAMILY, MONITORIDÆ.

2. *Psammosaurus scincus.* (Merrem. *Isis*, vi., p. 688.) The Warran.

This well-known huge Lizard we found near Engedi. The Arabs give the name of *Warran* both to this and the following species.

It is found all through North Africa, the Sahara, and Egypt.

3. *Monitor niloticus.* Geoffr. Rept. Egypte, i., 121, t. 3, f. 1.

The Nilotic Monitor inhabits the region to the south of the Dead Sea, and the Southern Judæan desert.

It is an inhabitant of Africa generally.

FAMILY, LACERTIDÆ.

4. *Lacerta viridis.* L. S. N. Petiv. Gaz., t. 95, f. 1. Green Lizard.

The Green Lizard is most abundant in every part of the country.

It is an inhabitant of all Central and Southern Europe, North Africa, Asia Minor, Transcaucasia, and North Persia.

5. *Lacerta strigata.* Eichw. Zool. Spec. iii., p. 189.

This species is very close to the Green Lizard, but always smaller. It is not so common as the former species.

It is recorded from Greece, Asia Minor, Persia, and Syria.

6. *Lacerta judaica.* Camerano. Atti. Accad. Sci., Torino, Bd. 13.

This species has been separated by its describer from *Z. muralis*, with which it had previously been confounded.

It appears to be generally spread over Palestine from the Lebanon to Jerusalem, but only in the upper and hilly country.

Beyond these limits it is only known from Cyprus.

7. *Lacerta lævis.* Gray. Ann. and Mag. N. H. i., p. 279.

Collected near Jerusalem and in the Plains of Jericho.

It is a South European species.

8. *Lacerta agilis.* Linn. Syst. Nat. i., p. 363. Sand Lizard.

Found in every part of the hill country and maritime plains.

It inhabits all the mountainous and hilly districts of Europe and the Caucasus.

9. *Zootoca taurica.* (Pall. Zool. Ross.-Asiat. iii., p. 30.)

Found in the Phœnician plain and all round the base of Lebanon.

It is a native also of Greece and some of its islands, of the Crimea, and of the Caucasus.

10. *Zootoca muralis.* (Laur. S. 61, 160, t. i., f. 4.)

Very common in the north of Palestine, but not observed by us in the south.

Found at Beyrout, Sidon, Tyre, and Lake Huleh.

The same Lizard, with slight variation (var. *fusca*, var. *neapolitana*), inhabits all South Europe and South-western Asia, including Armenia and North Persia.

11. *Zootoca tristrami.* Günther. P. Z. S., 1864, p. 491.

Plate XVI., Fig. 2.

This new species was collected by us in the Lebanon district.

It has not yet occurred elsewhere.

12. *Zootoca deserti.* Günther. P. Z. S., 1859, p. 470.

This Lizard was first discovered by me in an oasis of the African Sahara. We again collected it on the plains beneath Hermon and Lebanon.

13. *Acanthodactylus scutellatus.* Audouin. Descr. Egypte, i. 172. Suppl. t. i., f. 7.

This Egyptian and North African species has been found near Beyrout by Dr. Lortet.

14. *Acanthodactylus savignii.* Audouin. Descr. Egypte, i. 172. Suppl. t. i., f. 8.

Collected in various localities on the coast by Böttger and Schrader.

Its range extends from Palestine along the African coast to Algeria.

15. *Acanthodactylus boskianus.* (Daud. Rept. iii. 188.)

This species has a wider range than the last, being found not only in North Africa, but in Asia Minor. It does not appear to reach Northern Persia.

I have not myself found it, but it is among collections at Beyrout.

16. *Eremias guttulata.* (Licht. Doubl. 101.)

Found by Dr. Lortet near Beyrout.

It is an inhabitant of North Africa, from Algiers to Egypt.

17. *Mesalina pardalis.* (Licht. Doubl. 99.)

This Lizard was taken by me near Beersheba in the sandy plain.

It is very common in Egypt and A leria, extends through Arabia, and was found by Mr. Blanford throughout Persia.

18. *Ophiops* (*Ophisops*, err.) *elegans.* Ménétr. Cat. Rais., p. 63.

This is a very common species of Lizard everywhere except in the Jordan valley.

It extends throughout Asia Minor, the Taurid, and the whole of Persia.

19. *Ophiops schluëteri.* Böttger. Ber. Senck. Nat. Ges., 1879-80, p. 176.

This species, which differs considerably from *O. elegans,* has been described from specimens collected by Schrader near Beyrout and in Cyprus, where it appears common. It has not yet been recorded from other localities.

FAMILY, ZONURIDÆ.

20. *Pseudopus apoda.* (Pall. N. Com. Petrop. xix. 435, t. 9-10.)

I procured this Lizard on Mount Hermon. Dr. Lortet found it in other places as well.

It is not an African species, but extends from Istria through the mountain ranges of Turkey, Greece, and Asia Minor to Transcaucasia.

FAMILY, GYMNOPHTHALMIDÆ.

21. *Ablepharus pannonicus.* Licht. Doubl. 103.

Several specimens were taken near Caiffa by Böttger.

It inhabits Hungary, Rumelia, Greece, Cyprus, Asia Minor, and North Persia.

FAMILY, SCINCIDÆ.

22. *Scincus officinalis.* Laur. Syn. 55.

The Egyptian Skink is given by Böttger. It inhabits North-east Africa, *i.e.*, Egypt, Nubia, and Abyssinia, and in Asia extends into Arabia and Syria.

23. *Euprepes fellowsii.* Gray. Cat. Liz., p. 113.

This species, described by Dr. Gray from Xanthus, Asia Minor, was found by me in every part of the country, from Dan to Beersheba.

24. *Euprepes septemtæniatus.* Reuss. Mus. Senck. i. 47.

This Lizard, described originally from Abyssinia, has also been found in Arabia, Persia, the Caucasus, Syria, and the Lebanon.

25. *Euprepes savignii.* Audouin. Rept. Egypte, 117, t. 2, ff. 3, 4.

Occurs on the coast. It has not been noticed elsewhere in Asia, and is an African species.

26. *Euprepes vittatus.* (Oliv. Voy. ii. 58, t. 29, f. 1.)

This Skink has been obtained in considerable numbers from Beyrout by Böttger. It has an immense range for a Lizard, but strictly Ethiopian, being found in West Africa and the Cape, as well as in Egypt, whence it was first described.

27. *Eumeces pavimentatus.* Geoffr. Desc. Egypte, p. 135.

I found this Skink on rough ground near the Dead Sea. It was also collected on the coast by Dr. Lortet and others, and by Simon at Jerusalem.

It inhabits the North African countries from Morocco to Egypt, and Asia as far as the east of the Caspian and Baluchistan.

28. *Anguis fragilis.* Linn. Syst. Nat. i., p. 392. Blind Worm.

The familiar Blind Worm or Slow Worm was collected in Palestine by Müller.

It inhabits the whole of Europe, Transcaucasia, and North Persia.

The South Persian form has been separated by Dr. Anderson.

FAMILY, OPHIOMORIDÆ.

29. *Ophiomorus miliaris.* (Pall. Reis. ii., 718.)

I collected this species in various parts of the north. It extends from the Caucasus westwards into the Taurid, and is also found in Greece and Algeria.

FAMILY, SEPIDÆ.

30. *Gongylus ocellatus.* (Forsk. F. Arab. 13.)

This Lizard swarms in every part of the country, mountain, or deep valley, in dry places among stones. The varieties of colour are endless in this species.

It inhabits all the islands of the Mediterranean, Canaries, and Madeiras, all North Africa from Morocco to Egypt and Sennaar, Arabia, Syria, and Persia.

31. *Seps monodactylus.* Günther. P. Z. S., 1864, p. 491.

PLATE XVI., FIG. 4.

This new species I discovered first near Nazareth, and afterwards at Lake Huleh, and under Hermon. Böttger has since collected it at Jaffa and Caiffa.

It has not been noticed out of Palestine.

32. *Sphænops capistratus.* Wagl. Syst., p. 161.

Collected by Dr. Böttger at Jaffa.

It has a wide range in North Africa from Senegal to Egypt.

FAMILY, GECKONIDÆ.

33. *Ptyodactylus hasselquisti.* (Schneid. Amph. ii., 13.) The Gecko. Extremely abundant in every part of the country.

Its only other known habitat is Egypt.

34. *Hemidactylus verruculatus.* Cuv. R. A. ii. 54.

Found in every part of the country.

This Lizard is spread over all the countries bordering the Mediterranean, north and south; in Africa as far south as Senaar; and in Asia it inhabits Arabia Petrea, and, according to Duméril, Persia also.

35. *Stenodactylus guttatus.* Cuv. R. A. ii. 58.

Found in the Ghor, north of the Dead Sea.

It is an Egyptian and Arabian species.

36. *Platydactylus mauritanicus.* (Linn. Syst. Nat. i., p. 361.)

Found by Dr. Buch in Syria and Arabia, and is probably the species taken by Dr. Lortet near Tiberias.

It has been noted from all the countries bordering the Mediterranean.

37. *Gymnodactylus geckoïdes.* Spix. Braz. 17, t. 18, f. 1.

Found by me on Mount Carmel. It occurs in Greece and European Turkey and in Asia Minor, but has not yet been observed in Africa.

38. *Gymnodactylus kotschyi.* Steindachner. Sitz.-Ber. Akad. Wiss. Wien., vol. lxii., p. 329.

Discovered by Simon at Caiffa, and by Dr. Böttger at Beyrout.

It has been obtained in Persia, Cyprus, Asia Minor, and the Cyclades, and also is said to be from Egypt and Senegambia.

FAMILY, AGAMIDÆ.

39. *Trapelus sinaiticus.* (Heyden. Rüpp. Zool. N. Afr. vol. x., f. 3.)

PLATE XVI., FIG. 3.

Extremely abundant in the Jordan valley, and especially round the Dead Sea.

It is a native of the whole of Arabia, Egypt, and Senaar.

40. *Trapelus ruderatus.* (Oliv. Voy. Ottom. ii. 428, t. 29, f. 3.)

This species occurs in most parts of Palestine.

It extends from Algeria eastwards to Mesopotamia.

41. *Stellio cordylina.* (Laur. Syn. 47.)

This is the common Gecko of every part of the country, running about on rocks, walls, and trunks of trees throughout the summer.

This well-known species inhabits Greece and its islands, Asia Minor, and Armenia, Arabia and Egypt.

N.B.—I am satisfied that I have observed at least three other species of *Stellio,* or of genera closely allied to it, but have not preserved specimens.

42. *Uromastix spinipes.* Merrem. Tent. 56.

The Mastiguer inhabits the Southern desert of Judæa.

It is a native of North Africa, the Sahara, and Egypt.

43. *Uromastix ornatus.* Rüpp. Zool. N. Afr. i., t. 1.

I met with this North African species in the Southern Desert.

FAMILY, CHAMELEONIDÆ.

44. *Chameleo vulgaris.* Daud. Rept. iv. 181.

The Chameleon is very common throughout every part of the country, but flourishes especially in the Ghor and Gennesaret.

This well-known and interesting creature is found in Southern Spain, all through North Africa as far as the White Nile, and in Asia Minor, Syria and Arabia.

R. Mintern del. et lith.

Mintern Bro's. imp.

1. RHYNCOCALAMUS MELANOCEPHALUS 2. ZOOTOCA TRISTRAMI.
3. TRAPELUS SINAITA. 4. SEPS MONODACTYLUS.

ORDER, CROCODILIA.

FAMILY, CROCODILIDÆ.

Crocodilus vulgaris. Cuv. Oss. Foss. v. 42. The Crocodile.

It was long questioned whether it were possible that the Crocodile was still to be found in Palestine. That it had formerly existed there seemed evident, both from tradition and from the fact that a river rising in a swamp to the south of Mount Carmel, and entering the sea at the north of the Plain of Sharon, is known as the Zerka, or Crocodile River. It is mentioned by Pliny and Strabo, and Pococke, in the last century, speaks of its capture. I had often heard of it from the Arabs, who aver that it frequently steals their young kids when they go to water in the marshes; and I saw footprints in the mud, near the head marshes not far from Samaria, which left not the smallest doubt that a Crocodile of large size had been there very shortly before. The promise of a reward produced its effect, and very soon after a fine specimen was brought by my friends into Nazareth, in a state of such decay that only the bones and the head could be preserved. It measured 11 feet 6 inches in length, and the skull, which is before me as I write, is 19·5 inches long. When I look at my Crocodile's head, brought home by myself, and read the long disquisitions written in various languages as to the possibility of the Crocodile inhabiting Palestine, I feel that an ounce of fact is worth a ton of theory.

Mr. McGregor believes he saw a Crocodile in the Kishon, on the Plain of Acre. This is not impossible, though it has never been reported to be found to the north of Carmel. No doubt the Crocodiles in the Zerka must be very few in number, and on the verge of extinction.

Palestine is the only country beyond the limits of Africa where the Egyptian Crocodile is found, but it inhabits all the great rivers and the coast of the whole of that Continent, from the Nile to the Cape; and occurs along the whole west coast as well as in the upper waters of the feeders of the Niger.

ORDER, CHELONIA.

FAMILY, TESTUDINIDÆ.

1. *Testudo ibera.* Pall. Zoog. Ross.-As. iii., p. 18. Mauritanian Tortoise.

This is the common Tortoise of the Holy Land, and is found in every part of the country, quite irrespective of the nature of the soil, till we reach Hebron. The hill country of Judæa appears to be its southern limit, south of which and of the Dead Sea it does not occur.

It is the common Tortoise of Barbary, of the Caucasus, Asia Minor, Mesopotamia, Persia, and Syria. To the south of Palestine it is replaced by the following species.

2. *Testudo kleinmanni.* Lortet. Poissons de Syrie, p. 90. Kleinmann's Tortoise.

This species, which has generally been confounded with *T. marginata*, is the Tortoise of the region between Hebron and Beersheba, and of the Arabah, south of the Dead Sea. It inhabits the whole of the Sinaitic Peninsula, as far as Egypt, where also in sandy districts it is very common.

Kleinmann's Tortoise has no posterior tubercles on the thigh. The carapace is extremely convex, especially at the juncture of the posterior third with the two anterior thirds. The marginal plates are expanded, almost horizontal in the adults, nearly vertical in young specimens. The nuchal plate is small and pointed. The sub-caudal plate is rhomboïdal, forming behind a strongly marked angle, extending distinctly beyond the marginal plates. This plate is generally single, but sometimes there are traces of a suture, as in *T. græca.* Marginal plates, eleven; median plates, including the nuchal and the caudal, seven; lateral plates, four on each side.

Testudo marginata has been stated to be a native of Palestine; but probably Kleinmann's Tortoise has been mistaken for it. I formerly erroneously stated that *T. græca* was found, mistaking for it *T. ibera.*

Dr. Lortet is satisfied that neither *T. græca* nor *T. marginata* have as yet been found in Syria.

3. *Emys caspica.* (Gmel. Syst. Nat. i., p. 1041.) Dum. et Bib., vol. ii., p. 235. Terrapin.

The Terrapin swarms in all the streams and pools of Palestine. In the larger lakes it grows to a great size, and is a pest to the collector and sportsman, seizing and dragging under water any killed or wounded bird with extraordinary promptness.

The Terrapin is found in Greece, Asia Minor, Northern Syria, and the Caucasus, as well as all round the Caspian.

4. *Emys europœa.* (Dum. et Bib., vol. ii., p. 220.) European Terrapin.

This large Water Tortoise inhabits the lakes of Gennesaret and Huleh, where it attains a great size.

It is found throughout Southern and Eastern Europe, excepting Spain, and also in Asia Minor, and as far as the Caspian.

FAMILY, TRIONYCHIDÆ.

5. *Trionyx ægyptiacus.* Geoff. Descr. de l'Egypte. Egyptian Soft Tortoise.

This Tortoise has not yet been observed in the Jordan valley, but occurs in the Litany and the Nahr el Kelb (Böttger).

It is a native of the Nile.

FAMILY, CHELONIIDÆ.

6. *Chelonia caretta.* (Linn. Syst. Nat. i., p. 351.) = *Thalassochelys caouana.* Bonnatt. Loggerhead Turtle.

This Turtle is not uncommon on the coast. I have seen it brought in by fishermen at Sidon.

The Loggerhead is found throughout the southern side of the Mediterranean, and ranges through all the warmer Atlantic seas. It is well known in the West Indies.

7. *Chelone viridis.* Schneider. Allg. Naturg. d. Schildkr., p. 299.

A single example of this rare Turtle, taken off Beyrout, is in the Frankfurt Museum.

AMPHIBIA.

ORDER, URODELA.

FAMILY, SALAMANDRIDÆ.

1. *Triton vittatus.* (Gray. P. Z. S., 1858, p. 140.) Banded Newt.

Dr. F. Müller found this widely distributed Newt near Beyrout. It is an inhabitant of Europe and Egypt.

ORDER, ANURA.

FAMILY, BUFONIDÆ.

2. *Bufo viridis.* Laur. Syn. Rept., pp. 27 and 111, pl. 1. Green Toad.

This Toad swarms in multitudes in all moist places. There is a curious variety I obtained on the shores of the Dead Sea, with numerous spine-like, very large, and prominent tubercles. Other specimens from the same locality are smooth, or provided with flat tubercles only. It is most variable in coloration. I have found individuals quite unspotted; others are marbled, or with spots, larger or smaller.

Bufo viridis is found in South-east Europe, in all North Africa, including the Sahara; and in Western Asia as far as Persia. East of the Himalayas it is replaced by other species.

I follow Dr. Günther and Boulenger in uniting *B. pantherinus* and *B. variabilis* (both of which are stated to be found in Palestine) as races or varieties of *B. viridis.*

3. *Bufo regularis.* Reuss. Mus. Senck. i., p. 60.

This African Toad is also Arabian, found in the Sinaitic Peninsula and up to the Judæan wilderness south of Beersheba.

It extends throughout the whole of Africa, except Barbary.

FAMILY, BOMBINATORIDÆ.

4. *Pelobates cultripes.* (Cuv. Regne Anim.) Tschudi. Batr., p. 83.

I have not seen this frog myself. It was discovered in the Lebanon by Lataste, as stated by Böttger. Previously it has only been known from the south of France, Spain, and Portugal. Probably further research will show it to exist along the Mediterranean coasts.

FAMILY, RANIDÆ.

5. *Rana esculenta.* Linn. Syst. Nat. i., p. 357. Edible Frog.

No one who has ever spent a night under a tent within reach of water will question the amazing number of these frogs, deafening the weary traveller through the long night. In no other country have I seen the frog population so dense. The present appears to be the only species inhabiting Palestine. One specimen, collected by the Dead Sea, is covered with minute tubercles.

The Edible Frog is found all over Europe, except the British Isles, through all North Asia, including Japan, as far as the Himalayas; and in Africa it inhabits the Barbary coast and Egypt. The race inhabiting China and Japan has been distinguished as a variety (var. *japonica*).

FAMILY, HYLIDÆ.

6. *Hyla arborea.* (Linn. Syst. Nat. i., p. 357.) Tree Frog.

The Tree Frog is common in all parts of the country, being found alike in the Jordan valley, on the plains, and in the hills. In the Ghor and in the woods it sits on the foliage of trees, but on the treeless plains it is equally common on the leaves of the artichoke and the great umbellifers.

The Tree Frog inhabits all Central and Southern Europe, Barbary, and West Asia, including Asia Minor, the Caucasus, and Mesopotamia. It does not appear to extend beyond Persia.

The specimens from Syria, Asia Minor, and Persia, have been separated off as a distinct variety (var. *A. savignyi*), but the differences appear to be slight and rather variable.

FRESHWATER FISHES.

SUB-CLASS TELEOSTEI.

ORDER, ACANTHOPTERYGII.

FAMILY, BLENNIIDÆ.

1. *Blennius varus.* Risso. Ichthy. de Nice, p. 131. Lortet, pl. xviii., fig. 3.

D. 30, A. 21, V. 2, P. 13.

Found abundantly in the Lake of Galilee, but especially at the mouths of the warm streams flowing into it, Ain et Tîn, Ain Tabighah, and Wâdy Semakh.

This Blenny is found about the mouths of rivers in the Mediterranean, especially in thermal waters.

2. *Blennius lupulus.* Bonap. Faun. Ital. Pesc. and fig.

PLATE XIX. FIG. 3.

D. 29, A. 18.

Found by me in the Lake of Galilee, in the Kishon, in the streamlets in the Bay of Acre, and in the Nahr el Kelb.

Originally described by Bonaparte from rivulets in South Italy.

3. *Blennius vulgaris.* Pollini. Viagg. al Lag. di Garda. viii., p. 20, fig. 1.

D. 32, A. 20, V. 3, P. 14. Vert. 35—36.

Found in the Nahr el Bared and other streams flowing into the Mediterranean. It was originally described from the Italian lakes, in all the freshwaters of which country it seems to be found.

FAMILY, MUGILIDÆ.

4. *Mugil capito.* Cuv. and Val. Hist. Nat. de Poiss. xi., p. 36. Grey Mullet.

D. 4, $\frac{2}{7}$, A. $\frac{3}{9}$, V. $\frac{1}{5}$, P. 16. L. lat. 45. L. transv. 14.

The Grey Mullet is very plentiful in the Nahr Ibrahim (Adonis) and the Nahr el Kelb (Dog River) near Beyrout.

It inhabits the coasts of Europe, the Nile, and all the Mediterranean embouchures, and is found as far south as the Cape of Good Hope.

5. *Mugil curtus.* Yarrell. Br. Fish., 3rd edition, ii., p. 186.

D. 4, $\frac{1}{8}$, A. $\frac{3}{8}$, V. $\frac{1}{5}$, P. 11. L. lat. 38. L. transv. 12.

Found near the mouth of the Nahr el Kadischa (Lebanon) and the Nahr el Bared.

6. *Mugil octoradiatus.* Günther. Cat. of Fishes, vol. iii., p. 437.

D. 4, $\frac{1}{8}$, A. $\frac{3}{8}$, V. $\frac{1}{5}$, P. 18. L. lat. 42—44. L. transv. 14.

This species has been described by Dr. Günther from specimens obtained on the English coast. Its range, however, must be a wide one, as Dr. Lortet has received specimens from M. Blanche taken in the Nahr el Bahsas. This Mullet is well known in the market at Tripoli.

7. *Mugil auratus.* Risso. Ichthy., de Nice, p. 344.

D. $\frac{4}{9}$, A. $\frac{3}{9}$, V. $\frac{1}{5}$, P. 17. L. lat. 43. L. transv. 14.

The Golden Mullet is very common in all the rivers of the Syrian coast, and at certain times of the year ascends them to a considerable distance from the sea.

It is common throughout the Mediterranean, and is found off the Canary Islands and on the English coasts.

ORDER, ACANTHOPTERYGII PHARYNGOGNATHI.

FAMILY, CHROMIDÆ.

The Chromidæ are the most characteristic and abundant of all the amazing multitude of fishes with which the Lake of Galilee teems. No less than eight species are now known from its waters, five brought home by me in 1864, prior to which date no fishes from the Jordan valley had been identified; and to these, three new species have since been added by the researches of Dr. Lortet.

8. *Chromis niloticus.* (Hasselquist. Reise. in Palestina, p. 392.)

Plate XVIII. Fig. 1.

$$D. \frac{15}{12}, A. \frac{3}{9}, V. \frac{1}{5}, P. 14. \quad L. lat. 34.$$

This fish is one of the most abundant species in the whole of the Jordan basin, especially in smooth water and deep pools, though it is not in nearly such prodigious numbers as the following species. It abounds in the Lake Huleh, and in the Lake of Gennesaret, as well as in the river itself. It is equally common in the Nile, and in all its canals, and is known as *Bolti* in Egypt, and as *Moucht* by the fishermen of Tiberias.

It can easily be recognised from the following species by its blackish-grey colour, and by its caudal fin convex and not concave, as in *Ch. tiberiadis*, by the white spots on its dorsal fin, and by its forehead retiring instead of convex and prominent, as in its congener.

All these *Chromidæ* are frequently found with their eyes extracted, and their foreheads pierced by the Grebes, which prey on them, but they seem to thrive perfectly well in spite of this mutilation, and to flourish in a state of absolute blindness. Such specimens may often be seen in the market.

9. *Chromis tiberiadis.* Lortet. Poiss et Rept. du lac de Tiberiade, p. 37, pl. vi.

$$D. \frac{16}{13}, A. \frac{3}{10}, V. \frac{1}{5}, P. 13. \quad L. lat. 33—34.$$

This fish, peculiar to the Jordan and its affluents alone, is found in the most amazing numbers from the Lake Huleh to the head of the Dead

Pl. XVII.

R. Mintern del. et lith.

Mintern Bro's. imp.

1. CHROMIS ANDREA. 2. CHROMIS SIMONIS.

Sea. It is by far the most abundant of all the species in the lakes. I have seen them in shoals of over an acre in extent, so closely packed that it seemed impossible for them to move, and with their dorsal fins above the water, giving at a distance the appearance of a tremendous shower pattering on one spot of the surface of the glassy lake. They are taken both in boats and from the shore by nets run deftly round, and enclosing what one may call a solid mass at one swoop, and very often the net breaks. They are also taken in large quantities by poisoned crumbs thrown from the shore on to the surface of the water. By casting nets hundreds are often taken at once.

This species especially is carried down at the mouth of the Jordan by thousands into the Dead Sea. The fishes never get further than a few yards, when they become stupefied, and soon turn over on to their backs, while Cormorants and King-Fishers, perched on the snags or floating logs, gorge themselves without effort, and often heaps of putrifying carcases washed on the shore poison the atmosphere, and afford a plenteous feast to the ravens and vultures.

The *Chromis tiberiadis* is distinguished by the fishermen as *Mouchtlebet.*

10. *Chromis andreæ.* Günther. P. Z. S., 1864, p. 492.

$$\text{D. } \frac{15}{11}, \text{ A. } \frac{3}{9}, \text{ V. } \frac{1}{5}, \text{ P. } 15. \quad \text{L. lat. } 31.$$

PLATE XVII. FIG. 1.

This species, first described by Dr. Günther from specimens I procured in the Sea of Galilee in 1864, is not so common as the species already described, and seems to remain generally in the deep waters. It does not attain a great size, nor has it been found either by Dr. Lortet or myself in any other part of the course of the Jordan except in this lake. It is not distinguished by any special name among the Arab fishermen. Our largest specimen was 7½ inches long.

11. *Chromis simonis.* Günther. P. Z. S., 1864, p. 492.

$$\text{D. } \frac{15}{9}, \text{ A. } \frac{3}{8}, \text{ V. } \frac{1}{5}, \text{ P. } 12. \quad \text{L. lat. } 32.$$

PLATE XVII. FIG. 2.

This is another of the species peculiar to the Jordan system, and there, so far as we know, restricted to the little Lakes of Gennesaret and

Huleh. It was first described by Dr. Günther from our specimens collected in 1864.

Dr. Lortet has made some very interesting notes on this peculiar fish, which he found, as I also did, in the fountains adjoining the lake on the west side (Ain Mudawarah, Ain et Tîn, and Ain et Tâbighah). It is not distinguished by the fishermen with any special name, but included under *Moucht*.

Dr. Lortet has made most interesting observations on the propagation of this fish, which I venture, with his kind permission, to transcribe.

The spawn is of the size of No. 4 shot, of a rich deep green. The female deposits about two hundred eggs in a little excavation which she works out among the rushes and roots. When she has completed her labour, she appears exhausted, and remains motionless at a little distance. The male, on the contrary, appears much agitated, turns himself round the spawn, swimming constantly above them, and probably fecundates them at this moment. In a few minutes afterwards he takes the ova one after another into his mouth, and keeps them in the buccal cavity against his cheeks, which then appear swollen in an extraordinary manner. Some of them, however, escape through his gills. The ova, though they are not attached by any membrane, nor by any glutinous matter whatever, remain very securely in his mouth, and are never dropped while he is in the water. It is only when he is thrown out on the sand that, in the struggles of his death-agony, they fall out, many, however, remaining even then in his mouth.

In this novel hatching-oven the eggs, during several days, undergo all their metamorphoses. The little ones rapidly increase in size, and appear to be much incommoded in their narrow prison. They remain in great numbers, pressed one against another, like the grains of a ripe pomegranate. The mouth of the father-nurse now becomes so distended by his progeny that his jaws cannot meet. The cheeks are swollen, and the animal presents the strangest appearance. Some of the young, arrived at their perfect state, continue to live and develope among the folds of the bronchiæ. Others have their heads turned towards the mouth of the parent, and do not quit the sheltering cavity till they are about 4 inches long, and sufficiently active and nimble to escape their numerous enemies. It is difficult to understand how the male, who thus carries more than two hundred young for several weeks, can feed himself without swallowing along with his prey a great number of his fry. It is in the springs close to the lake that these fishes spawn.

A similar observation as to the breeding habits of a fish, probably of the genus *Chromis*, on the edge of the Lake Tanganyika, has been made by Dr. Livingstone in his Journal.

12. *Chromis flavii-josephi.* Lortet. Poiss. et Rept. du lac de Tiberiade, p. 43, pl. viii., fig. 2.

$$D. \frac{15}{8}, A. \frac{3}{7}, V. \frac{1}{5}, P. 12. \quad L. lat. 26.$$

This species, discovered by Dr. Lortet, has not been found by him in the Lake of Gennesaret itself, or in Lake Huleh; but only in the reaches of the Jordan between the two lakes, and in the basins of Ain Mudawarah and Ain et Tâbighah. It is a very small species, the largest specimen only just exceeding 4½ inches in length. It may be recognised at once by the regularly formed yellow spots on the anal fin, and is known to the fishermen as *Addadi*.

13. *Chromis microstomus.* Lortet. Poiss. et Rept. du lac de Tiberiade p. 41, pl. viii., fig. 1.

$$D. \frac{16}{12}, A. \frac{3}{9}, V. \frac{1}{5}, P. 14. \quad L. lat. 34.$$

This is one of the species for the discovery of which we are indebted to the invaluable exploration of Dr. Lortet.

It is very abundant in the Lake of Gennesaret, much less so in that of Huleh, and is rather rare in the Jordan. It is in great numbers in the basin of Ain Mudawarah and the fountains of Et Tîn and Et Tâbighah, all communicating with the lake. It is known to the fishermen as *Moucht Kart*. It seldom attains the length of 8 inches.

14. *Chromis magdalenæ.* Lortet. Poiss. et Rept. du lac de Tiberiade, p. 48, pl. ix., fig. 2.

$$D. \frac{15}{10}, A. \frac{3}{7}, V. \frac{1}{5}, P. 13. \quad L. lat. 32.$$

This new species of Dr. Lortet is by no means common in the Lakes of Gennesaret and Huleh, but, unlike the other species of *Chromidæ* found in the Jordan basin, is very abundant in the marshy lakes east of Damascus, into which the Barada (ancient Abana) and the Sabirany (ancient Pharpar) empty themselves.

It is not known whether the Arabs have a distinctive name for it. Dr. Lortet has observed that the male hatches its young in its mouth after the same fashion as *Chromis simonis*, and probably all the other species of the genus. Its spawn is much smaller than that of *Ch. simonis*, but of the same colour.

15. *Hemichromis sacra.* Günther. P. Z. S., 1864, p. 493. Lortet. Poiss. et Rept. du lac de Tiberiade, pl.

$$\text{D. } \frac{14}{9}, \text{ A. } \frac{3}{8}, \text{ V. } \frac{1}{4}, \text{ P. } 13. \quad \text{L. lat. } 32\text{—}34.$$

PLATE XVIII. FIG. 2.

Discovered by me in 1864, and described from our specimens by Dr. Günther. It lives among the bulrushes and flags in different parts of the Lake of Gennesaret, especially near the outlet from the lake, and is also to be found in the three well-known fountains of Mudawarah, Et Tîn, and Tâbighah, probably attracted by the warmth of the water, in which so many of these fishes seem to luxuriate. It has not been found either in Lake Huleh or in the stream of the Jordan itself. The habits of propagation, as observed by Dr. Lortet, are similar to those of *Ch. simonis*, described above. The eggs and young fry are to be found in the maw of the male in the month of June, and when they emerge from their shelter are about ·35 inch long. Dr. Lortet frequently observed a large number of fry already hatched, and suspended in a large bladder, while the other half of the eggs shewed no signs of development. The spawn is larger and of a darker colour than that of *Chromis simonis*.

The genus *Hemichromis* is, like *Chromis*, exclusively Ethiopian; and the occurrence in such variety of these African forms in the Jordan basin is one of the most significant links which attach the Palestine Fauna to the Ethiopian.

Pl. XVIII.

R. Mintern del. et lith. Mintern Bro's. imp.

1. CHROMIS NILOTICUS. 2. HEMICHROMIS SACER.

ORDER, PHYSOSTOMI.

FAMILY, SILURIDÆ.

16. *Clarias macracanthus.* Günther. Cat. of Fishes, v., p. 16. The Silurus.

$$D. 73, A. 55, V. \frac{1}{5}, P. \frac{1}{8}.$$

PLATE XIX. FIG. 1.

This most extraordinary fish is very abundant in the muddy bottoms, and wherever there are flags or papyrus, both in the Lakes of Gennesaret and Lake Huleh. It is also found in the Upper Nile, but differs from the common species of the Lower Nile, *C. anguillaris.*

It is spoken of by Josephus as the *Coracinus*, and he mentions it as inhabiting the fountain of Capernaum, and coming thither by a subterranean communication from the Nile. As it is equally abundant in all the three fountains on the west of the lake, this statement of the historian throws no light on the identification of the disputed site of Capernaum. Beyond these warm fountains, one of the most curious natural features of the lake, the Silurus does not appear to extend, probably because there are few muddy bottoms anywhere else in the course of the Jordan suited to its habits.

I first obtained this fish in Ain Mudawarah, in the month of March, and was surprised to find that some of the specimens exceeded three feet in length, a size very rarely attained by the inhabitants of the lake itself. Ain Mudawarah is a large circular basin of ancient masonry, thirty-six yards in diameter, about a mile and a half from the lake, and immediately under an intruding spur of the surrounding mountains. At the further end of the basin there boils up a magnificent spring of clear water, which fills the reservoir to the depth of from three to five feet, and at once forms a little stream, which meanders through the dense brush down to the lake. The fountain itself is almost buried in the oleanders and fig trees which overshadow and screen it on all sides. It seems strange that so vast a number of fish should live and attain such a great size in this little confined basin. The mystery was explained to me when crossing this little streamlet last year in the month of April, not more than four hundred yards from the lake. The stream was scarcely more than a thread of water, across which I stood astride. It was about a foot deep. One continuous file of *barbours*, as the Arab fishermen call the *Silurus*, was struggling up the stream, evidently on their way to spawn in the fountain, or in the mud holes near it. The fishes pressed on regardless of my presence, the snout of each touching the tail of the emigrant in front. In places the water was not sufficient to cover them ; in one or two places there was no water at all. Still in single file they pressed on, over land as well as in water. I took them out rapidly with my hands, and threw them to some distance. They squeaked and shrieked with a hissing sound, like a cat at bay, and rapidly floundered back to the streamlet, working their way rapidly among grass or over gravel. I selected six specimens, each over three feet long, and making a sack of my 'abeih, slung them on my horse, they still hissing and squeaking defiance, and carried them to our camp, near Semakh, three hours' distance. They were still alive and vigorous, and continued so for two days, when the survivors of the demands of the cook were remitted to the water We all considered them excellent eating, and far superior to any other fish of the lake. The flesh is firm and rich, like an eel's. The extraordinary migration, mentioned above, which seemed to surprise my Arab attendant (not being a fisherman) as much as myself, explains the presence of these great fish in the Round Fountain. Since writing the above I am interested to find that Dr. Lortet also notices the remarkable vocal powers of this fish.

FAMILY, CYPRINODONTIDÆ.

17. *Cyprinodon dispar.* Rüpp. Atl. Fische, p. 66, pl. 18.

D. 9, A. 10, V. 7, P. 16. L. lat. 26. L. transv. 9.

This tiny fish, rarely reaching a length of two inches, swarms by myriads in the little thermal and saline springs which fringe the shores of the Dead Sea. In the overcharged waters of the sea itself they perish. Rüppell first discovered the species in hot salt springs near the Red Sea. I found it in swarms in a brine spring near Jebel Usdum, at the south end of the Dead Sea, of the temperature of 91° Fahr; also at Ain Feshkhah, Ain Terabeh, Ain 'Sghir, and on the east side in various hot and sulphurous springs a little to the south of the mouth of the Callirrhoe, and at the mouth of the Wâdy Mojib (Arnon). In most of these springs, and the little lagoons round them, in which the fish live, the water is almost as dense and as salt as that of the Dead Sea, which in many cases during a gale overflows the lagoons. Yet the fish, which thrive and multiply in the one, perish as soon as they are placed in a jar of the other. In the little marshes full of *Salicornia fruticosa* at the mouth of the Wâdy Zuweirah, at the south-west corner of the Dead Sea, there are myriads of the fry of this fish, about an inch long. I caught them by hundreds in my handkerchief, yet, in trying to escape, none of them would ever attempt to enter the sea, though I might be holding my improvised net within a yard of it. Those we placed in a jar of the water of the salt spring were well and active after a night's confinement; those put into a jar of the Dead Sea water perished in a few minutes. M. Lortet has explained the cause, that though the amount of chloride of sodium is as great or greater in the water of the salt spring than in that of the Dead Sea, the amount of chloride of magnesium is much less.

The larvæ of the mosquitoes supply abundant food for the fish in these salt springs.

18. *Cyprinodon cypris.* (Heckel. Fische Syr., p. 140, pl. xix., f. 1.)

D. $\frac{2}{9}$, A. $\frac{2}{8}$, V. $\frac{1}{4}$. P. 14. L. lat. 26. L. transv. 8.

This species was first described from the Tigris. I found it in the Jordan, at Ain Feshkhah, by the Dead Sea, in the Jabbok, and in the Fountain of Nablous. Dr. Lortet also collected it at the pilgrims' bathing place in the Jordan, in the Wâdy Kelt, and in the Damascus lakes.

19. *Cyprinodon mento.* (Heckel in Russegger Reis. i., p. 1089, pl. 6, f. 4.)

D. 12, A. 11, V. 6. L. lat. 27. L. transv. $\frac{3}{4}$.

This fish was described from Mosul, on the Tigris. I found it in the little stream by Amman (Rabbath Ammon) which flows eastward and is lost in the desert.

20. *Cyprinodon sophiæ.* (Heckel in Russegger Reis. ii., 3, p. 267, pl. 22, f. 2.)

D. 11—12, A. 10—12, V. 6. L. lat. 26. L. transv. 7.

The type specimens of this species were from a salt spring near Persepolis. I found it, along with several other species, in the warm brackish spring of Ain Feshkhah, and in other similar springs down to Jebel Usdum.

FAMILY, CYPRINIDÆ.

21. *Discognathus lamta.* (Ham. Fish. Gang., pp. 343, 393.)

D. 11, A. 8, V. 8, P. 13. L. lat. 35. L. transv. 5.

PLATE XIX. FIG. 5.

This little Carp is very abundant in the Jabbok, the Arnon, and in all the little affluents of the Jordan on the east side. Dr. Lortet found it also in the Lake of Gennesaret.

It has a wide range, unlike all the species we have hitherto recorded, which are African, and never extend into Asia; this is a strictly Asiatic, and especially Indian species, its range reaching from Syria to Assam, and probably further east.

Under this species Dr. Günther (Cat. Fishes vii., p. 69) includes *Discognathus rufus.* Heckel in Russegger Reis. i., p. 1071, pl. 8, f. 2.

PLATE XIX. FIG. 4.

This variety is found abundantly in the streamlets of Gilead.

22. *Capoeta damascina.* (Cuv. and Val. xvi., p. 314, pl. 482.)

D. 11, A. 8, V. 12, P. 21. L. lat. 76. L. transv. $\frac{14-17}{16}$. L. vert. $\frac{26}{19}$.

This Cyprinoid is abundant, not only in the Jordan and all its affluents, but in every little stream flowing into the Mediterranean. It is common all over Syria and Asia Minor. It attains the length of 13 inches. It is carried down into the Dead Sea in great numbers, and perishes at once, strewing the north shore.

All the species of *Capoeta* are called *Hefafi* by the fishermen of Tiberias.

R. Mintern del. et lith.

Mintern Bro's imp.

1. COBITES INSIGNIS 2. COBITES GALILÆUS. 3. DISCOGNATHUS RUFUS.
4. BLENIUS LUPULUS. 5. CLARIAS MACRANTHUS.

23. *Capoeta syriaca.* (Cuv. and Val. xvii., p. 407, pl. 514.)

D. 12, A. 7, V. 10, P. 18. L. lat. 72—80.

This species is very common in the Lake of Gennesaret, and is also found in the Jordan. The type specimen was from Abraham's River, at the foot of Mount Sinai. It has also been found in the Euphrates at Birajik.

24. *Capoeta fratercula.* (Heckel. Fische Syr., p. 69, pl. v., f. 2.)

D. 13, A. 8, V. 9, P. 17. L. lat. 70—72. L. trans. 13.

This fish has not been yet found in the Jordan system, but it abounds in all the mountain-streams of Lebanon. At the Algerian village of Deichûn, near Safed in Galilee, there is a large fountain full of this species. These fish are looked upon by the Arabs as sacred to Mohammed, and they will on no account allow anyone to take them. A little to the north of Tripoli also, at the shrine of Sheikh el Bedawi, is a copious spring, with a large basin and streams flowing from it, choked with these fishes, which seem piled in layers, with hardly space to move. They are an object of veneration, and are always fed by the worshippers. They follow in masses any visitor as he walks by the edge, gaping for food; but we were not allowed to take one, though a little lower down I secured a number.

This fish ascends the mountain-streams of Lebanon to spawn, and leaps the cascades like a salmon. It is excellent eating, and its flesh is a pale pink colour.

The Arabs call it *Semakh nahri.*

25. *Capoeta socialis.* (Heckel. Fische Syr., p. 115, pl. xv., f. 2.)

D. 12, A. 8—9, V. 10, P. 16. L. lat. 67. L. transv. 14.

This species, which attains the length of 12 inches, is very abundant in Lakes Huleh and Gennesaret, and down the whole course of the Jordan, to which it seems confined. It is nearly allied to *C. damascina.*

26. *Capoeta amir.* (Heckel in Russegger Reis. ii., 3, p. 258.)

D. 12, A. 8, V. 10, P. 14. L. lat. 72. L. transv. 13.

This fish occurs in the northern streams flowing into the Mediterranean.

27. *Capoeta sauvagei.* Lortet. Poiss. et Rept. du lac de Tiberiade, p. 56, pl. 13, f. 2.

D. 9, A. 7, V. 5, P. 15. L. lat. 33.

This species, recently described by Dr. Lortet, was discovered by him in the Lake of Gennesaret, where it is evidently very rare, and only taken at a great depth. Its extreme length is 4 inches. It is most brilliantly coloured, its back being a brilliant blue, the belly golden yellow, the cheeks bright green, and the fins a silvery yellow.

28. *Barbus canis.* Cuv. and Val. xvi., p. 186, pl. 468.

D. 13, A. 8, V. 10, P. 15. L. lat. 33. L. transv. 4.

PLATE XX. FIG. 1.

This is one of the most abundant of the many species which swarm in the Lake of Gennesaret and in the Jordan, to which system it is peculiar, not being found in any of the rivers entering the Mediterranean.

I have seen thousands of these fishes in the Jordan, when an army of locusts has been attempting to cross the river, standing almost upright in the stream, with their heads partially out of the water, and their mouths wide open, devouring the locusts with inconceivable rapidity.

This Barbel attains the length of 19 inches.

29. *Barbus beddomii.* Günther. Cat. Fishes, vol. vii., p. 110.

D. 13, A. 8. L. lat. 28. L. transv. $5\frac{1}{2}$—5.

This Barbel was taken in the Lake of Galilee by Mr. Beddome.

The type is in the British Museum. No other specimens are known.

It is 4 inches long.

30. *Barbus longiceps.* Cuv. and Val. xvi., p. 179, pl. 467.

D. 11, A. 8, V. 9, P. 18. L. lat. 54—55. L. transv. $\frac{11}{13}$. L. vert. $\frac{25}{19}$.

PLATE XX. FIG. 2.

This is one of the most abundant of the many abundant species in the Lake of Galilee. It is also one of the best kinds for the table, and attains a length of 19 inches. It is not found above the lake, but is common throughout the course of the Lower Jordan, though but attaining the

1.

2.

R. Mintern del. et lith.

Mintern Bro's. imp.

1. BARBUS CANIS. 2. BARBUS LONGICEPS.

size it does in the lake, and is the last of the fish tribe to succumb to the poisonous influences of the Dead Sea, to which it is carried down in hundreds.

The fishermen call it *Escheri.*

It is peculiar to the lake and the Jordan.

31. *Phoxinellus libani.* Lortet. Poiss. et Rept. du lac de Tiberiade, p. 66, pl. xi., f. 4.

D. 41, A. 9, V. 8, P. 14 L. lat. 48.

This curious little fish, belonging to the genus *Leuciscus* of Günther, was discovered by Dr. Lortet in the little lake of Yammûneh, a mountain tarn above Ainâta in Lebanon, well known to visitors to the Cedars from Baalbek, and 4,800 feet above the sea. These little fishes, apparently the only inhabitants of the lake, at the season when the little streamlets of the tarn are fullest, crowd into them, and form an important article of commerce for the villagers.

This fish rarely reaches 2½ inches in length, generally less than 2 inches.

32. *Leuciscus zeregi.* (Heckel in Russeg. Reis. i., p. 1063, pl. 6, f. 3.)

D. 10, A. 9, V. 7. L. lat. 57—66.

We know nothing of this little fish, further than that Mr. Beddome found it in the Lake of Galilee, his specimen being in the British Museum, and that it has also been found in the Lake of Antioch.

It is 2½ inches long.

33. *Leuciscus lepidus.* (Heckel in Russeg. Reis. i., p. 1079, pl. 10, f. 2.)

D. 11, A. 13, V. 8, P. 14. L. lat. 48. L. transv. 7.

A native of the Tigris, from whence it was originally described, this fish has since been found in the Nahr el Arab, near Lattakieh. It may therefore probably occur also in the other rivers of the coast. Its length is about 4 inches.

It is known by the Arabs as *El Baraan.*

34. *Leuciscus tricolor.* Lortet. Poiss. et Rept. du lac de Tiberiade, p. 68, pl. xii. f. 2.

D. 10, A. 12, V. 8, P. 13. L. lat. 60.

This little species, rarely exceeding $3\frac{1}{4}$ inches in length, is the characteristic species of the lakes of Damascus into which the rivers from the north and east of Hermon drain and evaporate.

This is the ordinary sprat of the Damascus market.

35. *Rhodeus syriacus.* Lortet. Poiss. et Rept. du lac de Tiberiade, p. 70, pl. 12, f. 3.

D. 10, A. 8, V. 7, P. 14. L. lat. 48.

This little fish, very nearly allied to the Bitterling of Germany and to similar Chinese species, is the inhabitant of the fountains and streams of Ba'albek. It is also known from the Damascus lakes.

36. *Alburnus sellal.* Heckel in Russegger Reis. i., p. 1082, pl. 11, f. 4.

D. 10, A. 15, V. 9, P. 14. L. lat. 70,

This species, originally described from the Orontes, has been found by Dr. Lortet in the Lake of Galilee. It is a small species, averaging 6 inches in length.

37. *Alburnus vignoni.* Lortet. Poiss. et Rept. du lac de Tiberiade, p. 72, pl. 16, f. 3.

D. 9, A. 13, V. 8, P. 14. L. lat. 56.

This species, very brilliantly coloured, metallic blue above, with silvery sides and yellow abdomen, its fins tipped with purple, was discovered by Dr. Lortet in the Damascus lakes, but it has not yet been found in any of the waters of the Jordan system.

38. *Acanthobrama centisquama.* Heckel in Russegger Reis. i., p. 1074, pl. 9, f. 1.

D. 11, A. 22, V. 9. L. lat. 100. L. transv. 20.

This species of the Barada and the Damascus Lakes also occurs in the upper affluents of the Jordan, but has not yet been observed in the lower waters of that system.

39. *Nemachilus tigris.* (Heckel in Russegger Reis. i., p. 1088, pl. 12, f. 4.)

D. 10, A. 7, V. 6, P. 10.

This species occurs in the affluents of the Kadisha, as well as in the lakes of Damascus. It has also been found near Aleppo.

40. *Nemachilus galilæus.* (Günther, P. Z. S. 1864, p. 493.)

D. 12, A. 8, V. 7.

This species was found in the Lake of Galilee by Mr. Beddome. It is scaleless, and with the caudal fin truncate. It has not been elsewhere met with. The type is 3 inches long.

41. *Nemachilus insignis.* (Heckel. Fische Syr., p. 97, pl. 12, f. 3.)

D. 10, A. 7, V. 7.

PLATE XIX. FIG. 2.

This diminutive fish was found by me in great numbers in the warm brackish streamlets flowing into the north-west part of the Dead Sea, also in a spring between Jacob's Well and Nablus. Dr. Lortet also took it in the Wâdy Kelt, near Jericho.

42. *Nemachilus leontina.* Lortet. Poiss. et Rept. du lac de Tiberiade, p. 73, pl. 18, f. 1.

D. 9, A. 7, V. 5, P. 11.

This little fish, three inches in length, was discovered by Dr. Lortet in the Lake of Galilee, where, however, it is not common. It is not known from any other locality.

FAMILY, MURÆNIDÆ.

43. *Anguilla vulgaris.* Turton. Brit. Faun., p. 87. The Eel. Arab. انكليس, *Anklis.*

The Eel of Palestine was distinguished by Kaup as *A. microptera*, but both Dr. Günther and Lortet agree in identifying it with our common species.

I found it in the Kishon, the Wâdy Kurn, and the Nahr el Kelb, but neither Dr. Lortet nor myself have met with it in any part of the Jordan system. It is most abundant in the Lake of Antioch, where it reaches the length of four and a half feet, and to judge by the market, appears to form the staple food of the inhabitants.

TERRESTRIAL AND FLUVIATILE MOLLUSCA.

THE character of the Molluscan Fauna of Palestine partakes, as might have been expected, of the same variety which marks the other branches of its Fauna and Flora. There are, however, fewer exceptions to its general character as a part of the Mediterranean basin, and fewer traces of the admixture of African and Indian forms. Northern types, especially of the genus *Clausilia*, are frequent in the Lebanon and on its southern spurs in Galilee. The Molluscan Fauna of the maritime plains and the coast possesses no features distinct from those of Lower Egypt and Asia Minor. The shells of the central region are scarce and not generally interesting; while on the borders of the Jordan valley and in the southern wilderness we meet with very distinct groups of *Helix* and of *Bulimus*, chiefly of species peculiar, or common in some few cases to the Arabian desert.

The fluviatile Mollusca are of a type very much more tropical in its character than that of the terrestrial shells. There are here but few species similar to those of the east of Europe. Most of the species are identical with, or similar to, those of the Nile and of the Euphrates; and some of the genus *Melanopsis*, and no less than sixteen *Unios*, are peculiar to the Jordan or its feeders. It seems probable that the inhabitants of the waters were better able to sustain the cold of the glacial epoch than the mollusks of the land; and from the post-tertiary remains found by the Dead Sea we may infer that the species now existing have been transmitted from a period antecedent to the glacial; while the more boreal forms introduced at that epoch have maintained their existence in the colder districts of Northern Palestine to the exclusion of the southern species, which have not succeeded in re-establishing themselves. The beautiful group *Achatina*, requiring a degree of moisture not generally found in Palestine, is only represented by a few insignificant and almost microscopic species.

The Molluscan Fauna of this country has been less neglected than other branches of its natural history. Olivier first published a few species through Férussac in 1821. Ehrenberg added many more, of which eighteen were described as new. Boissier published his list in the 'Zeitschrift für Malakologie' in 1847. Bourguinat published and figured in 1853 the collection made by M. de Saulcy; and Dr. Roth, in his 'Molluscorum Species,' in his 'Spicilegium Molluscorum,' 1855, and 'Coquilles Terrestres et Fluviatiles,' edited by A. Mousson, 1861, has supplied us with a catalogue far more complete and exhaustive than any of his predecessors.

But all of these contributions to our knowledge of the Molluscan Fauna shrink into insignificance when compared with the magnificent work of Mr. Arnould Locard, on the Fluviatile Molluscs of the Lake of Galilee: 'Malacologie des Lacs de Tibériade, d'Antioche et d'Homs,' 4to., Lyon, 1883. This work, superbly illustrated, gives, with the fullest details, descriptions of all the known Molluscs of the Lake of Galilee, and to it I am indebted for the opportunity of enumerating a great part of the Unios of that unique inland sea, which contains no less than eighteen species confined to the waters of the Jordan valley. One of the most remarkable, and as yet unexplained, features of the phenomena of this unique depression is that many of the species are found at depths of 25, 50, and even 100 fathoms; and some of them only at that depth. Most of these deepwater species are small in comparison with the allied species found in the deeper waters of other European and Asiatic rivers; yet the thickness of the shells proves that there is no want of calcareous matter for their formation. The richness and brilliancy of the nacreous lining of most of these shells surpasses the colouring of any European species.

1. *Limax phœniciacus.* Bourg. Test. Nov., p. 9, 1852.

Very common in the maritime plains from Beyrout to Jaffa, and in the valleys which abut on them. Not observed in the interior. Easily distinguished from *L. agrestis* by its larger size and its crowded black spots. It is very slightly wrinkled, and reaches a length of upwards of two inches.

2. *Limax berytensis.* Bourg. Test. Nov., p. 10, 1852.

In the same localities as the preceding, but by no means so plentiful. It is of much smaller dimensions, and may be at once distinguished by its deep black colour, and its mantle, placed not in front, but almost on the centre of its back.

3. *Limax tenellus.* Müll. Verm. Terr. et Fluv. Hist. ii., p. 11.

I found several specimens of a slug in moist valleys south of the Lebanon, which I can in no way distinguish from the European species.

4. *Limax variegatus.* Drap. Tabl. Moll., p. 103.

5. *Parmacella moquini.* Bourg. Amén. Malac. ii., p. 139.

Plain of Sharon.

6. *Daudebardia saulcyi.* (Bourg. Test. Noviss., p. 10.) (=*D. syriaca.* Roth.)

I dug up four fine specimens of this interesting species in the Wâdy Kurn, near the Plain of Acre.

7. *Daudebardia gaillardoti.* Bourg. Rev. et Mag. Zool., 1855, p. 326.

Plain of Phœnicia.

8. *Succinea pfeifferi.* Rossm. Iconogr. i., p. 92, f. 46.

Among reeds near Beyrout.

9. *Succinea globosa.* Tristram. P.Z.S., 1865, p. 531.

Long. 14, diam. 10, alt. 8½ mill.

This beautiful and most peculiar species was obtained by me on papyrus-stems in the marshes of Huleh (waters of Merom), in the Upper Jordan. In the rotundity of its form and the diaphanous redness of its coloration, it is widely removed from any other of the group which I have seen. The animal is very large for the shell.

10. *Helix sancta.* Bourg. Test. Noviss. Or., p. 15.

Near Jerusalem. Mousson considers this only a giant variety of *H. cellaria.* The differences, however, appear constant, both in colour and convexity.

11. *Helix nitellina.* Bourg. Test. Noviss. Or., p. 16.

Scarce throughout the country.

12. *Helix protensa.* Fér. Tabl. Syst., p. 40, No. 207.

Nablus.

13. *Helix cellaria.* Müll. Verm. Terr. et Fluv. Hist. ii., p. 38.

In the north only.

14. *Helix jebusitica.* Roth. in Malak. Bl., 1855, p. 24.

Near Jerusalem, Sarepta, and Nazareth. Easily distinguishable from *H. sancta* by its less regular and less delicate striation, and by its much larger umbilicus—and from the following species by its rounded umbilicus and the less rapid increase of its whorls.

15. *Helix æquata.* Mouss. Coq. Bell. Or., p. 16, 55.

Only in the north, near the coast.

16. *Helix camelina.* Bourg. Test. Noviss. Or., p. 14.

Near Nazareth, Jericho, and Jerusalem.

17. *Helix hierosolymitana.* Bourg. Test. Noviss. Or., p. 13.

Not uncommon close to Jerusalem; not met with elsewhere.

18. *Helix pulchella.* Müll. Verm. Terr. et Fluv. Hist. ii., p. 30.

We found a single specimen of this world-wide species under a stone in the Plain of Acre.

19. *Helix conspurcata.* Drap. Tabl. Moll., p. 93.

On the coast near Sidon.

20. *Helix erdelii.* Roth. in Pfr. Mon. i., p. 205.

Near Jerusalem.

21. *Helix joppensis.* Roth. in Schmidt. Stylomm., p. 29; var. *multinotata.* Mouss.

Ditto, var. *subkrynichiana.* Mouss.

22. *Helix simulata.* Fér. Tabl. Syst., p. 45. Prodr., p. 289.

23. *Helix syriaca.* Ehrenb. Symb. Phys. Pfeiff. Mon. Hel. i., p. 131.

One of the most abundant shells in every part of the country.

24. *Helix rufilabris.* Jeffreys. Syn. Moll. Linn. Trans. xvi., p. 509.

Very common everywhere.

25. *Helix montis-carmeli.* Tristram. P. Z. S., 1865, p. 532.

Diam. maj. 8, min. 7, alt. 4 mill.

Two adult and several young specimens of this very distinct and pretty little shell were collected by us on Mount Carmel. It seems to bear no affinity to any other species in the country; but it is somewhat like *H. partita*, Pfr., from Ceylon, which, however, is umbilicated.

26. *Helix berytensis.* Fér. Prodr., p. 260.

Generally distributed in small numbers through the country.

27. *Helix lenticula.* Fér. Tabl. Syst., p. 41.

Near the coast.

28. *Helix nummus.* Ehrenb. (=*H. hedenborgi*, Pfr. Hel. Viv. i., p. 209).

Very abundant in the Nahr el Kelb, near Beyrout.

29. *Helix genezerethana.* Mouss. Coq. Voy. Roth., 1861, p. 28.

Perhaps a large variety of *H. nummus.*

30. *Helix pratensis.* Pfr. P. Z. S., 1845, p. 132.

Galilee.

31. *Helix obstructa.* Fér. Tabl. Syst., p. 69.

Phœnician plain.

32. *Helix solitudinis.* Bourg. Test. Noviss. Or., p. 15.

Cœle Syria.

33. *Helix bargesiana.* Bourg. Rev. et Mag. Zool., 1854, p. 15.

Northern plains.

34. *Helix pisana.* Müll. Verm. Terr. et Fluv. ii., p. 60, No. 255.

Plentiful along the coast, to which it is strictly confined. Specimens from the north are very richly coloured, while from the district near Gaza they are blanched and colourless in life.

35. *Helix cæspitum.* Drap. Hist. Moll., p. 92.

On the coast and the hills near it, in the north of Palestine.

36. *Helix variabilis.* Drap. Tabl. Moll., p. 73.

Very common on Mount Carmel, and with many variations of colour and size. The eastern specimens seem generally to be smaller than those of Europe. Probably several of the species not recognised by us may be referred to varieties of this widely spread and most variable shell. M. de Saulcy does not appear to have met with it, but perhaps distinguished it under some other name.

37. *Helix lineata.* Oliv. Zool. Adriat., p. 77.

Found on the hills along the coast. From the study of a long series of intermediate varieties, I should feel disposed to diminish very greatly the number of described species of this variable group.

38. *Helix intersecta.* Poiret. Coq. Flùv. et Terr. de l'Aisn., p. 80. (=*H. langloisiana*, Bourg. ?)

Common near Jerusalem.

39. *Helix hierochuntina.* Roth. Malak. Bl., 1855, p. 24.

Takes the place of the preceding species in the Jordan valley. It is at once distinguished by its red peristome and flattened spire.

40. *Helix turbinata.* Jan. Mantiss., p. 2.
Scarce on the coast.

41. *Helix neglecta.* Drap. Hist. Moll., p. 108.

42. *Helix arenosa.* (Beck. Ind. Moll., p. 14.)
Nablus.

43. *Helix apicina.* Lam. Anim. S. Vert. vi., ii., p. 93.
In the north, on the dry rocks near the coast.

44. *Helix campestris.* Ziegl. Mus. Rossmull. viii., p. 34.
Found on the high plateau of Moab and Eastern Gilead.

45. *Helix protea.* Ziegl. Rossmäster, Hist. Moll. viii., p. 34.
Common and variable from the coast to the southern deserts. I have many specimens corresponding to *H. langloisiana* of Bourguinat, which appears to be only a strongly marked desert and blanched variety of the present species.

46. *Helix amanda.* Rossm. Icon. vii., p. 10.
Jerusalem.

47. *Helix improbata.* Mouss. Coq. Voy. Roth., 1861, p. 11.
Jerusalem.

48. *Helix crispulata.* Mouss. Coq. Voy. Roth., 1861, p. 12.
Jerusalem, rare.

49. *Helix neglecta.* Drap. Hist. Moll., p. 108.

50. *Helix syrensis.* Pfr. Symb. ad Hist. Hel. iii., p. 69.
Lebanon.

51. *Helix vestalis.* Parr. Pf. Symb. i., p. 40.
Abundant in a few localities.

52. *Helix tuberculosa.* Conrad. in Lynch. Offi. Report, p. 229.
Erroneously identified by Bourguinat with *H. despreauxii* from the Canaries. This is the most peculiar and interesting *Helix* in Palestine,

and is found only sparingly in very restricted localities in the highlands west and south-west of the Dead Sea.

53. *Helix ledereri.* Pfr. in Malak. Bl. iii., 1856, p. 43.

In a few places on the coast, on sand-banks.

54. *Helix seetzeni.* Koch. Zeitschr. fur Malak., 1847, p. 14.

In immense numbers over the southern deserts, where it is the food of Sea-Gulls.

55. *Helix arabica.* Terver. Cat., p. 14.

Very scarce, and only south of the Dead Sea, taking the place of the preceding species.

56. *Helix candidissima.* Drap. Tabl. Moll., p. 75.

Very common.

Var. *hierochuntina*, Boiss., granulated at the apex.

Var. β, extremely glossy, and less than one-third the size of African specimens.

57. *Helix desertorum.* Forsk. Ehr. Symb. Phys.

Southern desert.

58. *Helix fimbriata.* Bourg. Test. Noviss. Or., p. 11.

Found in a few restricted localities north and west of the Dead Sea.

59. *Helix prophetarum.* Bourg. Test. Noviss. Or., p. 12.

Scattered in several localities west and south of the Dead Sea, near Sebbeh and Jebel Usdum.

60. *Helix boissieri.* Charp. Zeitschr. fur Malak., 1847, p. 133.

This fine example of a desert species, with its thick cretaceous shell, its solid contracted mouth and black interior, is widely dispersed in different localities over the Judæan desert, but not so generally as *H. seetzeni.*

61. *Helix filia.* Mouss. Coq. Voy. Roth., 1861, p. 26.

This beautiful desert species has strong affinities both with *H. prophetarum* and *H. boissieri.* It is extremely scarce, and is found only in a few localities near the Dead Sea.

62. *Helix cariosa.* Oliv. Voy. ii., p. 221, pl. 31, f. 4.

Extremely abundant in the mountain districts of Western Palestine; not observed in the east. The three varieties, (1) *amphicyrta,* (2) *nazarensis,* (3) *crassocarina,* are easily recognisable. The third is the prevailing type in the north, distinguished by its depressed spire and broad flattened keel. About Nazareth it gives way to the second variety, rounder, with the keel more compressed, but still the spire depressed. Specimens about Jerusalem and Carmel partake of the character of the first variety, with elevated spire; while at Hebron, the southern limit we observed for this shell, the northern form *crassocarina* reappears unchanged in the slightest particulars.

63. *Helix guttata.* Oliv. Voy. ii., p. 334.

Dead Sea.

64. *Helix eremophila.* Boiss. Reeve. Conch. Ic., vii., No. 956.

Southern Desert.

65. *Helix cæsareana.* Parr. Mouss. Coq. Or., p. 34, 44.

Abundant in the plain of Sharon and about the Sea of Galilee. The specimens from Gennesaret are much larger and more richly marked than those from Judæa.

66. *Helix spiriplana.* Oliv. Voy. Lev. i., p. 415, pl. 17, f. 7. (=*H. guttata,* Bourg.)

Generally distributed, but not numerous, in the higher grounds of Southern Palestine, and not found in the same localities as the preceding.

67. *Helix masadæ.* Tristram. P. Z. S., 1865, p. 535.

Diam. maj. 30, min. 25, alt. 14 mill.

Apert. diam. maj. 13, min. 11 mill.

Found on Sebbeh, the ancient Masada, and the most barren and sterile mountains from thence to Jebel Usdum, the salt-mountain. The deep and regular striation of this shell distinguishes it at once from *H. spiriplana*, for a small variety of which (such as that which Conrad has described under the name *H. lithophaga*) it might otherwise be mistaken.

68. *Helix aspersa.* Müll. Verm. Terr. et Fluv. Hist. ii., p. 59.

Very common in the gardens of Tyre, Sidon, Beyrout, Jaffa, and all places on the coast. We did not meet with it inland. It reaches a very large size—quite equal to the specimens from Algeria, and far surpassing those of the Greek islands. This as well as all the following species and *H. cæsareana* are collected and sold in the markets for food.

69. *Helix cavata.* Mouss. Coq. Bell. Or., p. 21.

Common in the interior; not plentiful near the coast.

70. *Helix prasinata.* Roth in Malak. Bl., 1855, p. 31.

We did not find this species ourselves; but I possess three specimens given me at Jerusalem by my lamented friend, its discoverer, Dr. Roth.

71. *Helix lucorum.* Linn. Syst. Nat., p. 1247.
Lebanon.

72. *Helix figulina.* Parr. in Rossm. Icon. ix., p. 9.
Dry Plains.

73. *Helix ligata.* Müll. Verm. Terr. et Fluv. Hist. ii., p. 58.
In the Lebanon.

74. *Helix grisea.* Linn. Syst. Nat., p. 693.
Between Nablus and the Jordan.

75. *Helix pachya.* Bourg. Rev. et Mag. Zool., 1860, p. 162.
Near the Lake of Gennesaret, and north of Beyrout.

76. *Helix engaddensis.* Bourg. Test. Noviss. Or., p. 11.
In the wilderness of Judæa.

These eight species appear to me to be very closely allied, the most important differences being in the aperture, which is almost circular in *H. cavata* (a species closely allied to *H. figulina*), and is oval and elongated in

H. prasinata and *H. engaddensis.* The differences in size and colour are certainly very great; yet I am inclined to believe that they are attributable rather to climate and locality, and that further research will embrace all of them in two or at most three species. In the immense series we collected, it is difficult anywhere to draw a satisfactory line.

77. *Helix vermiculata.* Müll. Verm. Terr. et Fluv. Hist. ii., p. 20.

A dwarf form of this widely spread and variable shell occurs between Beyrout and Tripoli. In Northern Syria it is as large as in North Africa.

78. *Bulimus acutus.* (Müll. Verm. Terr. et Fluv. Hist. ii., p. 100.)

Common on the sandy banks near the shore between Beyrout and Sidon; scarcer to the southward.

79. *Bulimus decollatus.* Linn. Syst. Nat. i., p. 773.

Found by us sparingly in the plain of Sharon. This is, so far as I am aware, the most eastern locality hitherto noticed for this shell. I cannot altogether agree with the remark of Bourguinat, that *B. decollatus* does not vary in the east and west, excepting in size, those from the east being considerably larger than from the west. I possess an extensive series collected by myself in every country bordering on the Mediterranean, from Spain and Morocco to Asia Minor, Cyprus, and Syria. The specimens from Algeria and Tunis are very much larger than any on the northern side, reaching the length of 2¼ inches without the rejected portion of the apex. The Spanish specimens are much more obtuse, and with fewer whorls, than those from countries further east; and the further we proceed eastward, the longer and the more attenuated do we find the shell, till in Cyprus and Palestine it reaches its extreme attenuation, though not approaching African specimens in size. It does not appear to occur in Egypt.

80. *Bulimus fasciolatus.* Oliv. Voy. Lev. i., p. 416, pl. 17, f. 5 (var. *eburneus*).

Scarce, in the neighbourhood of the Wâdy Kelt, near Jericho.

81. *Bulimus candelaris.* Pfr. P. Z. S. 1846, p. 40.

Maritime Plains.

82. *Bulimus labrosus.* Oliv. Voy. Lev. ii. 222, p. 31, f. 10.

The finest and most characteristic shell of Palestine. It is found generally concealed in small fissures of the limestone rocks, sometimes under stones, throughout the whole of Western Palestine, as far as the edge of the Ghor or Jordan valley, but not beyond. It is most abundant near the coast, where it attains its greatest size. A very small variety is found in the southern wilderness.

83. *Bulimus carneus.* Pfr. Phil. Abbild. ii., pl. 4, f. 5.

This beautiful shell takes the place of the preceding species in the basin of the Dead Sea towards the south; but we never found it north of Engedi, nor on the east side. It is most plentiful about the famed rock of Masada, the modern Sebbeh. We brought a considerable number home alive, which are now depositing their eggs, and feeding on succulent plants. It is impossible, after observing a large series, to have any hesitation in separating *B. carneus* specifically from *B. labrosus.* The elongated form, the mouth proportionally less than half the size of the other species, and circular instead of being extended towards the right, the solidity of the peristome, and the callosity largely extended over the last whorl, at once distinguish every specimen; nor have I ever detected any intermediate forms.

The typical *B. carneus* of Dr. Pfeiffer is from Lycia. I have not been able to compare my specimens with the type, though they appear to coincide exactly with the diagnosis and the figure. It is possible that our Dea Sea species may be distinct.

84. *Bulimus alepi.* Fér. Prodr., p. 418.

Generally diffused, but scarce in number of individuals. Collected near Jerusalem and by the Dead Sea.

85. *Bulimus syriacus.* Pfr. Symb. iii., p. 88.

Extremely abundant in certain localities of the Lebanon.

86. *Bulimus sidoniensis.* Charp. Reeve. Conch. Ic., v., pl. lxiii., No. 433.

In the plain of Phœnicia and the neighbouring hills.

87. *Bulimus ehrenbergi.* Pfr. P. Z. S., 1846, p. 113.

Erroneously identified by Bourguinat with *B. obesatus*, Webb and Berthelot, from the Canaries. Frequent throughout the wooded hills and under brushwood in Western Palestine. The rich olive-green epidermis of the living shell seems to have escaped the notice of its describers.

88. *Bulimus forskalii.* Beck. Ind., p. 68.

Southern desert. This species partakes much of the character of *Pupa*.

89. *Bulimus uriæ.* Tristram. P. Z. S., 1865, p. 537.

Long. tota 15, lat. 7; apert. long. $5\frac{1}{2}$, lat. $4\frac{1}{2}$ mill.

The Wâdy of Amman (Rabbath Ammon).

This *Bulimus*, the Transjordanic representative of *B. attenuatus*, is intermediate in character between it and *B. pupa* of Greece and Algeria. From the latter it may be distinguished at once by its olive-green colour and by its suddenly expanding fifth whorl, which gives it a peculiar obese appearance. From the former it is distinguished by the sixth and seventh whorls increasing instead of contracting.

90. *Bulimus* (*Chondrus*) *triticeus.* Rossm. Ic. iii., p. 89.

Near Jerusalem.

91. *Bulimus* (*Chondrus*) *sulcidens.* Mouss. Pfr. Mon. vi., p. 71.

The Bukâa.

92. *Bulimus* (*Chondrus*) *tricuspidatus.* Küster, p. 62, pl. 8, f. 5, 6.

Not uncommon near Beyrout, and by the Dead Sea.

93. *Bulimus* (*Chondrus*) *septemdentatus.* Roth. Diss., p. 19, pl. 2, f. 2.

Common throughout the whole country, and subject to great variations in size. The mouth is frequently six-toothed, and sometimes only five-toothed.

94. *Bulimus* (*Chondrus*) *ovularis.* Oliv. Voy. i., p. 225, pl. 17, f. 12.

Common. For the distinctions between this and the last species, see Mousson, Coq. p. 46.

95. *Bulimus* (*Chondrus*) *bidens.* Kryn. Bull. Mosc. vi., p. 401.
Anti-Lebanon.

96. *Bulimus lamelliferus.* Rossm. Icon. iii., 17, 1859, p. 95.

97. *Bulimus* (*Chondrus*) *saulcyi.* Bourg. Test. Noviss. Or., p. 18.
About the plain of Gennesaret and the Dead Sea. Confined, apparently, to the Jordan valley. Like *B. ovularis*, but invariably sinistral, and found in distinct localities.

98. *Bulimus* (*Chrondrus*) *nucifragus.* Parr. Pfr. Monog. ii., p. 135.
Scarce; found at Jaffa and near Jerusalem.

99. *Bulimus benjamiticus.* Benson. Ann. and Mag., N. H., 3rd Ser., 1859, p. 393.
Hills of Benjamin.

100. *Pupa delesserti.* Bourg. Test. Noviss. Or., p. 17.
Scarce, in the Anti-Lebanon.

101. *Pupa saulcyi.* Bourg. Test. Noviss. Or., p. 19.
Two specimens found near Tyre, in the hills.

102. *Pupa rhodia.* Roth. Diss., p. 19, pl. 2, f. 4.
Scarce near Jerusalem. Very abundant on a rock near the Lake of Gennesaret, but extremely local.

103. *Pupa granum.* Drap. Tabl. Moll., p. 50.
Near Sidon, in the plain of Phœnicia.

104. *Pupa scyphus.* Pfr. Zeitschr. f. Mal., 1848, p. 7.
A single dead specimen in Lebanon.

105. *Pupa chondriformis.* Mousson. Coq. Voy. Roth. 1861, p. 49, Jerusalem. Scarce.

106. *Pupa libanotica.* Tristram. P. Z. S., 1865, p. 538.
Long. tota 11, lat. $4\frac{1}{2}$ mill.
Found at Ainat, in the Lebanon.

107. *Pupa orientalis.* Parr. Pfr. in Malak. Bl. viii., 1861, p. 168.
Nazareth.

108. *Pupa raymondi.* Bourg. Rev. et Mag. Zool. xv., 1863, p. 259.
Anti-Lebanon.

109. *Pupa michonii.* Bourg. Test. Noviss. Or., p. 19.
One dead specimen found near Nazareth.

110. *Pupa hebraica.* Tristram. P. Z. S., 1865, p. 539.
Long. $2\frac{3}{4}$, lat. $1\frac{1}{2}$ mill.
Found in a tomb near Jericho. The beautiful and regular tranverse ridges on the whorls, as seen through a magnifying-glass, at once distinguish this from every other species of *Pupa.*

111. *Clausilia mœsta.* Fér. Pr. 539.
Near Jaffa, near Beyrout, and occasionally in the hills behind the plain of Phœnicia.

112. *Clausilia strangulata.* Fér. Pro. 516.
Plentiful in the ravine of the Nahr el Kelb, Lebanon.

113. *Clausilia saulcyi.* Bourg. Cat. Rais. Moll. Or., p. 50.
Only found by us at the Ladder of Tyre. Collected by M. de Saulcy near Jerusalem.

114. *Clausilia delesserti.* Bourg. Cat. Rais., p. 47.
In the Nahr el Kelb, in damp caves ; scarce.

115. *Clausilia albersi.* Charp. Journ. Conch., 1852, p. 374.
In the valley of the Kadisha, Lebanon.

116. *Clausilia boissieri.* Charp. Zeitschr. f. Malak., 1847, p. 142.
Excessively abundant near the Nahr el Kelb and on the rocks near Beyrout. Found abundantly on rocks, a few yards from the spray of the sea. It reaches a larger size here, and the peristome is more expanded than in specimens from Crete and other parts of Greece.

117. *Clausilia genezerethana.* Tristr., P. Z. S., 1865, p. 539.

Long. 20½, diam. 3 mill.

Found only on rocks near the plain of Gennesaret.

118. *Clausilia oxystoma.* Rossm. Ic. x., p. 19, f. 625.

Eastern slopes of Lebanon.

119. *Clausilia hierosolymitana.* Bourg. Rev. et Mag. Zool., 1868, p. 428.

Jerusalem.

120. *Clausilia corpenlenta.* Pfr. in Zeitschrift f. Malak., 1848, p. 7.

Beyrout.

121. *Clausilia cedretorum.* Bourg. Rev. et Mag. Zool. xv., 1863, p. 109.

Lebanon.

122. *Clausilia cylindrelliformis.* Bourg. Rev. et Mag. Zool., 1855, p. 330.

Lebanon.

123. *Clausilia bitorquata.* Rossm. in Malak. Bl. iv., 1857, p. 38.

Lebanon, near Tripoli.

124. *Clausilia medlycotti.* Tristram. P. Z. S., 1865, p. 540.

Long mill. 19, diam. 3½; apert. long. 3, lat. 2½ mill.

This most beautiful *Clausilia*, which I have great pleasure in dedicating to my friend and fellow-traveller, Sir W. C. P. Medlycott, Bart., was found by us only in one place, but in considerable plenty, in the hills behind Surafend (Sarepta). It may at once be distinguished from all others by the boldness of its sculpture, and by its very deep and distinct, though sometimes irregular, ridges.

125. *Clausilia sidonia.* Parr. Mal., 1848, p. 10.

Mughdooshy, near Sidon.

126. *Clausilia bicarinata.* Rossm. x., p. 17, f. 620.
Northern hills.

127. *Clausilia sancta.* Bourg. Rev. et Mag. Zool., 1868, p. 427.
Beyrout.

128. *Clausilia vesicalis.* Frév. Rossm. Malak. Bl. iv., 1857, p. 38.
Beyrout.

129. *Clausilia dextrorsa.* Böttg. Claus. Stud., 1877, p. 46.
Northern hills, Lebanon.

130. *Clausilia ehrenbergi.* Roth. Malak. Bl., 1855, p. 44.
Beyrout, in gardens.

131. *Clausilia fauciata.* Parr. Rossm. Malak. Bl. iv., 1857, p. 39.
Beyrout. Lower Lebanon, in cliffs.

132. *Clausilia phœniciaca.* Bourg. Rev. et Mag. Zool., 1868, p. 425.
Nahr el Kelb.

133. *Clausilia porrecta.* Frév. Rossm. Malak. Bl. iv., 1857, p. 39.
Lebanon, near the sea.

134. *Clausilia davidiana.* Bourg. Rev. et Mag. Zool., 1868, p. 376.
Nahr el Kelb valley, Lower Lebanon.

135. *Clausilia prophetarum.* Bourg. Rev. et Mag. Zool., 1868, p. 378.
Near Beyrout.

136. *Clausilia raymondi.* Bourg. Rev. et Mag. Zool. xv., 1863, p. 110.
Nahr el Kelb.

137. *Clausilia filumna.* Pfr. Malak. Bl. xiii., 1866, p. 151.
Lebanon.

138. *Clausilia pleuroptychia.* Bött. Jahrb. Mal. Ges. v., p. 291.
The Leontes valley.

139. *Clausilia galeata.* Rossm. x., p. 17, f. 621.
Near Baalbec.

140. *Clausilia dutaillyana.* Bourg. Rev. et Mag. Zool., 1868, p. 424.
Western Lebanon.

141. *Clausilia nervosa.* Parr. Schmidt. Clausil., p. 102.
Lebanon and Anti-Lebanon.

142. *Tornatellina* (Beck) *hierosolymarum.* Roth. Malak. Bl., 1855, p. 39.
Scarce, in tombs in various parts of the country.

143. *Glandina* (*Cæcilianella*) *tumulorum*, var. *judaica*, Bourg. iv., 625.
In tombs at Jerusalem.

144. *Glandina* (*Cæcilianella*) *liesvillei.* Bourg. Rev. et Mag. Zool., 1856, p. 385.

145. *Planorbis hebraicus.* Bourg. Test. Noviss. Or., p. 23.
Ain Mellaheh, near Lake Huleh.

146. *Planorbis piscinarum.* Bourg. Test. Noviss. Or., p. 22.
Near Zebdany, in Cœle Syria.

147. *Planorbis vortex.* L. Syst. Nat.

148. *Planorbis alexandrinus.* Roth. Moll., pl. ii., f. 8.

149. *Planorbis saulcyi.* (Bourg. Voy. Mer Morte, p. 68).

150. *Limnæa tenera.* Parr. Reeve. Conchol. Icon., vol. xviii., Limnæa, pl. xiv., sp. 96.
Near the Lake Huleh.

151. *Limnæa syriaca?* Mouss.

Near Baalbec.

152. *Limnæa truncatula.* (Mull. Verm. Hist. ii., p. 130.)

Lake Huleh.

153. *Cyclostoma olivieri.* Chemn. ed. nov., p. 156, pl. 21, f. 20.

Very common in the neighbourhood of the plains of Phœnicia and Acre, but not met with further south or east.

154. *Cyclostoma elegans.* Müll. Verm. Terr. et Fluv. Hist. ii., p. 137.

155. *Bithinia saulcyi.* Bourg. Voy. Mer Morte, p. 63.

The Bukâà.

156. *Bithinia gaillardoti.* Bourg.

157. *Bithinia hawadieriana.* Bourg. Voy. Mer Morte, p. 63.

Lake Huleh.

158. *Bithinia moquiniana.* Bourg.

159. *Bithinia hebræorum.* Bourg.

Ain Fijeh, and other fountains in the Bukâà; very common.

160. *Bithinia* (*Paludina*) *phialensis.* Conrad.

Birket er Ram (Lake Phiala).

161. *Bithinia rubens.* Menke. Synopsis, p. 134.

Lake Huleh.

There are several other species of minute *Paludinidæ*, which I have not been able to determine.

162. *Melania tuberculata.* Müll. Verm. Terr. et Fluv., p. 191.

Occurs living in various streams, and semi-fossil in great numbers on the marl-deposits by the Dead Sea. By the shores of the Lake of Galilee dead and bleached specimens are very common.

163. *Melania rothiana.* Mouss. Coq. Rec. Roth., p. 61.

We obtained several dead specimens of this shell by the Sea of Galilee; I am more than doubtful of its specific value, believing it to be merely an elongated form of *M. tuberculata;* but M. Lortet, a much better judge, considers it distinct.

164. *Melania gemmulata.* Reeve. Conch. Icon., vol. xii., Melania, pl. xiii., sp. 86.

In the Nahr el Kelb. Always a deep brown-black, and differing from *M. tuberculata* in the absence of the longitudinal ridges and tubercles on the spire.

165. *Melania rubro-punctata.* Tristram. P. Z. S., 1865, p. 541.

Long. 21, diam. 5 mill.; altera 17 long., $3\frac{1}{2}$ diam.

Lives buried in the sand, in fountains near the Dead Sea.

Had I not consulted more experienced naturalists than myself, I should have felt disposed to have included this as a delicate and very beautiful variety of the variable *M. tuberculata.* The distinctive characters are the extreme smallness of the aperture and the sudden termination of the longitudinal sculpture, which does not extend to the lower whorls.

166. *Melania judaica.* Roth. Malak. Bl. ii., p. 53.

167. *Melanopsis buccinoïdea.* (Oliv., Voy. I., p. 297.)

Very abundant in almost all the streams of Palestine, and found sub-fossil in the old marl-deposits by the shores of the Dead Sea. There is a distinct variety peculiar to almost every district.

Var. A, from the Nahr el Kelb, near Beyrout, is horn-colour, with three dark brown bands.

Var. B, from streams near Engedi and other streams flowing into the Dead Sea, is much larger than any other specimens I have seen, and may be at once recognised by a compression on the right side of the peristome, near the columella. It is rarely black, but of a rich brown colour, and the inside of the mouth a pale purple. It may be hereafter separated as a distinct species.

Var. C, from the waters of Merom and the Lake of Galilee, is very large, almost approaching the specimens of Engedi in size, but black, more inflated and obtuse, and with a rich deep purple colour inside the mouth. Those from the Kishon are similar, but smaller.

168. *Melanopsis ammonis.* Tristram. P. Z. S., 1865, p. 542.

Long. 25, diam. 10; apert. long. 7, larg. 5 mill.

I was at first inclined to place this shell as a variety of *M. prærosa;* but its more elegant and elongated shape, the smallness of its mouth, and the traces of longitudinal ridges, appear to me sufficient to justify its separation. Found only in streams at Heshbon and Ammon, east of Jordan, where the other species does not occur.

169. *Melanopsis saulcyi.* Bourg. Voy. Mer Morte, p. 66.

In a few restricted localities; chiefly at Ain Sultan, Jericho.

170. *Melanopsis ferrusaci.* Roth. Moll. Spec., 1839, p. 24, pl. ii., f. 10.

A species of Asia Minor and Northern Syria, but also found here and there in Lebanon.

171. *Melanopsis doriæ.* Issel. Moll. Persiæ, 1865, p. 16, pl. i., f. 7, 8.

Found in running water near Beyrout. Originally described from Persia.

172. *Melanopsis variabilis.* V. de Busch. in Phil. Abbild., 1847, p. 175, pl. iv., ff. 7, 8, 10.

Also found in streams near Beyrout.

173. *Melanopsis brevis.* Parr. in Mouss. Coq. Bellardi, 1854, p. 51.

In the Litany River.

174. *Melanopsis costellata.* Fér. Monog. Melanop., p. 28.

In the Kishon. The differences between this and *M. saulcyi* are clearly pointed out by Bourguinat. This species is less fusiform, more inflated, does not increase regularly, and its last whorl is three times the size of the others united; while that of . *Msaulcyi* is not more than once and a half as large.

175. *Melanopsis costata.* Oliv., Voy. ii., p. 294.

Very abundant in the Huleh (waters of Merom), the Lake of Galilee, and the Upper Jordan. In immense quantities in a sub-fossil state round the Dead Sea.

176. *Melanopsis jordanica.* Roth. Moll. Spec., p. 25, pl. ii., f. 12, 13.

Peculiar, so far as we could ascertain, to the Lake of Galilee and the Jordan below it.

Rossmässler, as well as Roth at an earlier period, considered this to be a variety of the preceding species. There is, however, a striking difference, not only in the shape and coloration, but in the habit of the living animal. *M. costata* is always found adhering to the stems and the under surface of the leaves of aquatic plants; while the obtuse and striped form, *M. jordanica,* adheres only to rocks and stones. *M. costata* we never met with south of the entrance to the Lake of Galilee in a living state, nor *M. jordanica* to the north of it.

177. *Melanopsis eremita.* Tristram. P. Z. S., 1865, p. 542.

Long. 16, diam. 6½ mill.; apert. long. 5, lat. 3½ mill.

Collected only in the little stream of the Wâdy Um Bagkek, between Sebbeh and Jebel Usdum, at the south-west corner of the Dead Sea, where it was very abundant.

This beautiful and very small species of *Melanopsis* may be at once recognised by its peculiarly brilliant gloss. It may be remarked, that in the same region which supplies the smallest of its group, the common *Melanopsis prærosa* attains its greatest magnitude.

Besides the species here enumerated, the following have been named by Bourguinat, but no descriptions published. They are unknown to me, and are:

178. *Melanopsis prophetarum.*

Elisha's Fountain. Beyrout. Lake of Antioch.

179. *Melanopsis ovum.*

Lake of Galilee.

180. *Neritina syriaca.*

181. *Neritina jordani.* Buttler. Sowerby. Conch. Ill., f. 49.

Found in the Jordan and its two lakes, Huleh and Galilee, but chiefly under the leaves of water-lilies in the Huleh. We never discovered it living in the lower course of the Jordan.

182. *Neritina michonii.* Bourg. Test. Nov., p. 25.

Abundant in almost every stream and spring throughout the whole of Palestine, east and west. It attains its greatest size in the thermal springs of the Ghor.

183. *Neritina bellardi.* Mouss. Coq. Bellardi, p. 52, pl. i., f. 11.

I have not had the opportunity of examining a type specimen; but, from the diagnosis, I believe this species to be that which is found in the Jabbok and its affluents. It is certainly different from *N. michonii*, as may be at once recognised by an examination of its operculum.

184. *Corbicula saulcyi.* Bourg. Moll. Nouv. Litig., p. 315, pl. xlv., f. 6—9.

Long. 36, lat. 35, diam. 16·5 mill.

This shell is at once distinguished from the next species by the coarse and irregular striations which furrow the whole surface of the valves. These in *C. fluminalis* are fine and regular.

Found in the Jordan. Dead valves by the shore of the Lake of Galilee.

185. *Corbicula fluminalis.* Müll. Verm. Fluv. Hist. ii., p. 205.

This species is not uncommon in the Jordan and Sea of Galilee. It is very variable in size, as may be seen from the measurements of two specimens from the lake.

Long. 17—27, lat. 20—28, diam. 16—21 mill.

186. *Corbicula syriaca.* Bourg. Locard. Malac. Lac Tiberiade, p. 29, pl. xxii, f. 22—24.

Long. 19—24, lat. 17—21, diam. 11—16 mill.

Found sparingly in the Lake of Galilee. Very common in the Lakes of Antioch and Hums.

187. *Cyrena crassula.* Mouss. Reeve. Conch. Icon. Cyrena, sp. 72.

Adonis river. Litany river, etc. Jaffa.

188. *Cyclas casertanum.* Poli. Test. Sicil. i., 65.

Wâdy Kadisha, Lebanon.

189. *Unio terminalis.* Bourg. Voy. Mer Morte, p. 76, pl. 3, f. 4.

This is the common *Unio* of the Lake of Galilee, and I found it also in the Litany (Leontes) river. The *Unio jordanicus*, Bourg., seems to be only a thinner and shorter variety of *U. terminalis.* I have obtained so many intermediate specimens that it appears to be impossible to separate the two. *U. terminalis* also seems to be identical with the *U. dignatus*, Lea, from the Tigris, as I find on comparison of type specimens in Mr. Cuming's collection.

190. *Unio delesserti.* Bourg. Voy. Mer Morte, Moll., p. 77.

In the Zerka or Crocodile river, in the Plain of Sharon. Found by M. de Saulcy near Jaffa, in the same plain.

191. *Unio michonii.* Bourg. Voy. Mer Morte, p. 74.

In the Zerka or Crocodile river. Figured by Bourguinat from the neighbouring streams of Jaffa. Appears to be but a variety of the *U. marginalis*, Lamarck, a variable species found in India, Mauritius, and China.

192. *Unio saulcyi.* Bourg. Voy. Mer Morte, p. 74, pl. iii., f. 1—3.

From the Kishon. Found by M. de Saulcy in the streams near Jaffa. This species is very like *U. mosalensis*, Lea, from the Tigris, if indeed it be not the same species.

193. *Unio simonis.* Tristram. P. Z. S., 1865, p. 544.

Long. 66, lat. 44, diam. 32 mill.

This shell is found in the Jordan, the Sea of Galilee (where it reaches its greatest dimensions), the Orontes, and the Leontes (Litany). Its rotundity, thickness, solidity, and the brilliant rosy tint of its nacreous interior distinguish it at once from every other species. The massive solidity of the young shells is very remarkable. The rosy tint is equally brilliant in all the specimens I have seen.

194. *Unio episcopalis*. Tristram. P. Z. S., 1865, p. 544.

Long. 98, lat. 56, diam. 35 mill.

This, the prince of Oriental *Unionidæ*, is not uncommon in the Orontes. I found a dead valve by the Leontes, but did not meet with it in the Lake of Galilee. From its brilliant purple hue, which is preserved in the most worn valves, from its size, its jet-black epidermis, and the peculiar compression, it is a remarkable and isolated species. I can find no *Unio* in the collection of Mr. Cuming which at all resembles it.

195. *Unio tripolitanus*. Bourg. Voy. Mer Morte, p. 75, pl. iv., f. 10—12.

Near Tripoli.

196. *Unio requieni*. Mich. Suppl., pl. xvi.

197. *Unio rhomboïdeus*. Schröter. Fluss. Conch., p. 186.

198. *Unio rothi*. Bourguinat. Moll. Nouv. Litig., 1865, p. 133, pl. xx.

Long. 46—57, lat. 30—39, diam. 21—23 mill.
In the Lake of Galilee and in the Jordan.

199. *Unio luynesi*. Bourg. Locard. Malac. Lac Tiberiade, p. 11.
Long. 53, lat. 28, diam. 23 mill.
In the Jordan.

200. *Unio galilæi*. Locard. Malac. Lac Tiberiade, p. 12, pl. xx., f. 10—12.

Long. 35, lat. 28, diam. 23 mill.
In the Lake of Galilee, but rare.

201. *Unio timius*. Bourg. Locard. Malac. Lac Tiberiade, p. 13, pl. xx., f. 13, 14.

Long. 18, lat. 14, diam. 8 mill.
In the Jordan. Rare.

202. *Unio raymondi.* Bourg. Locard. Malac. Lac Tiberiade, p. 14.

Long. 65—69, lat. 32—35, diam. 21—24 mill.

The most delicate, light, and translucid of all the *Unios* of the Lake of Galilee, where alone it is found.

203. *Unio tristrami.* Locard. Malac. Lac Tiberiade, p. 15, pl. xx., f. 15, 16.

Long. 59, lat. 31, diam. 24 mill.

This peculiar form, in some respects like *U. euphraticus*, Bourg., has been found only in the Lake of Galilee, where Dr. Lortet discovered it.

204. *Unio pietri.* Locard. Malac. Lac Tiberiade, p. 16, pl. xx., f. 17—19.

Long. 50—57, lat. 28—31, diam. 20—22 mill.

Discovered by Dr. Lortet in the Lake of Galilee. It had been previously taken by me in some quantity, but not discriminated from former species.

205. *Unio ellipsoideus.* Bourg. Locard. Malac. Lac Tiberiade, p. 17, pl. xxi., f. 1—3.

Long. 55, lat. 32, diam. 23 mill.

In the Lake of Galilee.

206. *Unio jordanicus.* Bourg. Amœn. Malacol. I., p. 167, pl. xvi., f. 1—4.

Long. 55—60, lat. 30—32, diam. 23—26 mill.

Very common in the Jordan. Less abundant in the Lake of Galilee.

207. *Unio genezerethanus.* Letourneux. Locard. Malac. Lac Tiberiade, p 19, pl. xxi., f. 4—6.

Long. 56, lat. 34, diam. 23 mill.

Lake of Galilee.

208. *Unio grelloisianus.* Bourg. Amœn. Malac. I., p. 165, pl. xvii., f. 1—4.

Long. 30, lat. 18, diam. 15 mill.

This small *Unio* was discovered by De Saulcy in the Jordan.

209. *Unio lorteti.* Locard. Malac. Lac Tiberiade, p. 21, pl. xxi., f. 7—12.

Long. 58—60, lat. 30—32, diam. 23—24 mill.

This rather variable shell is common in the Lake of Galilee, and a larger form of it also exists in the Lake of Antioch.

210. *Unio tiberiadensis.* Letourneux. Locard. Malac. Lac Tiberiade, p. 22, pl. xxi., f. 13—15.

Long. 57, lat. 30, diam. 23 mill.

Rare. In the Lake of Galilee.

211. *Unio prosacrus.* Bourg. Locard. Malac. Lac Tiberiade, p. 25, pl. xxi., f. 16, 17.

Long. 47—55, lat. 25—30, diam. 18—23 mill.

This is the most cuneiform of all the *Unios* of the Lake of Galilee, where alone it is found, and is there rather common.

212. *Unio lunulifer.* Bourg. Amœn. Malacol. I., p. 166, pl. xvii., f. 5—8.

Long. 49, lat. 31, diam. 22 mill.

Collected in the Jordan by De Saulcy.

213. *Unio zabulonicus.* Bourg. Locard. Malac. Lac Tiberiade, p. 26, pl. xxii., f. 11—13.

Long. 50, lat. 31, diam. 22 mill.

In the Lake of Galilee.

FLORA OF PALESTINE.

THE catalogue of the Flora has been compiled both from my own Herbarium, comprising 1,400 species, formed by Mr. B. T. Lowne, who accompanied me as botanist in my expedition of 1863—4, and from the various additions made by me in two subsequent journeys. The Herbariums of the late W. Amherst Hayne, Esq., and of Rev. H. E. Fox and Rev. W. Linton, have contributed several additions. The M.S. catalogue of the Flora of Palestine, compiled by Mr. Hanbury and Sir Jos. Hooker, now in the Herbarium at Kew, has been most kindly placed at my service by Professor Oliver, F.R.S., to whom I am under the greatest obligation for his generous assistance while revising my list by the aid of the Kew Herbarium. Very many species have been added on the authority of these catalogues, though unfortunately many of the older collectors have omitted to mark the precise locality. But by far the most important materials for the compilation of this catalogue are to be found in the unrivalled work of M. Edmond Boissier, '*Flora Orientalis*,' of which I have not hesitated unsparingly to avail myself. For carefulness of detail in description, and for accuracy of statement as to geographical area, no botanical writer has ever surpassed M. Boissier. In his later work, M. Boissier has suppressed many of his own species described in his 'Diagnosis' and other earlier works, including them as local or climatic varieties. I have generally followed his later decision; but in some instances, where the distinctive characters are easily recognisable, and the species have been treated as distinct by several subsequent writers, I have for convenience of reference retained the earlier synonymy.

It is a question what should be the geographical limits of the Palestine Flora. I have given a liberal interpretation, and have looked on 'the Land of Promise' rather than 'the Land of Possession' as our area—*i.e.*, have comprised the whole country from the Wâdy el Arish and Petra up to the Orontes valley at Hamath as within our limits. If a plant occurs north of Petra, or by the El Arish, we may be pretty certain that it will be found in the southern wilderness and plateau; and no botanist has yet exhaustively worked Southern Judæa. With the exception of the district close to Gaza, it has only been skimmed; and the following pages will show how many species belonging to the Arabian Flora rest their claims to insertion on Gaza alone, as their only known Palestine habitat. So northwards the Bukâà supplies a great number of species, which we may be certain will be found, when looked for, further south. To the eastward, again, we have a great volcanic region, which has not been even skimmed by the botanical collector, besides the whole of Gilead, possessing by far the richest Flora of the whole region, and which has never been worked except by a hurried and passing traveller. It is to the east of Jordan that we must look for our principal future additions to the Flora of Palestine; and there lies a rich field for any enterprising young botanist who is not afraid of Bedouin.

I have endeavoured to give the geographical area of each species as briefly as possible. The letter 'P.' after the locality given implies that the species is, so far as known, peculiar to Palestine. Where a Palestine locality is given and no 'P.' follows, the plant inhabits other parts of the Oriental region.

The Oriental region is taken to comprise all the countries bordering on the Mediterranean from Greece and Egypt eastwards to the frontiers of India—*i.e.*, Greece, Southern Turkey, Lower Egypt, Arabia Petræa, Asia Minor, Syria, Mesopotamia, Armenia, Persia, and Afghanistan.

Whenever a species extends beyond these limits, its further extension is given after the word 'Area.'

Thus:

'*Rhus coriaria.* L. Sp. 379.

'Generally distributed. Area, Mediterranean region'—

implies that it is found generally in Palestine, in the Oriental region given above, and also in the countries bordering on the Mediterranean west of Greece and Egypt.

PLANTÆ VASCULARES.

CLASS, DICOTYLEDONEÆ.
SUB-CLASS, THALAMIFLORÆ.

ORDER, RANUNCULACEÆ.

1. *Clematis cirrhosa.* L. Sp. 766.

On the coast and hills east and west of Jordan. Area, Spain, Barbary, South Italy.

2. *Clematis orientalis.* L. Sp. 765.

In the north. Area, South Siberia, North-west India.

3. *Clematis flammula.* L. Sp. 766.

On the coast and hills. Area, Central and Southern Europe, North Africa.

4. *Clematis vitalba.* L. Sp. 766. Traveller's Joy. Arab. شراج, *Scheradj.*

On the coast and hills. Area, Central and Southern Europe, North Africa.

5. *Clematis recta.* L. Sp. 767.

Gilead. Area, Southern Europe.

6. *Thalictrum orientale.* Boiss. Ann. Sc. Nat., 1841, p. 349.

Lebanon.

7. *Anemone coronaria.* L. Sp. 760.

In every part of the country in profusion, almost invariably the red variety, yellow, blue, and purple occurring very early. The most gorgeously painted, the most conspicuous in spring, the most universally spread of all the floral treasures of the Holy Land; if any one plant can claim pre-eminence among the wondrous richness of bloom which clothes the Land of Israel in spring, it is the anemone, and therefore it is on this we fix, as the most probable 'lily of the field' of our Lord's discourse.

Area, the Mediterranean region.

8. *Anemone blanda.* Schott. and Ky. Œst. Bot. Woch., 1854, p. 129.

Lebanon and Hermon. Area, Eastern Mediterranean.

9. *Adonis palæstina.* Boiss. Diagn. Ser. i., viii., p. 1.

In every part of the country—most abundant. P.

10. *Adonis autumnalis.* L. Sp. 771.

Under Lebanon. Area, Central and Southern Europe.

11. *Adonis microcarpa.* De Cand. Syst. i., p. 223.

Gilead. Area, Spain, Italy, North Africa.

12. *Adonis æstivalis.* L. Sp. 771. Pheasant's Eye.

Abundant on the plains. Moab. Area, Central and Southern Europe, North Africa, Western Himalayas.

13. *Adonis dentata.* Del. Fl. Eg., tab. v., p. 1.

On the sandy plains. Area, North Africa.

14. *Adonis flammea.* Jacq. Austr., tab. 335.

In Lebanon. Area, Central Europe.

15. *Myosurus minimus.* L. Sp. 407. Mousetail.

Anti-Lebanon, Moab. Area, Europe, North Africa, North America.

16. *Ranunculus aquatilis.* L. Sp. 781, *ex parte.* Water Ranunculus.

In pools near the coast in various forms and varieties. Area, the temperate Northern Hemisphere.

Var. *heterophyllus.* Near Acre.

17. *Ranunculus trichophyllus.* Chaix in Vill. Delph. i., p. 335.

In the valley of the Upper Jordan. Area, Europe, North America.

18. *Ranunculus calthæfolius.* (*Ficaria.*) (Jordan Obs. vi., p. 2.)

General, coast and hills. Area, South Europe, North Africa.

19. *Ranunculus ficarioïdes.* Bory et Chaub. Flor. Pelop., p. 55, pl. xvi., p. 2.

Lebanon.

20. *Ranunculus crymophilus.* Boiss. et Hohm. Diagn. Ser. i. viii., p. 6.

Snow-line of Lebanon. (W. A. Hayne.)

21. *Ranunculus myosuroïdes.* Boiss. and Ky. Pl. Syr. Exs., 1855.

On the snow-line of Lebanon and Anti-Lebanon, 6,500 feet. P.

22. *Ranunculus orientalis.* L. Sp. 781.

On the hills in the north. Area, North Africa.

23. *Ranunculus damascenus.* Boiss. et Gaill. Diagn. Ser. ii., vi., p. 5.

In fields in the north.

24. *Ranunculus muricatus.* L. Sp. 780.

South Judæa. Area, Mediterranean region, North-west India.

25. *Ranunculus philonotis.* Retz Obs. vi., p. 3. =*R. hirsutus.* Ait.

Under Lebanon. Area, Europe, North Africa.

26. *Ranunculus tuberculatus.* Kit.

27. *Ranunculus cuneatus.* Boiss. Diagn. Ser. i., viii., p. 2.
Lebanon. P.

28. *Ranunculus asiaticus.* L. Sp. 777.
All over the country east and west.
Ditto, var. *Tenuilobus.* Boiss. Gilead.

29. *Ranunculus chærophyllus.* De Cand. Fl. Fr. iv., p. 900.
Northern Palestine. Area, South Europe, North Africa.

30. *Ranunculus sprunerianus.* Boiss. Diagn. Ser. i., i., p. 64.
The Anti-Lebanon.

31. *Ranunculus myriophyllus.* Schrad. Journ., 1799, p. 424.
On the northern plains, and Moab.

32. *Ranunculus hierosolymitanus.* Boiss. Flor. Or. i., p. 36.
Through all the hill country. Differs slightly from the last species. P.

33. *Ranunculus eriophyllus.* Koch. Linn. xix., p. 46.
Marshes in Cœle-Syria. Area, Mediterranean region.

34. *Ranunculus lateriflorus.* De Cand. Syst. i., p. 252.
By pools in Gilead. Area, coast of Spain, South Italy, Sicily.

35. *Ranunculus demissus.* De Cand. Syst. i., p. 275.
On the summits of Lebanon and Hermon, 8,000—9,000 feet. Area, Sierra Nevada, Spain.

36. *Ranunculus comatocarpus.* F. and M.

37. *Ranunculus cassius.* Boiss. Ann. Sc. Nat., 1841, p. 354.
Lebanon and Anti-Lebanon.

38. *Ranunculus constantinopolitanus.* Urv. Enum., p. 64.
Ditto, var. *Palæstinus.* Boiss. In Northern Palestine.

39. *Ranunculus sceleratus.* L. Sp. 776.

In wet places in Northern Palestine. Area, all Europe, North Africa, India, North America.

40. *Ranunculus brachycarpus.* C. A. Mey.

41. *Ranunculus ophioglossifolius.* Vill. Delph. iv , p. 732, pl. 49.

The northern plains. Area, Mediterranean region.

42. *Ranunculus glaberifolius.* De Cand. Syst. i., p. 254.

Marshes of the Kishon. Area, Mediterranean region.

43. *Ranunculus chius.* De Cand. Syst. i., p. 299.

Lebanon. Area, Sicily, Dalmatia.

44. *Ranunculus trachycarpus.* F. and M. Ind. iii., Petrop., p. 46.

Hermon, Moab, and all the hill-country.

45. *Ranunculus cornutus.* De Cand. Syst. i., p. 300; and Ditto, var. *Rhyncocarpus.* Boiss. Fields in the north.

46. *Ranunculus arvensis.* L. Sp. 780. Corn Crowfoot.

Universal. Area, Europe, North Africa, Himalayas.

47. *Ranunculus pinardi.* Boiss. Diagn. Ser. ii., v., p. 10.

The Anti-Lebanon. P.

48. *Ranunculus parviflorus.* L. Sp. 780.

Area, South Europe, North and West Africa, Canaries, North America.

49. *Ceratocephalus falcatus.* Pers. Syn. 341.

The hilly districts and Aroer. Area, South Europe.

50. *Helleborus orientalis.* Lam. Encycl. iii., p. 92.

In the north.

51. *Helleborus fœtidus.* L. Sp. 784.

Doubtfully indigenous. Area, Europe.

52. *Garidella unguicularis.* Lam. Ill., pl. 379, f. 2.

Throughout Palestine.

53. *Nigella arvensis.* L. Sp. 753. Var. *Divaricata.*

Generally distributed. Area, Central and Southern Europe, North Africa.

54. *Nigella deserti.* Boiss. Ann. Sci. Nat., 1841, p. 359.

Near Gaza, Gilead.

55. *Nigella sativa.* L. Sp. 753. Hebr., קצח. A.V., '*Fitches*' (Is. xxviii.. 25, 27).

Cultivated. Area, South Europe.

56. *Nigella orientalis.* L. Sp. 753.

Mount Tabor, Moab.

57. *Nigella damascena.* L. Sp. 753.

In fields. Area, Mediterranean region.

58. *Nigella aristata.* Sibth. Fl. Gr., pl. 510.

59. *Nigella stellaris.* Boiss. Diagn. Ser. i., viii., p. 8.

In the north.

60. *Nigella ciliaris.* De Cand. Syst. i., p. 327.

Through the coast and central districts.

61. *Delphinium axilliflorum.* De Cand. Prodr. i., p. 341. *Larkspur.*

Cultivated generally. Wild about Hermon.

62. *Delphinium oliganthum.* Auch. in Boiss., Pl. Or. An. Sci. Nat., 1841, p. 365.

The Eastern desert.

63. *Delphinium pusillum.* Labill. Syr. Dec. iv., pl. 2, f. 2.
About Hermon.

64. *Delphinium rigidum.* De Cand. Syst. i., p. 244.
The coast and hilly country. P.

65. *Delphinium deserti.* Boiss. Fl. Or. i., p. 83.
Near Gaza, in the desert.

66. *Delphinium aconiti.* Linn. Mant. 77.

67. *Delphinium halteratum.* Sibth. Fl. Græc. vi., pl. 107.
Area, South Italy.

68. *Delphinium ajacis.* L. Sp. 748. = *D. pubescens.* Gris.
Area, South Europe.

69. *Delphinium anthoroïdeum.* Boiss. Ann. Sc. Nat., 1841, p. 369.
Lebanon and Hermon.

70. *Delphinium peregrinum.* L. Sp. 749.
The hill-country. Area, Italy, Dalmatia, and Spain.

71. *Delphinium bovei.* Decaisne. Ann. Sc. Nat. Sec. Ser. iv. p. 356.
Desert near Gaza.

72. *Delphinium virgatum.* Poir. Sup. ii., p. 458.
The northern plains.

73. *Delphinium ithaburense.* Boiss. Diag. Ser. i., viii., p. 9.
Mount Tabor. P.

74. *Pæonia corallina.* Retz. Obs. iii. p. 32.
Rare in North Lebanon. Area, Central and Southern Europe.

ORDER II., BERBERIDEÆ.

1. *Bongardia chrysogonum.* (L. Sp. 447.)

Distinguished as two species, *B. rauwolfii* and *B. olivieri*, by C. A. Mey. Verz., p. 174.

Through the hilly districts.

2. *Leontice leontopetalum.* L. Sp. 448.

Coast and hills. Area, South Italy.

3. *Berberis vulgaris.* L. Sp. 471. Barberry. Arab. طرح, *Tarah.*

Lebanon. Area, Europe.

4. *Berberis cratægina.* De Cand. Syst. ii., p. 9.

5. *Berberis cretica.* L. Sp. 472.

Hermon and Lebanon, 5,000—7,000 feet.

ORDER III., NYMPHÆACEÆ.

1. *Nymphæa alba.* L. Sp. 729. White Water Lily.

In still water, especially Lake Huleh. Area, Europe and Siberia.

2. *Nuphar luteum.* (L. Sp. 729.) Yellow Water Lily.

Lake Huleh. Area, Europe and Siberia.

ORDER IV., PAPAVERACEÆ.

1. *Papaver libanoticum.* Boiss. Ann. Sc. Nat., 1841, p. 373.

Lebanon and Hermon. P.

2. *Papaver dubium.* L. Sp. 726. In fields. Area, Europe.

3. *Papaver umbonatum.* Boiss. Diagn. Ser. i., viii., p. 11.

Rocky plains of Southern Lebanon. P.

4. *Papaver rhœas.* L. Sp. 727. Field Poppy.
On the coast, and Dead Sea shore. Area, Europe, North Africa.

5. *Papaver syriacum.* Boiss. Diag. Ser. ii., vi., p. 8.
Plain of Tyre. P.

6. *Papaver polytrichum.* Boiss. and Ky. Diagn. Ser. ii., v., p. 14.
Northern Plains.

7. *Papaver somniferum.* L. Sp. 726. Opium Poppy.
Cultivated.

8. *Papaver hybridum.* L. Sp. 725. Rough Poppy.
General in fields. Area, Central and Southern Europe, North Africa.

9. *Papaver argemone.* L. Sp. 725.
In fields. Area, Central and Southern Europe, North Africa.

10. *Rœmeria hybrida.* (L. Sp. 724.)
In fields. Area, Mediterranean region.

11. *Rœmeria orientalis.* Boiss. Ann. Sci. Nat., 1841, p. 374.
Zara, east side of Dead Sea.

12. *Glaucium corniculatum.* (L. Sp. 724.) Horned Poppy.
Coast and hills. Area, Central and Southern Europe, North Africa.

13. *Glaucium arabicum.* Fres. Mus. Senck., p. 174, pl. 10.
Southern desert.

14. *Glaucium luteum.* Scop. Carn. i., p. 369.
On the coast. Area, Canaries, Central and Southern Europe.

15. *Glaucium fulvum.* Smith. Exot. Bot., p. 11.
Perhaps a variety of *G. luteum.*

16. *Glaucium leiocarpum.* Boiss. Fl. Or. i., p. 122.
Lebanon and Hermon.

17. *Glaucium vitellinum.* Boiss. et Buhse. Aufz., p. 12, f. 4.
Mount Hermon, 4,000—5,000 feet.

18. *Hypecoum procumbens.* L. Sp. 181.
Jordan valley. Wâdys in Gilead. Area, Mediterranean region.

19. *Hypecoum grandiflorum.* Benth. Catal. Pyr., p. 91.
Coast and hilly districts. Area, Spain, and Southern France.

20. *Hypecoum imberbe.* Sibth. Fl. Gr., pl. 156
Gaza, and other sandy districts.

21. *Hypecoum pendulum.* L. Sp. 181.
The Bukâà. Area, Central and Southern Europe.

ORDER V., FUMARIACÆ.

1. *Coryaalis rutæfolia.* (Sibth. Fl. Gr., pl. 867)
Lebanon and Hermon. Snow-line, 8,000 feet.

2. *Corydalis solıda.* (Smith. Engl. Bot., pl. 1471.)
Northern mountain region. Area, mountains of Central and Southern Europe, and Siberia.

3. *Corydalis libanotica.* Hochst. in Lorent. Wand., p. 350.
Neighbourhood of the Cedars of Lebanon. Area, mountains in Europe and Siberia.

Var. *C. modestum*, var. *C. pulchellum*, var. *C. purpurascens*, Schott. and Ky. As. Min. Bot. i., p. 359.

4. *Ceratocapnos palæstina.* Boiss. Diagn. Ser. i., viii., p. 12.
Walls and rocks. Coast and hill districts. P.

5. *Fumaria officinalis.* L. Sp. 984. Fumitory.
In the north; general. Area, Europe, North Africa, North Asia.

6. *Fumaria parviflora.* Laur. Enc., p. 567.
General. Area, Central and Southern Africa, North-west Himalayas.

7. *Fumaria asepala.* Boiss. Fl. Or. i., p. 135.
Lebanon.

8. *Fumaria micrantha.* Lag. el Hort. Matrit., 1816, p. 21.
Common on cultivated land, and Moab. Area, Central and Southern Europe, India.

9. *Fumaria anatolica.* Boiss. Diagn. Ser. i., viii., p. 14.
Hill region.

10. *Fumaria capreolata.* L. Sp. 985.
Coast region. Area, Western and Southern Europe.

11. *Fumaria thureti.* Boiss. Diagn. Ser. ii., i., p. 15.
Lebanon.

12. *Fumaria macrocarpa.* Parlat. Pl. Nov., 1842, p. 5. Var. *Oxyloba.*
Hill region.

13. *Fumaria oxyloba.* Boiss. Diagn. Ser. i., viii., p. 14.
Mounts Carmel and Gerizim. P.

14. *Fumaria judaica.* Boiss. Diagn. Ser. i., iii., p. 15.
Coast and hill region.

15. *Fumaria gaillardoti.* Boiss. Fl. Or. i., p. 139.
Fields on the coast. P.

ORDER VI.—CRUCIFERÆ.

1. *Chorispora syriaca.* Boiss. Ann. Sc. Nat., 1842, p. 384.
Universal.

2. *Morettia canescens.* Boiss. Diagn. Ser. i., viii., p. 17.
Desert east of Gilead.

3. *Matthiola albicaulis.* Boiss. Ann. Sc. Nat., 1842, p. 46.
Moab.

4. *Matthiola sinuata.* (L. Sp. 926.) Sea Stock.
Rocks near the coast. Area, Mediterranean region.

5. *Matthiola crassifolia.* Boiss. Diagn. Ser. ii., vi., p. 10.
Rocks on the coast. P.

6. *Matthiola damascena.* Boiss. Diagn. Ser. i., viii., p. 16.
Bare chalky hills, north and east. P.

7. *Matthiola arabica.* Boiss. Ann. Sc. Nat., 1842, p. 49.
In the eastern desert.

8. *Matthiola tricuspidata.* (L. Sp. 926.)
On the coast. Area, Mediterranean region.

9. *Matthiola aspera.* Boiss. Diagn. Ser. i., viii., p. 16.
Barren hills west of the Dead Sea. P.

10. *Matthiola oxyceras.* De Cand. Syst. ii, p. 173.
Round the Dead Sea, in the Ghor.

11. *Matthiola lunata.* Boiss. Fl. Or. i., p. 156.
Ghor, north end of Dead Sea, and Callirrhoe.

12. *Matthiola livida.* (Delil. Ill. Æg, No. 591.)
Barren plains in Moab, Wâdy Zerka, Wâdy Mojib. P.

13. *Eremobium lineare.* (Del. in Lab., p. 85, f. 3.)
Sandy desert south of Beersheba. Area, the Sahara.

14. *Farsetia agyptiaca.* Turra. Diss., p. 1, pl. 1.
South, east and west of the Dead Sea. Area, the Sahara.

15. *Farsetia ovalis.* Boiss. Diagn. Ser. i., viii., p. 32.
South end of Dead Sea.

16. *Farsetia incana.* (L. Sp. 978.)
Eastern desert.

17. *Cardamine hirsuta.* L. Sp. 915. Bitter Cress.
Lebanon. Area, Europe.

18. *Arabis auriculata.* Lam. Dict. i., p. 219.
Mount Gilead. Area, Southern Europe.

19. *Arabis verna.* (L. Sp. 928.)
Mountain districts. Area, Mediterranean region.

20. *Arabis montbretiana.* Boiss. Ann. Sc. Nat., 1842, p. 53.
Northern mountain region.

21. *Arabis hirsuta.* Scop. Carn. ii., p. 30.
Northern hills. Area, Europe.

22. *Arabis sagittata.* (Bechst. Amœn. It., p. 185.)
Mountain region. Area, Europe, Siberia.

23. *Arabis albida.* Stev. Cat. Hort. Gor., p. 51.
Lebanon. Area, Sicily.

24. *Arabis brevifolia.* De Cand. Syst. ii., p. 218.
Mountain regions under Lebanon and Hermon.

25. *Arabis billardieri.* De Cand. Syst. ii., p. 218.
Hermon, 6,000 feet.

26. *Nasturtium officinale.* R. Brown. H. Kew. iv., p. 109. Watercress.

General. Area, the temperate Northern Hemisphere.

27. *Nasturtium coronopifolium.* De Cand. Syst. ii., p. 224.

South end of Dead Sea.

28. *Nasturtium sylvestre.* (L. Sp. 916.) Creeping Watercress.

Mountain districts, 4,000 feet. Area, Northern and Central Europe and Asia.

29. *Nasturtium macrocarpum.* Boiss. Diagn. Ser. ii., viii., p. 18.

Under Hermon. P.

30. *Barbarea minor.* C. Koch. L. xix., p. 55.

Lebanon and Hermon snowline, 8,000 feet.

31. *Barbarea sicula.* Presl. Del. Pras. 17, Guss. Syn. ii., p. 180.

Area, South Italy.

32. *Cheiranthus cheiri.* L. Sp. 924.

On maritime cliffs, Lebanon. Area, Central and Southern Europe.

33. *Erysimum repandum.* L. Sp. ii., p. 923.

Northern and eastern mountain districts, Gilead. Area, East Europe, North Africa, North-west India.

34. *Erysimum smyrnæum.* Boiss. Diagn. Ser. ii., v., p. 23.

Under Hermon.

35. *Erysimum verrucosum.* Boiss. Diagn. Ser. ii., vi., p. 12.

Slopes of Hermon. P.

36. *Erysimum scabrum.* De Cand. Syst. ii., p. 505.

Lebanon.

37. *Erysimum goniocaulon.* Boiss. Diagn. Ser. ii., viii., p. 25.

Lebanon and Hermon district.

38. *Erysimum crassipes.* C. A. Mey. Hohm. Enum. Talysch., p. 141.

Round Hermon, Lebanon, and Gilead.

39. *Erysimum purpureum.* Auch. in Boiss., Ann. Sc. Nat., 1842, p. 42.

Round Hermon and Lebanon.

40. *Erysimum rupestre.* (Sib. Flor. Gr., pl. 633.)

Lebanon.

41. *Conringia orientalis.* (L. Sp. 931.)

Northern Palestine. Area, Europe.

42. *Conringia clavata.* Boiss. Ann. Sc. Nat., 1842, p. 84.

Northern mountain region.

43. *Sisymbrium pumilum.* Steph. in Willd. iii, p. 507.

In desert districts. Area, Southern and Eastern Russia.

44. *Sisymbrium schimperi.* Boiss. Ann. Sc. Nat., 1842, p. 76.

Eastern mountains of Moab.

45. *Sisymbrium sophia.* L. Sp. 922. Flixweed.

Ravine of the Arnon. Area, Europe.

46. *Sisymbrium columnæ.* L. Sp. 655.

Plains in the north. Area, Southern Europe.

47. *Sisymbrium pannonicum.* Jacq. Coll. i., 70.

Plains of the highlands.

48. *Sisymbrium nudum.* (Belay. Voy. Ic. Boiss. Ann. Sc. Nat., 1862, p. 54. Sub Arabide.)

49. *Sisymbrium pannonicum,* var. *rigidulum.* Jacq. Coll. i., 70.

Southern highlands. Area, Germany, Russia, West Thibet.

50. *Sisymbrium irio.* L. Sp. 921.

General; especially Jordan valley. Area, Europe, North Africa, North-west India.

51. *Sisymbrium damascenum.* Boiss. Diagn. Ser. ii., vi., p. 11.

Northern Palestine. P.

52. *Sisymbrium officinale.* (L. Sp. 922.) Hedge Mustard.

General. Area, Europe, North Africa.

53. *Sisymbrium runcinatum.* Lag. in De Cand. Syst. ii, p. 478.

Northern Palestine, Moab. Area, North Africa, Spain.

54. *Malcolmia pulchella.* (De Cand. Syst. ii., p. 455.)

Universal in the plains, maritime and sandy.

55. *Malcolmia pygmæa.* (Del. Ill. Æg., p. 19.)

Coast at Askalon.

56. *Malcolmia africana.* (L. Sp. 928.)

Eastern Desert of Moab. Area, Mediterranean region, and North-west India.

57. *Malcolmia maritima.* (L. Amœn. iv., p. 180.)

On the coast. Area, Mediterranean region.

58. *Malcolmia littorea.* Willd. Sp. Pl. iii., p. 521.

Area, Southern France, Spain.

59. *Malcolmia torulosa.* (Desf. Att. ii., p. 84, pl. 159.)

In the bare and desert regions north and south. Also var. *B. contortuplicata* and *V. leiocarpa.* Area, North Africa.

60. *Malcolmia chia.* (Lam. Dict. iii., p. 324.)

Coast and hills.

61. *Malcolmia crenulata.* (De Cand. Syst. ii., p. 456.)

Hill country and bare plains inland. Plains of Moab.

62. *Malcolmia conringiodes.* Boiss. Fl. Or. i., p. 230.
Lower slopes of Lebanon and Hermon. P.

63. *Hesperis kotschyana.* Fenzl. Pugill, p. 13.
Lebanon, higher parts.

64. *Hesperis pendula.* De Cand. Syst. ii., p. 457.
Plains of the Upper Jordan.

65. *Hesperis secundiflora.* Boiss. Diagn. Ser. i., i., p. 70.
On rocks. Gilead and Moab. Area, Dalmatia.

66. *Nasturtiopsis arabica.* Boiss. Fl. Or. i., p. 237.
Southern Desert of Judæa.

67. *Anchonium billardieri.* De Cand. Syst. ii., p. 578.
Lebanon and Anti-Lebanon, 6,000—8,000 feet.

68. *Sterigma sulphureum.* De Cand. Syst. ii., p. 212.
Northern barren plains.

69. *Aubrietia deltoidea.* (L. Sp. 908.)
Galilee, Anti-Lebanon.

70. *Aubrietia canescens.* Boiss. Flor. Or. i., p. 252.
Lebanon.

71. *Aubrietia libanotica.* Boiss. Diagn. Ser. i., viii., p. 32.
Lebanon, 8,000 feet. P.

72. *Ricotia lunaria.* De Cand. Syst. ii., 284.
In the Jordan basin. P.

73. *Fibigia clypeata.* (L. Sp. 909.)
Northern parts.

74. *Fibigia lunarioïdes.* (Reich. Bot. Mag., pl. 3087.)

75. *Fibigia macroptera.* (Ky. Pl. Exs., 1859.)
Northern mountains.

76. *Fibigia rostrata.* (Schenk. Pl. Spec., p. 42.)

Hill country, north and south, and Gilead and Moab.

77. *Fibigia eriocarpa.* (De Cand. Syst. ii., p. 288.)

Anti-Lebanon.

78. *Alyssum alpestre.* L. Mant. 92.

Lebanon and Anti-Lebanon, 6,000—9,000 feet. Area, Alps, Southern Russia, Siberia, Corsica—on hill tops.

79. *Alyssum argenteum.* Wittm. Summ. iv., p. 430. Var. *Chrysanthum.* Boiss. Fl. Or. i., 271.

Below Lebanon. Area, Italy, and South-eastern Europe.

80. *Alyssum cassium.* Boiss. Diagn. Ser. i., viii., p. 34.

Northern hills.

81. *Alyssum montanum.* L. Sp. 907.

Lebanon and Hermon. Area, Central and Southern Europe.

82. *Alyssum suffrutescens.* Boiss. in Bourg. Pl. Exs., Ann., 1860.

Summit of Hermon.

83. *Alyssum xanthocarpum.* Boiss. Ann. Sc. Nat., 1842, p. 154.

Hermon, 7,000 feet.

84. *Alyssum tetrastemon.* Boiss. Ann. Sc. Nat., 1842, p. 153.

Hill country of Galilee, and Lebanon.

85. *Alyssum szowitsianum.* F. and M. Ind. iv., Hort. Petrop.

Lebanon and Anti-Lebanon.

86. *Alyssum campestre.* L. Sp. 909.

Everywhere in cultivated ground. Area, Central and Southern Europe.

87. *Alyssum damascenum.* Boiss. Diagn. Spec. Ser. ii., vi., p. 18.

Cultivated ground in the north. P

88. *Alyssum micranthum.* C. A. Mey. Ind. i., Hort. Petrop., p. 22.
Northern plains and hills.

89. *Alyssum calycinum.* L. Sp. 908.
Jebel Khaisun, near Damascus. Area, Central and Southern Europe.

90. *Alyssum aureum.* (Fenzl. Pug., No. 44.) = *Meniocus grandiflorus.* Jaub.
Hill country, Jerusalem, etc.

91. *Alyssum meniocoides.* Boiss. Ann. Sc. Nat., 1842, p. 158.
The Hauran.

92. *Koniga maritima.* (L. Mant., 42.)
Coast, near Askalon. Area, Mediterranean region.

93. *Koniga lybica.* (Viv. Pl. Lib., p. 34, pl. 16, f. 1.)
Desert of Moab.

94. *Koniga arabica.* Boiss. Diagn. Ser. i., viii., p. 26.
Desert south of Beersheba.

95. *Draba oxycarpa.* Boiss. Diagn. Ser. i., viii., p. 28.
Snow-line of Lebanon and Hermon. P.

96. *Draba velutina.* Boiss. Diagn. Ser. ii., vi., p. 14.
Highest parts of Lebanon and Hermon.

97. *Draba vesicaria.* Desv. Journ. Bot. iii., p. 186.
Lebanon and Hermon, near the snow-line, 7,000—9,000 feet. P

98. *Draba aizoides.* L. Mant. 91.
Lebanon. Area, Europe.

99. *Erophila minima.* C. A. M. Ind. Cauc., p. 184.
Lebanon and Antibanon, Gilead.

100. *Erophila præcox.* (Stev. Mem. Mosq. iv., p. 269.)
Hauran. Area, Central Europe.

101. *Erophila vulgaris.* De Cand. Syst. ii., p. 356. = *Draba verna.* L. Sp. 896. Whitloe Grass.
Common. Area, Europe.

102. *Erophila setulosa.* Boiss. Diagn. Ser. ii., v., p. 31.
Fields near Sidon. P.

103. *Coluteocarpus reticulatus.* (Lam. Ill., pl. 559, f. 2.)
Lebanon and Hermon, 7,000—9,000 feet.

104. *Peltaria augustifolia.* De Cand. Syst. ii., p. 329.
Lebanon and Hermon, 3,000—6,000 feet.

105. *Clypeola jonthlaspi.* L. Sp. 910.
Hill-districts, Moab, Gilead. Area, Mediterranean region.

106. *Clypeola echinata.* De Cand. Syst. ii., p. 328.
Hill-country west and east of Jordan.

107. *Clypeola lappacea.* Boiss. Ann. Sc. Nat., 1842, p. 174.
Mountain region in the north.

108. *Camelina sylvestris.* Wallr. Sched. 347.
General. Area, Central and Southern Europe.

109. *Notoceras canariense.* R. Br. Kew. iv., p. 117.
Jericho. Area, Mediterranean, Canaries, and North-west India.

110. *Anastatica hierochuntina.* L. Sp. 895. Rose of Jericho.
Only on the shores of the Dead Sea. Area, Deserts of North Africa.

111. *Heldreichia kotschyi.* Boiss. Ann. Sc. Nat., 1842, p. 186.
Lebanon, higher zone.

112. *Biscutella columnæ.* Ten. Nap. iv., pl. 162.
Hill-districts and Moab highlands. Area, Mediterranean region.

113. *Biscutella apula.* Lam. Dict. iii., p. 618.
Hill-districts of the south, Jerusalem, etc. Area, Mediterranean.

114. *Biscutella ciliata.* Desf. Area, South Europe.

115. *Biscutella leiocarpa.* De Cand. Syst. ii.
Jordan basin, by Dead Sea.

116. *Thlaspi bellidifolium.* Griseb. Spic. Add., p. 505.
Lebanon.

117. *Thlaspi brevicaule.* Boiss. Diagn. Ser. ii., v., p. 40.
Lower slopes of Lebanon and Hermon. P.

118. *Thlaspi arvense.* L. Sp. 901. Penny-cress.
Area, Northern Hemisphere.

119. *Thlaspi microstylum.* Boiss. Diagn. Ser. i., viii., p. 38.
Lebanon. P.

120. *Thlaspi perfoliatum.* L. Sp. 902.
Wâdys of Moab and Gilead. Area, South Europe.

121. *Thlaspi natolicum.* Boiss. Ann. Sc. Nat., 1842, p. 180.
Coast, hill, and central districts.

122. *Carpoceras oxyceras.* Boiss. Diagn. Ser. i., viii., p. 38.
Galilean hills.

123. *Iberis odorata.* L. Sp. 906.
In the plains. Area, North Africa.

124. *Iberis aleppica.* Scop. (doubtful).

125. *Crenularia glaucescens.* (Boiss. in Tchih. As. Min. iii., p. 325.)
Lebanon.

126. *Hutchinsia petræa.* (L. Sp. 899.)
Bare hills, Moab. Area, Europe, West Africa.

127. *Capsella procumbens.* (L. Sp. 898.)
Philistian coast. Area, Mediterranean. Elsewhere introduced.

128. *Capsella bursapastoris.* L. Sp. 903. Shepherd's Purse.
General. Area, Temperate Northern Hemisphere.

129. *Æthionema oppositifolium.* Labill. Déc. v., p. 14, pl. 9, f. 2.
Top of Lebanon only. P.

130. *Æthionema coridifolium.* De Cand. Syst. ii., p. 561.
Bare slopes of Lebanon, 6,000—8,000 feet.

131. *Æthionema stylosum.* De Cand. Syst. ii., p. 562.
Top of Lebanon. P.

132. *Æthionema cordatum.* (Desf. Cor., pl. 52.)
Lebanon and Hermon, 6,000 feet.

133. *Æthionema cristatum.* De Cand. Syst. ii., p. 560.
Hill-country down to Jerusalem and Hebron.

134. *Æthionema heterocarpum.* J. Gay in F. and M. Ind. Sem. Petrop. iv.
Below Hermon.

135. *Æthionema buxbaumii.* (Fisch. in Horn. Hafn. Suppl., 71.)
On the hills and plains.

136. *Lepidium draba.* L. Sp., p. 645.
Plain of Esdraelon, Moab. Area, South Europe.

137. *Lepidium sativum.* L. Sp. 899. Cress.
Cultivated. Spontaneous on Mount Gilead. Area, South Russia.

138. *Lepidium spinescens.* De Cand. Syst. ii., p. 534.

139. *Lepidium spinosum.* L. Mant. 253.
Marshes near Banias.

140. *Lepidium cornutum.* Sibth. Prodr., No. 1491.
Coast and north.

141. *Lepidium chalepense.* L. Amœn. iv., p. 321.
In fields and lower ground.

142. *Lepidium latifolium.* L. Sp. 899.
In plains and near marshes, and under Hermon. Area, Europe, North Africa, Siberia, Thibet.

143. *Lepidium perfoliatum.* L. Sp. 897.
Northern plains. Area, Spain, Danube, Southern Russia.

144. *Senebiera coronopus.* (L. Sp. 904.)
Among stony rubbish everywhere, Gilead, etc. Area, Europe.

145. *Cakile maritima.* Scop. Carm., No. 844. Sea Rocket.
On the coast. Area, all Europe and North Africa.

146. *Erucaria aleppica.* Gært. Carp. ii., p. 298, pl. 143, f. 9, and var. *puberula.*
On barren hills, and Judæan wilderness, Mount Nebo. Area, South Spain.

147. *Erucaria microcarpa.* Boiss. Diagn. Ser. i., viii., p. 47.
Wilderness of Judæa, Moab.

148. *Erucaria crassifolia.* (Forsk. Æg., p. 118.)
South end of Dead Sea.

149. *Erucaria lineariloba.* Boiss. Ann. Sc. Nat., 1842 p. 390.
Eastern highlands.

150. *Hussonia uncata.* Boiss. Diagn. Ser. i., viii., p. 47.
Desert south of Beersheba. Area, African deserts.

151. *Euclidium syriacum.* (L. Sp. 895.)

Fields and scrub. Area, Danube valley, South Russia, North and West India.

152. *Ochthodium ægyptiacum.* (L. Sp. iii., p. 231.)

Generally distributed.

153. *Neslia paniculata.* L. Sp. 894.

Common in fields. Area, Europe, Northern Asia.

154. *Texiera glastifolia.* (De Cand. Syst. ii., p. 337.)

The Hauran, near Damascus.

155. *Isatis latisiliqua.* Stev. Mem. Mosq., 1812, iii., p. 267.

Lebanon and Hermon.

156. *Isatis tinctoria.* L. Sp. 936. Woad.

Plain of Gennesaret. Area, Central and Southern Europe.

157. *Isatis microcarpa.* Boiss. Ann. Sc. Nat., 1842, p. 201.

Desert east of Moab.

158. *Isatis aleppica.* Scop. Del. Insubr. ii., p. 31, pl. 10.

Hill and mountain regions east and west of Jordan.

159. *Schimpera arabica.* (Schimp. Pl. Arab. Exs. 244.)

Desert south of Gaza.

160. *Moricandia arvensis.* (L. Mant. 95.) (W. A. Hayne.) Area, South Europe, North Africa.

161. *Moricandia dumosa.* Boiss. Diagn. Ser. i., viii., p. 25.

Southern desert.

162. *Moricandia sinaica.* (Boiss. Ann. Sc. Nat., 1842, p. 85.)

Wâdys south and south-west of Dead Sea.

163. *Diplotaxis viminea.* (L. Sp. 919.)

Among stone-heaps everywhere. Area, Central and Southern Europe, Northern Africa.

164. *Diplotaxis harra.* (Forsk. Fl. Æg., 118.)
Judæan wilderness, and near Dead Sea.

165. *Diplotaxis erucoïdes.* (L. Amœn. iv., p. 322.)
Plains and Jordan valley, Eastern Ghor. Area, Mediterranean region.

166. *Diplotaxis acris.* (Forsk. Fl. Æg. Arab., p. 118.)
Southern desert.

167. *Hirschfeldia adpressa.* Mœnch. Meth. 261.
Fields and stony places east and west. Area, Mediterranean region.

168. *Brassica rapa.* L. Sp. 931. Rape-seed.
Cultivated. Area, Europe and Siberia.

169. *Brassica bracteolata.* F. and M. Linn. xii., Litt., p. 153.
Moab.

170. *Brassica napus.* L. Sp. 931. Turnip.
Cultivated. Area, Europe.

171. *Brassica deflexa.* Boiss. Ann. Sc. Nat., 1842, p. 87.
In fields in the north.

172. *Brassica tournefortii.* Gon. Ill., p. 44, pl. 20.
In sand on the coast. Area, Southern Spain and Southern Italy.

173. *Sinapis juncea.* L. Sp. 934.
Jericho.

174. *Sinapis nigra.* L. Sp. 933.

175. *Sinapis arvensis.* L. Sp. 933. Charlock.
Generally distributed. Area, Europe.

176. *Sinapis orientalis.* L. Amœn. iv., p. 280.
Plains and hills.

177. *Sinapis alba.* L. Sp. 834. Mustard.
General. Area, Central and Southern Europe.

178. *Eruca sativa.* Lam. Fl. Fr. ii., p. 496.
Generally distributed. Area, South Europe.

179. *Savignya ægyptiaca.* De Cand. Syst. ii., p. 283.
The southern desert.

180. *Carrhichtera vellæ.* De Cand. Syst. ii., p. 641.
The Lower Jordan valley. Plains of Moab. Area, Southern Spain, Sicily, North Africa.

181. *Enarthrocarpus arcuatus.* Labill. Déc. v., p. 4, pl. 2.
Northern Palestine.

182. *Enarthrocarpus strangulatus.* Boiss. Diagn. Ser. i., viii., p. 44.
South end of Dead Sea.

183. *Raphanus sativus.* L. Sp. 935. Radish.
Grown everywhere. Area, Temperate Northern Hemisphere, Old World.

184. *Raphanus pugioniformis.* Boiss. Diagn. Ser. i., viii., p. 46.
Plains of Galilee, from Tabor to Hermon. P.

185. *Raphanus raphanistrum.* L. Sp. 935. Wild Radish.
In fields. Area, Europe.

186. *Raphanus aucheri.* Boiss. Diagn. Ser. i., viii., p. 45.
Everywhere, coast, plains, by the Dead Sea, base of Hermon.

187. *Rapistrum rugosum.* (L. Sp. 893.)
Sea coast and Lebanon. Area, Central and Southern Europe.

188. *Didesmus rostratus.* Boiss. Fl. Or. i., p. 405.
Central hill district. P.

189. *Didesmus ægyptius.* (L. Sp. 895.)
Plain of Gennesaret.

190. *Crambe maritima.* L. Sp. 937. Sea-kale.
On the coast. Area, Europe, shores of Mediterranean and Atlantic.

191. *Crambe orientalis.* L. Sp. 937.
Northern plains.

192. *Crambe filiformis.* Jacq.
Plain of Gennesaret.

193. *Crambe hispanica.* L. Sp. 937.
Zerka Main. Area, Mediterranean region.

194. *Zilla myagroïdes.* Forsk. Fl. Æg. 121, Icon. tab. 17 *a.*
South-western end of the Dead Sea.

195. *Calepina corvini.* (All. Ped., No. 937.)
Common in all the central districts. Area, Central and South Europe.

ORDER VII., CAPPARIDEÆ.

1. *Cleome pentaphylla.* L. Sp. 938.
Plain of Sharon. Area, subtropical and tropical Old World.

2. *Cleome arabica.* L. Sp. 939.
Desert of Ain Beida, lower end of Dead Sea. Area, Algerian and Arabian deserts.

3. *Cleome trinervia.* Fresn. Mus. Senck., p. 177, pl. 11.
Round the Dead Sea on gravel, and south of Gaza.

4. *Cleome droserifolia.* Del. Fl. Æg., p. 317, pl. 36, f. 2.
Engedi, and elsewhere round the Dead Sea shore.

5. *Capparis sodada*. R. Br. in Oudn. and Clappert, p. 17.

On sand in the plains of Shittim, north-east end of the Dead Sea.

6. *Capparis spinosa*. L. Sp. 720. The Caper. Hebr. אֵזוֹב, Arab. ازوف, *Azuf*.

Grows on walls, and on the perpendicular faces of rocks in Jerusalem and elsewhere, probably introduced. Very common in parts of the Jordan valley.

Var. *Ægyptia*, common by the Dead Sea. The caper is the hyssop of Scripture, as is shown by the identity of the Hebrew and Arabic names. Area, Mediterranean regions.

7. *Capparis galeata*. Fresn. Mus. Senck. Beitr. Abyss., p. 111.

With hesitation I assign to this species a caper found on the cliffs of the Akabah, south of the Dead Sea.

8. *Cratæva gynandra*.

Hot sulphurous springs by the east side of the Dead Sea.

ORDER VIII., RESEDACEÆ.

1. *Ochradenus baccatus*. Del. Fl. Eg., p. 15, pl. 31, f. 1.

Round the Dead Sea. Area, Deserts of North-west India.

2. *Reseda alba*. L. Sp. 645 = *R. suffruticulosa*. L.

General. Area, Mediterranean region.

3. *Reseda propinqua*. R. Br. Obs. Oudn., p. 22.

Southern desert. Area, North-east Africa.
Ditto, var. *Eremophila*. Boiss. Diagn. Ser. i., viii., p. 88.
Southern deserts.

4. *Reseda arabica*. Boiss. Diagn. Ser., i., p. 6.

Southern desert. Area, North Africa.

5. *Reseda alopecurus*. Boiss. Diagn. Ser. i., viii., p. 55.

Cultivated. Maritime and other plains. P.

6. *Reseda orientalis.* Boiss. Fl. Or. i., p. 427.

Sandy fields on the coast. P.

7. *Reseda phyteuma.* L. Sp. 449.

Central districts. Area, Mediterranean region.

8. *Reseda odorata.* L. Sp. 646. Mignonette.

Plentiful in the Wâdy Zuweirah, at the south-west end of the Dead Sea. Not hitherto found wild in Palestine.

9. *Reseda lutea.* L. Sp. 645.

In fields and barren plains, and Gilead. Area, Europe, North Africa.

10. *Reseda muricata.* Presl. Bot. Bemerk., p. 8.

By the Dead Sea.

11. *Reseda pruinosa.* Del. Fl. Eg., p. 15.

South-west end of Dead Sea, Wâdy Zuweirah, etc.

12. *Reseda luteola.* L. Sp. 643. Weld.

Hills south-west of Dead Sea. Area, Europe, North Africa.

13. *Oligomeris subulata.* (Delil. Fl. Eg., p. 15.)

Southern Desert. Area, Canaries, across North Africa to North India.

ORDER IX., CISTINEÆ.

1. *Cistus villosus.* L. Sp. 736.

The hill districts east and west of Jordan, especially plentiful on Carmel. From the Cistus is collected the *Ladanum*, Heb. לט, Arab. اللاذن, *ladan*, a balsam once much esteemed in pharmacy. Area, Corsica, Italy, North Africa.

2. *Cistus incanus.* L. Sp. 757.

Classed by Boissier as a variety of the last species.

3. *Cistus creticus.* L. Sp., p. 738.

The commoner form on the southern hills.

4. *Cistus salviæfolius.* L. Sp. 738.

On the hills everywhere. Area, Mediterranean region.

5. *Helianthemum syriacum.* Boiss. Diagn. Ser. i., viii., p. 49.

Sub-alpine parts of Lebanon. Area, Spain, Portugal, South of France.

6. *Helianthemum umbellatum.* (L. Sp. 739.)

Sub-alpine Lebanon district. Area, Spain and France.

7. *Helianthemum guttatum.* (L. Sp. 741.)

Coast, and North, and Moab. Area, Central Europe, Mediterranean region.

8. *Helianthemum niloticum.* (L. Mant., p. 246.)

Hill districts. Area, Mediterranean region.

9. *Helianthemum salicifolium.* (L. Sp. 742.)

Barren hills, east and west. Area, Mediterranean region.

10. *Helianthemum ægyptiacum.* (L. Sp. 742.)

On the hills east and west. Area, Mediterranean region.

11. *Helianthemum kahiricum.* Delil. Fl. Eg., p. 93, pl. 31, f. 2.

Southern desert. Area, North African desert.

12. *Helianthemum lippii.* (L. Mant. 245.)

Sandy districts on the coast. Area, North Africa, Sicily.

13. *Helianthemum lavandulæfolium.* (Lam. Dict. ii., p. 25.)

Central Hill district. Area, Mediterranean region.

14. *Helianthemum vesicarium.* Boiss. Diagn. Ser. i., viii., p. 30.

On all the southern hills. Plentiful on Olivet. Area, North-east Africa.

15. *Helianthemum kotschyanum.* Boiss. Diagn. Ser. ii., i., p. 53.

Anti-Lebanon in the plain.

16. *Helianthemum ellipticum.* Desf. Fl. Atl. i., p. 418, pl. 107.

Plain of Acre. Area, Atlas range.

17. *Fumana arabica.* (L. Sp. 745.)

Mount Carmel. Area, Atlas range.

18. *Fumana glutinosa.* (L. Mant. 246.)

Generally distributed. Area, Mediterranean region.

19. *Fumana procumbens.* De Cand. Prodrom. i., p. 275.

Near Beyrout. Area, Central and Southern Europe.

ORDER X., VIOLARIEÆ.

1. *Viola spathulata.* Willd. in Ræm. and Schutt. iii., p. 353.

Lebanon.

2. *Viola libanotica.* Boiss. Diagn. Ser. i., viii., p. 52.

Lebanon, 7,000—9,500 feet.

3. *Viola sylvatica.* Fries. Mant. iv., p. 121.

Lebanon. Area, Europe, North Asia.

4. *Viola parvula.* Tineo. Pug. Var., p. 5.

Anti-Lebanon. Area, mountains of South Spain, Corsica, and Sicily.

5. *Viola biflora.* L. Sp. 1326.

Lebanon. Area, mountains of Northern and Central Europe, Siberia, and North America.

6. *Viola ebracteolata.* Fenzl. Ill. Taur., p. 49.

Lebanon, 5,000 feet.

7. *Viola odorata.* L. Sp. 1524.

Wooded hills of Central and Eastern Palestine. Area, Europe North Africa, Canaries, Siberia.

8. *Viola modesta.* Fenz. Ill. Pl. Taur., p. 48.

Northern wooded hills.

ORDER XI., POLYGALEÆ.

1. *Polygala monspeliaca.* L. Sp. 987.

Generally distributed. Area, Mediterranean region.

2. *Polygala supina.* Schreb. Dec., p. 19. pl. 10.

Galilean hills.

3. *Polygala anatolica.* Boiss. Diagn. Ser. ii., i., p. 57.

On the scrub-covered hills.

4. *Polygala vulgaris.* L. Sp. 986.

Lebanon and Anti-Lebanon, 6,000 feet. Area, Central and South Europe.

ORDER XII., SILENEÆ.

1. *Velezia rigida.* L. Sp. 474.

Plain of Gennesaret. Area, Mediterranean region.

2. *Velezia fasciculata.* Boiss. Diagn. Ser. i., viii., p. 92.

Lebanon.

3. *Dianthus armeria.* L. Sp. 586. Area, Europe.

4. *Dianthus multipunctatus.* Ser. in De Cand. Prodr. i., p. 362.

Central and Northern Palestine.

5. Var. *Subenervis.* Boiss. Fl. Or. i., p. 483.
'Bsherreh, Lebanon.

6. Var. *Gracilior* = *D. striatellus.* Fenz.
Lebanon.

7. *Dianthus polycladus.* Boiss. Diagn. Ser. i., p. 65.
Anti-Lebanon, low grounds.

8. *Dianthus pallens.* Sibth. Fl. Gr. iv., p. 87, pl. 399.
Ditto, var. *Oxylepis.* Deserts below Hermon.

9. *Dianthus superbus.* L. Sp. 589.
Northern mountains. Area, Central and Northern Europe, Siberia.

10. *Dianthus caryophyllus.* L. Sp. 210. Area, Central and South Europe.

11. *Dianthus nazaræus.* Clarke. It., iii., ex Spreng. Nen. Entd. iii., p. 161. (?)

12. *Dianthus polymorphus.* M. B. Taur. Cauc. i., p. 324.
Northern hills. Area, Hungary, South Russia.

13. *Dianthus judaicus.* Boiss. Diagn. Ser. i., viii., p. 66.
Hills surrounding Dead Sea. P.

14. *Dianthus libanotis.* Labill. Dec. i., p. 14, pl. 4.
Higher parts of Lebanon.

15. *Dianthus fimbriatus.* N. B. Taur. Cauc. i., p. 382.
Ditto, var. *Brevifolius* = *D. pogonopetalus.* Boiss.
Lebanon, near the Cedars.

16. *Dianthus pendulus.* Boiss. Diagn. Ser. ii., vi., p. 28.
On the face of precipices, Lebanon.

17. *Dianthus zonatus.* Fenz. Pug., No. 35.
Mount Nebo.

18. *Tunica syriaca.* Boiss. Diagn. Ser. i., viii., p. 63.

Northern Lebanon.

19. *Tunica pachygona.* Fisch. et Mey. Ind. S. Petrop. iv., No. 2255.

Eastern plains of Moab.

20. *Tunica saxifraga.* (L. Sp. 584.)

Plains and dry valleys under Lebanon. Area, Central and South Europe.

21. *Tunica arabica.* Boiss. Diagn. Ser. i., viii., p. 62.

Southern Desert.

22. *Tunica prolifera.* (L. Sp. 382.)

Plains of Esdraelon. Area, Northern and Central Europe.

23. *Saponaria vaccaria.* L. Sp. 585.

Galilee, Gilead. Area, Central and Southern Europe, Siberia.

24. *Saponaria prostrata.* Willd. Enum. 465.

Bare hillsides.

25. *Saponaria oxyodonta.* (Boiss. Diagn. Ser. ii., i., p. 68.)

Common in cornfields everywhere.

26. *Saponaria mesogitana.* Boiss. Diagn. Ser. i., i., p. 16.

Northern hills.

27. *Saponaria pulvinaris.* Boiss. Diagn. Ser. i., viii., p. 74.

Higher parts of Lebanon and Hermon.

28. *Saponaria syriaca.* Boiss. Diagn. Ser. i., viii., p. 73.

Central districts.

29. *Saponaria orientalis.* L. Sp. 585.

Plains and wilderness of Judæa.

30. *Ankyropetalum cœlesyriacum.* Boiss. Diagn. Ser. i., viii., 59.

Plain of the Bukâa. P.

31. *Gypsophila rokejeka.* Del. Fl. Eg., p. 282, pl. 29, f. 1.
Wilderness of Judæa towards Dead Sea, ruins in Moab.

32. *Gypsophila libanotica.* Boiss. Diagn. Ser. i., i., p. 12.
Cedars of Lebanon, Hermon.

33. *Gypsophila damascena.* Boiss. Diagn. Ser. i., viii., p. 57.
Barren hills, Anti-Lebanon, etc.

34. *Gypsophila viscosa.* Murr. Comm. Gœtt., 1783, p. 9, pl. 3.
Fields in the northern plains.

35. *Gypsophila frankenioides.* Boiss. Diagn. Ser. i., i., p. 10. Var. *libanotica.*
Rocky places on the eastern slopes of Lebanon.

36. *Gypsophila hirsuta.* (Labill. Dec. Syr. iv., p. 9.) Var. *mollis.*
Fissures of limestone rocks. Lebanon, 5,000—6,000 feet.

37. Var. *Alpina.* Highest parts of Lebanon.

38. Var. *Barradensis.* Fissures of rocks, Souk Wâdy Barada.

39. Var. *Filicaulis.* Barren hills about Damascus.

40. *Silene coniflora.* Otth. in De Cand. Prodr. i., p. 371.
Cultivated land. General.

41. *Silene conoidea.* L. Sp. 598. Var. *Obcordata.*
Stony fields, north and east of Jordan, Moab, etc. Area, Spain, South France, North-west India.

42. *Silene reuteriana.* Boiss. Diagn. Ser. ii., v., p. 54.
Sandy districts, Lebanon. R.

43. *Silene armeria.* L. Sp. 601.
On the coast. Area, South Europe.

44. *Silene muscipula.* L. Sp. 601.
Coast. Area, South Europe, North Africa.

45. *Silene behen.* L. Sp. 599.
Coast, plains. Area, South Italy, Sardinia.

46. *Silene cretica.* L. Sp. 601.
Plains. Area, South Spain, Italy, Dalmatia.

47. *Silene crassipes.* Fenz. Pug., No. 23.
On the coast.

48. *Silene gonocalyx.* Boiss. Diagn. Ser. i., viii., p. 38.
Mountain and hill regions. General.

49. *Silene racemosa.* Otth. in De Cand. Prodr. i., p. 384. Var. *Sibthorpiana.*
Central districts.

50. *Silene dichotoma.* Ehr. Beitr. 7, p. 144.
Galilee, Gilead. Area, Danube, South Russia.

51. *Silene hispida.* Desf. Atl. i., 248. =*S. vespertina.* Retz.
Beds of watercourses. East plains of Moab.

52. *Silene gallica.* L. Sp. 595.
Coast, plains, and hills. Area, Central and Southern Europe, North Africa.

53. *Silene villosa.* Forsk. Desc. Cent. iii., p. 88.
Near the Dead Sea. Area, North African deserts.

54. *Silene setacea.* Viv. Fl. Lib., p. 23, pl. 12, f. 2.
Ghor, and salt plain south of Dead Sea. Area, North-east Africa.

55. *Silene siderophila.* Boiss. Diagn. Ser. ii., vi., p. 34.
Subalpine Lebanon. P.

56. *Silene damascena.* Boiss. Diagn. Ser. ii., vi., p. 34.
Mountain and hill districts. General. P.

57. *Silene palæstina.* Boiss. Diagn. Ser. i. viii., p. 80.
Desert near Gaza. P.

58. *Silene nocturna.* L. Sp. 595.
Coast and plains. Area, Mediterranean region.

59. *Silene bipartita.* Desf. Atl. i., p. 352, pl. 100.
Sea coast. Area, Mediterranean region.

60. Ditto, var. *Eriocaulon.* Boiss.
Phœnician plain. P.

61. *Silene oliveriana.* Otth. in De Cand. Prodr. i., p. 373.
Wilderness of Judæa towards Dead Sea, and sandy plains by coast.

62. *Silene rubella.* L. Sp. 600.
North plains. Area, Mediterranean region.

63. *Silene sedoides.* Jacq. Coll. Suppl., pl. 14, f. 1.
Coast. Area, South France, South Italy.

64. *Silene fuscata.* Link. in Brot. Fl. Lus. ii., p. 187.
Fields on the coast. Area, Portugal, South Spain, South Italy, North Africa.

65. *Silene atocion.* Murr. Syst., p. 421.
Coast and hills, near Gilead. Area, interior of Algeria.

66. *Silene kotschyi.* Boiss. Diagn. Ser. i., i., p. 40.
Under Lebanon and Hermon.

67. *Silene cassia.* Boiss. Diagn. Ser. i., viii., p. 78.
Wooded hills, Galilee.

68. *Silene juncea.* Sibth. Fl. Gr. v., p. 15, pl. 421.
On the coast.

69. *Silene chætodonta.* Boiss. Diagn. Ser. i., i., p. 39. Var. *Modesta.* Boiss.
Phœnician plain.

70. *Silene stenobotrys.* Boiss. Fl. Or. i., p. 611.
In rocks at Rascheya.

71. *Silene spergulifolia.* (Desf. Cor. Tourn., p. 73.)
Subalpine regions.

72. *Silene pruinosa.* Boiss. Diagn. Ser. i., viii., p. 85.
Anti-Lebanon, 7,000 feet.

73. *Silene subulata.* Boiss. Diagn. Ser. i., viii., p. 33.
High summits of Lebanon.

74. *Silene odontopetala.* Fenz. Pug., No. 28.
Rocks in Lebanon, 4,000—9,000 feet, and on top of Hermon.

75. *Silene inflata.* Sm. Brit., p. 467.
Coast and central regions, Jordan valley. Area, Europe, North Africa, Liberia.

76. *Silene physalodes.* Boiss. Diagn. Ser. i., viii., p. 83.
North, near Banias. P.

77. *Silene italica.* (L. Sp. 593.)
Mountains and hills. Area, Mediterranean region.

78. *Silene longipetala.* Venten. Hort. Cals., p. 83, pl. 83. Var. *Purpurascens.*
Coast, Judæan wilderness, Jordan valley, Gilead.

79. *Silene makmeliana.* Boiss. Diagn. Ser. i., viii., p. 89.
Lebanon and Anti-Lebanon. P.

80. *Silene libanotica.* Boiss. Diagn. Ser. i., viii., p. 89.
Rocky places, Lebanon up to 7,000 feet, Hebron, etc. P.

81. *Silene gigantea.* L. Sp. 598.
Lebanon.

82. *Silene brevipes.* Paine. Pal. Expl. Soc., No. 3, p. 98.
Mount Nebo. P.

83. *Silene grisea.* Boiss. Diagn. Ser. i., viii., p. 88.
Limestone rocks below the Cedars, Lebanon. P.

84. *Silene succulenta.* Forsk. Descr., p. 89.
On the coast.

85. *Melandrium pratense.* Rœhl. Deutsch. Fl. i., p. 254.
Lebanon up to 5,000 feet. Area, Northern and Central Europe, Siberia.

86. *Melandrium eriocalycinum.* Boiss. Diagn. Ser. ii., i., p. 78.
Lebanon to the summit.

ORDER XIII., ALSINEÆ.

1. *Sagina apetala.* L. Mant. 559.
Coast of Palestine. Area, Europe, the Canaries.

2. *Buffonia macrosperma.* Gay in Gre. and God. Fl. Fr. i., p. 248.
Anti-Lebanon. Area, Central and Southern Europe.

3. *Alsine procumbens.* (Vahl. Symb. i., 50, pl. 33.)
Sandy desert south of Gaza. Area, Mediterranean region.

4. *Alsine rupestris.* (Labill. Dec. iv., p. 8, pl. 4, f. 1.)
Highest parts of Lebanon and Hermon. P.

5. *Alsine juniperina.* Fenzl. Alsin., p. 18.
Hills and mountain districts up to 8,000 feet.

6. *Alsine libanotica.* Boiss. Diagn. Ser. i., viii., p. 98.
Bare alpine heights of Lebanon. P.

7. *Alsine meyeri.* Boiss. Diagn. Ser. i., viii., p. 96.
Lebanon, 4,000—6,000 feet.

8. *Alsine billardieri.* Boiss. Diagn. Ser. i., viii., p. 95.
Mount Gilead.

9. *Alsine smithii.* Fenz. Alsin., p. 57.
Lebanon.

10. *Alsine decipiens.* Fenz. Pug., No. 35.
Carmel and Central Palestine, Gilead.

11. *Alsine tenuifolia.* (L. Sp. 607.)
Lebanon. Area, Europe, Siberia, North Africa.

12. Ditto., var. *Macropetala.* Boiss. Diagn. Ser. i., i., p. 45.
Sandy plains, Phœnicia.

13. Ditto, var. *Lydia.* Boiss. Diagn. Ser. ii., i., p. 86.
Galilean hills, on the coast.

14. *Alsine thymifolia.* (Sibth. Fl. Gr. i., p. 305, pl. 441.)
Ditto, var. *Syriaca.* On the coast, near Beyrout.

15. *Alsine picta.* (Sibth. Prodr. Fl. Gr., p. 304.)
Sandy soils, general.

16. Ditto, var. *Sinaica.*
Desert south of the Dead Sea, ravine of the Arnon.

17. *Queria hispanica.* Lœfl. It. Hisp., p. 48.
Below Hermon. Area, Spain, Portugal.

18. *Arenaria graveolens.* Schreb. Nov. Act. Cur. iii., p. 478.
Northern mountain region.

19. *Arenaria cassia.* Boiss. Diagn. Ser. i., viii., p. 101.
Northern hills.

20. *Arenaria serpyllifolia.* L. Sp. 606.
Lebanon. Area, Europe, North Africa, Siberia.

21. Ditto, var. *Leptoclados.* Rchb. Cent. xv., pl. 32.
General in sandy places. Area, Europe, North Africa.

22. *Arenaria tremula.* Boiss. Diagn. Ser. i., viii., p. 101.
Northern hills. P.

23. *Stellaria media.* (L. Sp. 389.)
General. Area, almost world-wide.

24. *Holosteum umbellatum.* L. Sp. 130.
General in plains and hills alike. Area, Central and South Europe, North Africa.

25. *Holosteum liniflorum.* Stev. Ex. F. and M. Ind. Petrop. iv., p. 10.
In sandy fields.

26. *Cerastiüm anomalum.* W. K. Pl. Rar. i., p. 21, pl. 22.
The Bukâà.

27. *Cerastium kotschyi.* Boiss. Fl. Or. i., p. 715.
Anti-Lebanon, 5,000 feet. P.

28. *Cerastium illyricum.* Arduin. Sp. ii., p. 26.
Higher parts of Lebanon.

29. *Cerastium dichotomum.* L. Sp. 628.
Northern hills to Lebanon and Hermon. Area, Spain, North Africa.

30. *Cerastium inflatum.* Desf. Cat. Hort. Par., p. 462.
Moab.

31. *Cerastium viscosum.* L. Sp. 627.
Cultivated and moist ground, general. Area, almost world-wide.

32. *Cerastium fragillimum.* Boiss. Diagn. Ser. i., i., p. 54.
Higher parts of Lebanon and Anti-Lebanon.

33. *Malachium aquaticum.* (L. Sp. 29.)
Central Palestine, Nablus. Area, Europe, West Siberia, North-west India.

34. *Spergula arvensis.* L. Sp. 630.
Fields. Area, Europe, Siberia, India, Abyssinia.

35. *Spergula pentandra.* L. Sp. 630.
Gaza. Area, Europe, North Africa, Canaries.

36. *Spergularia rubra.* (Wahl. Ups. 151.)

The Bukâà and Hasbany valley. Area, Europe, Siberia, North Africa, India, North America.

37. *Spergularia media.* (Wahl. Fl. Goth., p. 45.)

Coast near Gaza. Area, as above.

38. *Spergularia marginata.* (De Cand. Fl. Fr. iv., p. 793.)

Salt plain south of Dead Sea. Area, as above.

ORDER XIV., PARONYCHIEÆ.

1. *Robbairea prostrata.* (Forsk. Dec., p. 207.) Arab., ربير, *Robbyr.*

Hill-country of Judæa and Southern Desert. Area, Oases of the Sahara.

2. *Polycarpon tetraphyllum.* (L. Sp. 89.)

Ghor, by Jericho. Area, Central and South Europe, Africa.

3. *Polycarpon arabicum.* Boiss. Diagn. Ser. i., x., p. 13.

Southern Desert.

4. *Polycarpæa fragilis.* Del. Fl. Eg., p. 241, pl. 24, f. 1.

Desert beyond Gaza. Area, deserts of Algeria.

5. *Herniaria cinerea.* De Cand. Fl. Fr. v., p. 375.

Very common. Area, Mediterranean region.

6. *Herniaria hirsuta.* L. Sp. 317.

On the coast and hills. Area, Central and South Europe.

7. *Herniaria incana.* Lam. Dict. iii., p. 124.

Galilee. Area, South Europe.

8. *Herniaria hemistemon.* J. Gay. Duch. Rev. Bot., 1846, p. 371.

Shores of the Dead Sea.

9. *Paronychia sinaica.* Fres. Mus. Senck. i., p. 180. Var., *Flavescens.* Boiss.

Southern Desert, near Beersheba.

10. *Paronychia argentea.* Lam. Fl. Fr. iii., p. 230.
Beersheba and south-west and east of Dead Sea. Area, Mediterranean region.

11. *Paronychia arabica.* (L. Mant. 51.)
Gaza. Much more desert habitat than the last.

12. *Gymnocarpum fruticosum.* Pers. Ench. i., p. 636.
Southern Judæa. Area, interior North Africa.

13. *Scleranthus annuus.* L. Sp. 580.
Jordan valley. Area, Europe, North Africa, Abyssinia, Western Siberia.

14. *Habrosia spinuliflora.* (Ser. in De Cand. Prodr. i., p. 406.)
The Hauran.

15. *Pteranthus echinatus.* Desf. Fl. Atl. i., p. 144.
Wilderness of Judæa. Area, North Africa, Malta.

ORDER XV., MOLLUGINEÆ.

1. *Telephium imperati.* L. Sp. 388.
Lebanon and Anti-Lebanon. Area, Central and South Europe.

2. *Telephium sphærospermum.* Boiss. Diagn. Ser. i., iv., p. 12.
The Southern Desert.

3. *Glinus lotoïdes.* L. Sp. 663.
On the coast. Area, Mediterranean region.

4. *Glinus dictamnoides.* L. Mant. ii., p. 243.
Near Rascheya, under Hermon. Area, India.

ORDER XVI., PORTULACEÆ.

1. *Portulaca oleracea.* L. Sp. 638.
In the plains. Area, Temperate and Tropical North Zones.

ORDER XVII., TAMARISCINEÆ.

1. *Reaumuria palæstina.* Boiss. Diagn. Ser. i., x., p. 10.

Barren marl hills round the Dead Sea. P.

2. *Tamarix syriaca.* Boiss. Fl. Or. i., p. 767. Heb. אֵשֶׁל, Arab. هطاف, *Ghataf.*

Side of streams near Sidon. P.

3. *Tamarix tetragyna.* Ehr. Linn. ii., p. 257.

Coast near Mount Carmel.

4. *Tamarix jordanis.* Boiss. Fl. Or. i., p. 771.

On the banks of the Jordan. P.

5. *Tamarix nilotica.* Ehr. Sched. Herb. Berol.

Southern desert.

6. *Tamarix pallasii.* Desv. Ann. Sc. Nat. iv., p. 349.

On the coast of the Phœnician plain. Area, Moldavia, South Russia, Turkestan.

ORDER XVIII., FRANKENIACEÆ.

1. *Frankenia pulverulenta.* L. Sp. 474.

On the coast. Area, Mediterranean region, Senegal, Cape of Good Hope.

2. *Frankenia hirsuta.* L. Sp. 474.

On the coast. Area, Central and Southern Europe, North and South Africa, Western Siberia.

ORDER XIX., ELATINEÆ.

1. *Elatine campylosperma.* Seub. Mon. Elat., p. 17.

On the Philistian plain. Area, South France, Sardinia, North Africa.

ORDER XX., HYPERICINEÆ.

1. *Triadenia russeggeri.* Fenz. Pug., No. 18.
Among ruins.

2. *Hypericum hircinum.* L. Sp. 1103.
Common in the north and on the coast. Area, south coasts of Europe.

3. *Hypericum nanum.* Poir. Suppl. Dict. iii., p. 699.
Clinging to rocks in the hills and mountains.
Ditto var. *Prostratum.* Ain Fijeh, Anti-Lebanon.

4. *Hypericum serpyllifolium.* Lam. Dict. iv., p. 176.
Galilæan hills among brushwood.

5. *Hypericum cuneatum.* Poir. Suppl. Dict. iii., p. 699.
Lebanon, 4,000 feet.

6. *Hypericum scabrum.* L. Amœn. iv., p. 287.
Galilee, Lebanon, and Anti-Lebanon.

7. *Hypericum confertum.* Chois. Monog. Hyper. p. 55, pl. 8. Var. *Stenobotrys.* Boiss.

8. *Hypericum hyssopifolium.* Vill. Dauph. iii., p. 505, pl. 44.
Lebanon and Anti-Lebanon. Area, South Europe, Western Siberia.

9. *Hypericum helianthemoides.* (Spach. Suit. Buff. v., p. 379.)
Lower hills, Galilee and Anti-Lebanon.

10. *Hypericum tetrapterum.* Fries. Nov. p. 236.
Hills, Northern Palestine. Area, Europe, North Africa.

11. *Hypericum crispum.* L. Mant., p. 106.
Generally distributed. Area, Mediterranean region.

12. *Hypericum lanuginosum.* Lam. Dict. iv., p. 171.
In all parts of the country, except the Jordan valley.

13. *Hypericum perforatum.* L. Sp. 1105.
Coasts and mountain region. Area, Europe, North Africa, Siberia.

14. *Hypericum cassium.* Boiss. Diagn. Ser. i., viii., p. 111.

ORDER XXI., MALVACEÆ.

1. *Malva sylvestris.* L. Sp. 969.
Plain of Sharon, Moab. Area, Europe, North Africa, Siberia.

2. *Malva nicæensis.* All. Ped. ii., p. 40.
Common. Coast and central districts. Area, South Europe.

3. *Malva rotundifolia.* L. Sp. 969.
Fields in the north, Moab. Area, Central Europe.

4. *Malva parviflora.* L. Sp. 960.
Among brushwood on hills, central district. Area, Southern Europe.

5. *Malva oxyloba.* Boiss. Diagn. Ser. i., viii., p. 109.
Plain of Esdraelon, and elsewhere. P.

6. *Lavatera cretica.* L. Sp. 973.
Plains, coast and inland. Area, Mediterranean region, Canaries.

7. *Lavatera punctata.* All. Auct., p. 26.
Northern Palestine, Lebanon. Area, Mediterranean region.

8. *Lavatera unguiculata.* Desf. Arb. i., p. 471.

9. *Lavatera thuringiaca.* L. Sp. 972.
Area, Central Europe.

10. *Lavatera trimestris.* L. Sp. 974.
Coast, and Jordan valley. Area, Southern Europe.

11. *Althæa hirsuta.* L. Sp. 966.
Carmel, Galilee, Nebo. Area, Central and Southern Europe.

12. *Althæa officinalis.* L. Sp. 966.

Northern plains. Area, Central and Southern Europe, West Siberia.

13. *Alcea acaulis.* (Cav. Diss. ii., pl. 27, f. 3.)

Coast, hill and mountain regions, east and west.

14. *Alcea rufescens.* (Boiss. Diagn. Ser. ii., i., p. 102.)

Banias, under Hermon.

15. *Alcea lavateræflora.* (De Cand. Prodr. i., p. 437.)

Hill country, north to south.

16. *Alcea rosea.* L. Sp. 966. Var. *Sibthorpii.*

Hills above Lake Huleh. Area, South Italy, Dalmatia.

17. *Alcea ficifolia.* L. Hort. Cliff., 348.

Area, South Russia, Dalmatia.

18. *Alcea setosa.* (Boiss. Diagn. Ser. i., vii., p. 107.)

Hills, Central Palestine. P.

19. *Alcea apterocarpa.* (Fenz. Cat. Hort. Vindob., 1858.)

Lebanon, Mount Nebo.

20. *Alcea kurdica.* (Schlecht. Linn. xvii., p. 127.)

Rocks, Anti-Lebanon.

21. *Malvella sherardiana.* (L. Sp. 1675.)

Generally distributed. Area, Spain.

22. *Abutilon fruticosum.* Guill. et Perr. Fl. Sénég. i., p. 70.

By the Dead Sea. Area, Senegal, Nubia, Abyssinia, Arabia Deserta.

23. *Abutilon muticum.* (Dell. Ill. Eg., No. 633.)

Engedi, Ghor, Safieh (by Dead Sea). Area, Senegal, Nubia, Arabia Deserta.

24. *Hibiscus syriacus.* L. Sp. 978.

Doubtful if now found, except in Northern Syria.

25. *Hibiscus trionum.* L. Sp. 981.
General. Area, Central Europe.

26. *Hibiscus cannabinus.* L. Sp. 979. Hemp.
Cultivated.

27. *Hibiscus esculentus.* L. Sp. 980.
Cultivated.

28. *Gossypium herbaceum.* L. Sp. 975. Cotton.
Cultivated.

ORDER XXII., STERCULIACEÆ.

None.

ORDER XXIII., TILIACEÆ.

1. *Corchorus olitorius.* L. Sp. 746.
Maritime plains. Area, all tropical countries.

2. *Corchorus trilocularis.* L. Mant. 77. Jew's Mallow.
The Ghor, north end of Dead Sea. Area, tropical Africa and Asia.

ORDER XXIV., LINEÆ.

1. *Linum gallicum.* L. Sp. 401.
Maritime plains. Area, Mediterranean region, Abyssinia.

2. *Linum corymbulosum.* Rchb. Fl. Excurs., p. 834. = *L. aureum.* De Cand. Prodr. i., 423.
Mount Carmel. Area, Italy, Dalmatia, Abyssinia.

3. *Linum strictum.* L. Sp. 400.
Mount Carmel. Area, Mediterranean region, Abyssinia.

4. *Linum nodiflorum.* L. Sp. 401.
Hill country. Area, Mediterranean region.

5. *Linum toxicum.* Boiss. Fl. Or. i., p. 854.
Hermon top, 9,500 feet. P.

6. *Linum flavum.* L. Sp. 399.

7. *Linum orientale.* Boiss. Fl. Or. i., p. 855.
Universally abundant.

8. *Linum syriacum.* Boiss. Fl. Or. i., p. 856.
Galilee, on limestone rocks.

9. *Linum usitatissimum.* L. Sp. 397. Common Flax.
Near Damascus. Cultivated everywhere.

10. *Linum pubescens.* Russ. Alep. ii., p. 268.
Generally distributed east and west, except in Jordan valley.

11. *Linum humile.* Mill. Dict., No. 2.
Under Hermon.

12. *Linum angustifolium.* Huds. Fl. Angl., 134.
General. Area, Central and South Europe, Canaries.

13. *Linum carnosulum.* Boiss. Diagn. Ser. i., viii., p. 104.
Lebanon, 8,000 feet. P.

14. *Linum hirsutum.* L. Sp. 398.
Gennesaret. Area, East Germany, Hungary, South Russia.

ORDER XXV., OXALIDEÆ.

Oxalis corniculata. L. Sp. 623.
In cultivated plains in the south. Area, world-wide.

ORDER XXVI., BALSAMINEÆ.

Not observed.

ORDER XXVII., GERANIACEÆ.

1. *Geranium subcaulescens.* De Cand. Prodr. i., p. 640.
Lebanon, 6,000—8,000 feet. Area, Dalmatia.

2. *Geranium tuberosum.* L. Sp. 953.
Universally distributed. Area, Mediterranean region.

3. *Geranium libanoticum.* Boiss. Fl. Or. i., p. 877.
Higher parts of Lebanon, from 4,000 feet.

4. *Geranium asphodeloides.* Willd. Sp. iii., p. 704.
Lebanon and Anti-Lebanon, lower parts. Area, South Italy.

5. *Geranium pusillum.* L. Sp. 957.
The Hauran, Jordan valley. Area, Europe, North Africa.

6. *Geranium columbinum.* L. Sp. 956.
Wooded districts. Area, Europe, North Africa, Siberia.

7. *Geranium ibericum.* Cav. Diss. iv., p. 209.
Northern mountainous districts.

8. *Geranium rotundifolium.* L. Sp. 957.
General, except Jordan valley. Area, Europe, North Africa, Siberia.

9. *Geranium dissectum.* L. Sp. 956.
Plains and hills. Area, Europe, North Africa, Canaries, Siberia, North America.

10. *Geranium molle.* L. Sp. 955.
Hills, plains, and Lebanon. Area, Europe, North Africa, Canaries.

11. *Geranium sylvaticum.* L. Sp. 933.
Lebanon. Area, Europe, Siberia.

12. *Geranium purpureum.* Vill. Dauph. iii., p. 174, pl. 40.
North and Lebanon. Area, Mediterranean region.

13. *Geranium lucidum.* L. Sp. 955.

General, except Jordan valley. Area, Central and Southern Europe, North Africa, West Siberia.

14. *Geranium robertianum.* L. Sp. 956. Herb Robert. Area, Central and Southern Europe, Siberia.

15. *Erodium trichomanefolium.* L'Hér. in De Cand. Prodr. i., p. 645.

Higher parts of Lebanon. Area, Southern Spain.

16. Var. *Albiflorum.* Boiss.

Hermon.

17. *Erodium romanum.* (L. Sp. 951.)

Lower parts of Lebanon and Anti-Lebanon. Area, Mediterranean, Europe.

18. *Erodium moschatum.* (L. Sp. 951.)

General. Area, Europe, Abyssinia, Canaries.

19. *Erodium ciconium.* (L. Sp. 952.)

Central and Southern Palestine. Area, South Europe.

20. *Erodium gruinum.* (L. Sp. 952.)

Plain of Esdraelon, Galilee, Gilead, and Jordan valley. Area, Sicily.

21. *Erodium botrys.* (Cavan. Diss. iv., pl. 90, f. 2.)

Plains of Jordan valley. South Europe, North Africa, Madeira, Canaries.

22. *Erodium laciniatum.* Cavan. Diss. iv., p. 228, pl. 113, f. 3.

Plain of Phœnicia. Area, Mediterranean region.

23. *Erodium cicutarium.* (L. Sp. 95.)

Hills and plains of Moab. Area, Europe, North Africa, Siberia.

24. *Erodium malacoïdes.* (L. Sp. 952.)

General, especially Jordan valley. Area, Mediterranean region.

25. *Erodium hirtum.* (Forsk. Desc., p. 123.)
Southern deserts. Area, Desert of Algeria and Tunis.

26. *Erodium gaillardoti.* Boiss. Diagn. Ser. ii., vi., p. 61.
Northern parts. P.

27. *Erodium glaucophyllum.* Ait. ii., Kew. ii., p. 416.
Slopes of Mount Gilead.

28. *Monsonia nivea.* (Decaisn. Fl. Sin., p. 61.)
Desert south of Gaza. Area, oases of Sahara.

29. *Bierbersteinia multifida.* De Cand. Prodr. i., p. 708.
Lebanon and Hermon, 8,000 feet.

ORDER XXVIII., ZYGOPHYLLEÆ.

1. *Tribulus terrestris.* L. Sp. 554.
Round the Dead Sea. Area, Mediterranean region.

2. *Tribulus alatus.* Del. Eg. Illustr., No. 438.
Southern Desert. Area, Arabia and Scinde.

3. *Fagonia glutinosa.* Del. Eg., p. 86, pl. 26, f. 3.
Desert south and west of Dead Sea.

4. *Fagonia kahirina.* Boiss. Diagn. Ser. i., viii., p. 122.
South-west end of Dead Sea. Area, Algerian deserts.

5. *Fagonia sinaica.* Boiss. Diagn. Ser. i., i., p. 61.
Rocky hills south-east of Dead Sea.

6. *Fagonia myriacantha.* Boiss. Diagn. Ser. i., viii., p. 123.
Desert south of Beersheba.

7. *Fagonia olivieri.* De Cand. Prodr. i., p. 704.
Northern plains, Bukâà.

8. *Fagonia mollis.* Del. Eg., p. 86, pl. 28, f. 2. = *F. cistoides.* Del.
Wâdy Akabah, south of Dead Sea. North-west of Moab.

9. *Fagonia grandiflora.* Boiss. Diagn. Ser. i., viii., p. 121.
Western slopes of Judæa, towards Dead Sea. P.

10. *Fagonia arabica.* L. Sp. 553.
Wâdy Zuweirah, south-west end of Dead Sea.

11. *Zygophyllum dumosum.* Boiss. Diagn. Ser. i., viii., p. 125.
South end of Dead Sea, desert south of Beersheba. P.

12. *Zygophyllum fabago.* L. Sp. 551.
Dry plains in the north. Area, Spain, Sardinia, North Africa.

13. *Zygophyllum album.* L. Dec. i., pl. 8.
Southern Desert, Ain Gades, Callirrhoe, Moab. Area, Spain.

14. *Zygophyllum coccineum.* L. Sp. 551.
Barren plains north and south of Dead Sea. Area, tropical Arabia, Scinde.

15. *Nitraria tridentata.* Desf. Atl. i., p. 372.
Wâdys south and south-east of Dead Sea. Area, Sahara, Senegal.

16. *Peganum harmala.* L. Sp. 638.
By shores of the Dead Sea. Area, Mediterranean region and sub-tropical Arabia.

ORDER XXIX., RUTACEÆ.

1. *Dictamnus froxinella.* Pers. Syn. i., p. 464.
Northern woods. Area, Europe, Siberia, North-west Himalayas.

2. *Ruta chalepensis.* L. Mant. i., p. 69.
Ditto, var. *Bracteosa.* De Cand.
Generally distributed. Area, the Mediterranean region, and tropical Arabia.

3. *Haplophyllum fruticulosum.* (Labill. Dec. i., p. 13, pl. 4.) Bare hills in the north.

4. *Haplophyllum buxbaumii.* (Poir. Encycl. vi., p. 356.) General in barren places. Area, Tunis.

5. Var. *Stenophyllum.* Boiss. Lebanon and Anti-Lebanon.

6. *Haplophyllum corymbulosum.* Boiss. Diagn. Ser. ii., i., p. 116. Jericho. P.

7. *Haplophyllum tuberculatum.* (Forsk. Descr., p. 86.) Southern desert; Ghor of the Seisaban. Area, Sahara, Nubia, tropical Arabia.

8. *Haplophyllum sylvaticum.* Boiss. Diagn. Ser. i., viii., p. 126. Hills in the northern parts.

9. *Haplophyllum longifolium.* Boiss. Diagn. Ser. i., viii., p. 127. Ravines of Dead Sea. P.

ORDER XXX., CORIARIEÆ.

1. *Coriaria myrtifolia.* L. Sp. 1467. Area, Southern Europe, Northern Africa.

ORDER XXXI., SIMARUBEÆ.

1. *Balanites ægyptiaca.* Del. Ill. Eg., p. 263, pl. 28, f. 1. Ghor. Round Dead Sea. Area, Nubia, Abyssinia, Arabia.

ORDER XXXII., SAPINDACEÆ.

1. *Acer pseudoplatanus.* L. Sp. 1495. Sycamore or Maple. Mountains of Gilead. Area, Central and Southern Europe.

2. *Acer hyrcanum.* F. and Mey. Ind. iv., Hort. Petrop., No. 9. Var. *Reygassii.* P.

On Lebanon.

3. *Acer creticum.* L. Sp. 1497.
Mountains of Bashan and Gilead.

4. *Acer monspessulanum.* L. Sp. 1497.
Lebanon; about 4,000 feet. Area, Mediterranean region.

5. Var. *Microphyllum.* Boiss. Fl. Or. i., p. 951.
Northern and western slopes of Hermon. Eastern lower slopes of Lebanon.

6. *Acer syriacum.* Boiss. Diagn. Ser. ii., v., p. 72.
Many of the upper Lebanon valleys. P.

7. *Staphylæa pinnata.* L. Sp. 386.
Northern and sub-alpine woods. Area, South and South-east Europe.

ORDER XXXIII., MELIACEÆ.

1. *Melia azedarach.* L. Sp. 550.
Cultivated.

ORDER XXXIV., AMPELIDEÆ.

1. *Vitis vinifera.* L. Sp. 293. Heb. גֶּפֶן, Arab. كرم, *Karm.* The Vine.
Cultivated everywhere.

2. *Vitis orientalis.* (Lam. Ill., p. 332, pl. 84, f. 2.)
Among ruins in the Hauran.

PLANTÆ VASCULARES.

CLASS, DICOTYLEDONEÆ.

SUB-CLASS, CALYCIFLORÆ, POLYPETALÆ.

ORDER XXXV., BURSERACEÆ.

1. *Balsamodendron opobalsamum.* (Forsk. Descr., p. 79.) Balm of Gilead. Heb. צֳרִי. Arab. ابرشام, *Abusham.*

In ancient times, and probably down to the date of the Crusades, the Balm of Gilead was cultivated about Jericho. But it is now lost, and was introduced from Arabia or Nubia, in both of which countries it is still found indigenous.

ORDER XXXVI., TEREBINTHACEÆ.

1. *Rhus cotinus.* L. Sp. 383.

Anti-Lebanon. Area, Central and Southern Europe.

2. *Rhus coriaria.* L. Sp. 379.

Generally distributed. Area, Mediterranean region.

3. *Rhus oxyacanthoides.* Dum. Cour. Bot. ed. 3, iii., p. 568. = *R. syriaca,* Boiss.

The Southern desert, and the base of Lebanon, behind Sidon. Area, North Africa, Sicily, the Nubian coast.

4. *Pistachia vera.* L. Sp. 1454. Heb. בָּטְנִים, Gen. xliii., 11. Arab, بطم, *But'm.*

Cultivated everywhere.

5. *Pistachia palæstina.* Boiss. Diagn. Ser. i., ix., p. i. Heb. אֵלָה, A. V., 'Teil tree,' 'Oak.'

On the lower slopes of the hills throughout the country.

This tree has generally been identified with *P. terebinthus*, L. Sp. 1455, but separated by Boissier, who does not acknowledge the occurrence of a second species in the country.

6. *Pistachia mutica.* Fisch. and Mey. Hoh. Talysch. Enum., p. 102.

Frequent. Probably introduced.

7. *Pistachia lentiscus.* L. Sp. 1455.

Most abundant on the plains and lower slopes of the hills. Area, Mediterranean region.

8. *Pistachia terebinthus.* L. Sp. 1455.

This tree, continually spoken of by travellers as the Terebinth, has always been referred to this Mediterranean species. But, as stated above, if distinct from *P. palæstina*, it probably does not occur.

ORDER XXXVII., CELASTRINEÆ.

1. *Evonymus europæus.* L. Sp. 286, var. *a.* Spindle-tree.

Northern Lebanon (doubtful). Area, Europe, Western Siberia.

ORDER XXXVIII., RHAMNEÆ.

1. *Paliurus aculeatus.* Lam. Ill., tab. 210. Christ's Thorn, A. V., 'Briers.' Hebr. שָׁמִיר. Arab. صعور, *Samûr.*

Covers the rocky hills over the whole country. Area, Southern Europe.

2. *Zizyphus vulgaris.* Lam. Dict. iii., p. 316.

Not uncommon in the warmer parts. Area, Southern Europe.

3. *Zizyphus spina-christi.* (L. Sp. 282.) Heb. נַעֲצוּץ. Arab. نبك, *Nubk.*

Extremely abundant in the whole Jordan valley, and in the warmer parts of the maritime plains. Area, Sahara, Nubia, Abyssinia, tropical Arabia, North-west India.

4. *Zizyphus lotus.* (L. Sp. 281.)

By the coast and in the Jordan valley. Area, Spain, Sicily, North Africa, Arabia.

5. *Rhamnus alaternus.* L. Sp. 281.

General. Area, Mediterranean region.

6. *Rhamnus punctata.* Boiss. Diagn. Ser. i., ii., p. 4.

Northern districts, maritime and mountain.

7. *Rhamnus grandifolia.* F. and M. Hoh. Talysch. Enum., p. 99.

Near the sources of the Jordan.

8. *Rhamnus prunifolia.* Sibth. and Sm. Prodr. i., p. 157.

Northern mountains.

9. *Rhamnus cornifolia.* Boiss. Diagn. Ser. i., ii., p. 3.

Northern mountains.

10. *Rhamnus palæstina.* Boiss. Diagn. Ser. ii., i., p. 119. Probably Heb. אָטָד, Judg. ix. 14, 15. A.V. 'Bramble.'

On the rocks in all parts of the country, east and west. P.

11. *Rhamnus petiolaris.* Boiss. Diagn. Ser. xi., v., p. 75.

In the northern districts.

12. *Rhamnus libanotica.* Boiss. Diagn. Ser. i., ii., p. 119.

Lebanon and Anti-Lebanon.

13. *Rhamnus oleoides.* L. Sp. 279.

Hills round Gennesaret. Area, Spain, South France, Sicily, Sardinia.

14. *Rhamnus kurdica.* Boiss. Diagn. Ser. i., ii., p. 3.
Under Lebanon and Anti-Lebanon.

15. *Rhamnus dahurica.* Pall.
Lake Huleh. Area, Caspian region, Persia.

16. *Rhamnus sibthorpiana.* De Cand. Prodr. ii., p. 25.

ORDER XXXIX., MORINGEÆ.

1. *Moringa aptera.* Gærtn. Fr., p. 315.
In the lower Jordan valley, and round the Dead Sea. Area, Nubia and Arabia.

ORDER XL., LEGUMINOSÆ.

1. *Anagyris fœtida.* L. Sp. 534.
Ravines near the coast, Moab, Judæan hills. Area, Mediterranean region and Arabia.

2. *Crotolaria ægyptiaca.* Benth. Hort. Lond. Journ. ii., p. 473.
Southern desert.

3. *Lupinus pilosus.* L. Sp. 1015.
In the north and Jordan valley alike.

4. *Lupinus digitatus.* Forsk. Æg. Arab., p. 131.
Cultivated.

5. *Lupinus palæstinus.* Boiss. Diagn. Ser. i., ix., p. 9.
Southern plains near Gaza, etc. P.

6. *Lupinus hirsutus.* L. Sp. 1015.
Near the coast. Area, Mediterranean region.

7. *Lupinus augustifolius.* L. Sp. 1015.
General. Area, Mediterranean region.

8. *Lupinus reticulatus.* Desv. Ann. Bot. iii., p. 100, and var. *Philisteus.* Boiss.

Plain of Philistia. Area, South France, Spain, Sicily.

9. *Lupinus termis.* Forsk. Æg. Arab., p. 131.

In the plains. Cultivated in Italy.

10. *Lotononis dichotoma.* (Del. Fl. Æg. Ill., No. 717.)

Southern desert, shores of Dead Sea.

11. *Argyrolobium crotalarioïdes.* Jaub. et Sp. Ill. Or. i., p. 114, pl. 59.

Northern hills and Cœle Syria.

12. *Argyrolobium uniflorum.* (Decaisn. Fl. Sin., p. 41.)

In deserts and barren grounds from north to south. Area, North Sahara.

13. *Adenocarpus divaricatus.* (L'Her. Stirp., 184.) Var. *Græcus* = *A. commutatus.* Guss.

Southern Lebanon, Galilee. Area, Spain, South France, South Italy, Sicily.

14. *Calycotome villosa.* (Vahl. Symb. ii., p. 80.)

Coast, southern wilderness, Jordan valley, Gilead. Area, Mediterranean region.

15. *Spartium junceum.* L. Sp. 995.

Central districts, upland plains, and sub-alpine region. Area, Mediterranean region and Canaries.

16. *Retama rœtam.* (Forsk. Æg. Arab., p. 214.) Heb. רֹתֶם, A. V. 'Juniper,' 1 Kings xix. 4, 5, etc., Arab. ريطم, *Rœtem.*

On all the hills in desert rocky parts, especially about the Dead Sea. Most plentiful in Gilead and Moab. Area, North-east Africa.

This is one of the exquisitely beautiful plants of the country. The gauzy delicate pink-and-white hues of a whole hill-side covered with shrub in blossom, as I have seen it in Gilead, is unsurpassed even by the apple-blossom of an English orchard.

17. *Genista acanthoclada.* De Cand. Prodr. ii., 146.
Northern hills.

18. *Genista sphacelata.* Decaisn. Ann. Sc. Nat., 1835, p. 360.
Central parts, Carmel, Gilead. P.

19. *Genista anatolica.* Boiss. Diagn. Ser. i., ii., p. 8.
Northern hills.

20. *Genista albida.* Willd. Sp. iii., p. 942.
Mountains of Gilead and Hauran. Area, South Russia.

21. *Genista patula.* M. B. Taur. Cauc. ii., p. 148.
Northern hills of Cœle Syria.

22. *Genista libanotica.* Boiss. Diagn. Ser. i., ix., p. 3.
Lebanon and Anti-Lebanon, 7,000 feet. P.

23. *Gonocytisus pterocladus.* (Boiss. Diagn. Ser. i., ii., p. 10.)
Dry sunny spots on the coast and on Lebanon.

24. *Cytisus cassius.* Boiss. Diagn. Ser. i., ix., p. 5.
Northern hills, Carmel.

25. *Cytisus syriacus.* (Boiss. Diagn. Ser. ii., ii.)
Lebanon. P.

26. *Cytisus orientalis.* Loisel. Duharn. Arb., p. 136.
Northern hills.

27. *Cytisus drepanolobus.* Boiss. Diagn. Ser. i., ix., p. 6.
Northern hill region.

28. *Ononis antiquorum.* L. Sp. 1006.
Ghor, east side of Jordan. Area, Mediterranean region.

29. *Ononis leiosperma.* Boiss. Fl. Or. ii., p. 57.

By the roadside in chalky ground. Common.

30. *Ononis columnæ.* All. Ped. i., p. 318, pl. 20, f. 3.

Lebanon. Area, Central and Southern Europe, North Africa.

31. *Ononis adenotricha.* Boiss. Diagn. Ser. i., ii., p. 14.

Lebanon and Anti-Lebanon.

32. *Ononis natrix.* L. Sp. 1008.

Galilee, Lebanon, Anti-Lebanon, Gilead, Moab. Area, Southern Europe.

33. Ditto, var. *Tomentosa.* About Jerusalem and Central Palestine, Eastern Gilead.

34. Ditto, var. *Stenophylla.* On the maritime plains. Arab. وزبا, *Wezba.*

35. *Ononis vaginalis.* Vahl. Symb. i., p. 53.

Barren and sandy plains under Anti-Lebanon. Area, Canaries, Cyrenaica.

36. *Ononis biflora.* Desf. Atl. ii., p. 143.

Crest of Mount Gilead. Area, Spain, Sardinia, Sicily, North Africa.

37. *Ononis ornithopodioïdes.* L. Sp. 1009.

Mount Carmel, Moab, etc. Area, Mediterranean region.

38. *Ononis sicula.* Guss. Prodr. Sic. ii., p. 387.

Southern desert. Area, Spain, Sicily, North Africa.

39. *Ononis breviflora.* De Cand. Prodr. ii., p. 160.

All the central region. Area, Mediterranean region.

40. *Ononis reclinata.* L. Sp. 1011.

Coast and plains, Dead Sea slopes. Area, Mediterranean region, Canaries, Abyssinia.

41. *Ononis pubescens.* L. Mant. 267.
General. Area, Mediterranean region.

42. *Ononis variegata.* L. Sp. 1008.
Maritime plains. Area, Mediterranean region.

43. *Ononis hirta.* Desf. Hort. Par. ex Poir. Suppl. i., p. 741.
Maritime and central plains. Area, South Spain.

44. *Ononis serrata.* Forsk. Æg. Arab., p. 130.
Coast near Gaza. Area, Mediterranean region.

45. Ditto, var. *Major.* Coast near Beyrout.

46. *Ononis phyllocephala.* Boiss. Fl. Or. ii., p. 63. P.
Boissier very clearly points out the differences between this and the preceding species.

47. *Ononis mitissima.* L. Sp. 1007.
General. Area, Mediterranean region, Madeiras, Canaries.

48. *Ononis alopecuroides.* L. Sp. 1008.
Esdraelon, Galilee, Lebanon. Area, Mediterranean region.

49. *Trigonella fænum-græcum.* L. Sp. 1095.
Callirrhoe, Moab. Area, Mediterranean region, chiefly cultivated ; Abyssinia.

50. *Trigonella berythæa.* Boiss. Diagn. Ser. ii., ii., p. 10.
Plain of Phœnicia. P.

51. *Trigonella cassia.* Boiss. Diagn. Ser. i., ix., p. 13.
Lower Lebanon.

52. *Trigonella astroites.* Fisch. et Mey. Ind. i., Petrop., 1835. = *T. sinuata.* Boiss.
Anti-Lebanon, Rasheya.

53. *Trigonella spinosa.* L. Sp. 1094.
General.

54. *Trigonella crassipes.* Boiss. Diagn. Ser. i., ii., p. 23.
Northern district.

55. *Trigonella monspeliaca.* L. Spec. 1095.
General. Area, Mediterranean region.

56. *Trigonella minima.* Paine. Pal. Expl. Soc., pt. 3, p. 101.
Heights of Mount Gilead. P.

57. *Trigonella monantha.* C. A. Mey. Ind. Cauc., p. 137.
In the north.
Ditto, var. *Brachycarpa.* Anti-Lebanon.

58. *Trigonella strangulata.* Boiss. Diagn. Ser. i., ix., p. 17.
Anti-Lebanon. P.

59. *Trigonella cœle-syriaca.* Boiss. Diagn. Ser. i., ix., p. 19.
The Bukâà and northern district.

60. Ditto, var. *Gaillardoti.* Plain of Phœnicia.

61. *Trigonella hierosolymitana.* Boiss. Diagn. Ser. i., ix., p. 15.
General, east and west.

62. *Trigonella sibthorpii.* Boiss. Diagn. Ser. i., ix., p. 14.
Northern district.

63. *Trigonella cylindracea.* Desv. Journ., 1814, i., p. 77.
Maritime plains.

64. *Trigonella filipes.* Boiss. Diagn. Ser. i., ix., p. 16.
Lebanon and Anti-Lebanon.

65. *Trigonella lilacina.* Boiss. Diagn. Ser. i., ii., p. 17.
Carmel and Galilee. P.

66. *Trigonella velutina.* Boiss. Diagn. Ser. i., ii., p. 18.
Anti-Lebanon.

67. *Trigonella laciniata.* L. Sp. 1095.
Marshes of Sharon.

68. *Trigonella hamosa.* L. Sp. 1094.
East side of Dead Sea.

69. *Trigonella maritima.* Del. in. Poir. Encyl. v., 361.
Maritime plains. Area, Sardinia, Sicily, Tunis.

70. *Trigonella stellata.* Forsk. Æg. Arab., p. 140.
Southern Desert, near Dead Sea. Area, the Sahara.

71. *Trigonella spicata.* Sm. Prodr. Fl. Gr. ii., p. 108.
Galilee, Phœnicia, and Lebanon.

72. *Trigonella glomerata.* Ledeb. Fl. Ross. i., p. 521.
Northern hills.

73. *Trigonella arabica.* Del. in Laborde, Ar. Petr., p. 86, f. 5 = *T. pecten.* Schenk.
Deserts near Gaza.

74. *Trigonella radiata.* (L. Sp. 1096.)
Northern districts, Gilead.

75. *Trigonella corniculata.* L. Sp. 1094.
Area, South Europe.

76. *Medicago marina.* L. Sp. 1097.
Maritime plains. Area, Mediterranean region.

77. *Medicago scutellata.* All. Ped., No. 1155.
General, in the north. Area, Mediterranean region.

78. *Medicago blancheana.* Boiss. Diagn. Ser. ii., v., p. 75.
In the plains.

79. *Medicago rotata.* Boiss. Diagn. Ser. i., ii., p. 24.
Universally distributed.

80. *Medicago elegans.* Jacq. in Willd. Sp. iii., p. 408.
Coast and central districts. Area, Corsica, Sardinia, South Italy.

81. *Medicago littoralis.* Rohde. in Lois. Not., p. 118.
Maritime plains. Area, Mediterranean region and Canaries.
Ditto, var. *Subinermis.* Phœnician plain. Area, as above.

82. *Medicago tribuloides.* Desr. in Lam. Enc. iii., p. 635.
Southern desert. Area, Mediterranean region.

83. *Medicago turbinata.* Willd. Sp. iii., p. 1409. Area, Mediterranean region.

84. *Medicago tuberculata.* Willd. Sp. iii., p. 1410.
Moab, Southern Judæa, and the desert. Area, Mediterranean region.

85. *Medicago gerardi.* Willd. Sp. iii., p. 1415.
General. Area, Mediterranean region.

86. *Medicago coronata.* Lam. Dict. iii., p. 634.
General. Area, Mediterranean region.

87. *Medicago galilæa.* Boiss. Diagn. Ser. i., ix., p. 10.
Hills and central plains.

88. *Medicago denticulata.* Willd. Sp. iii., p. 1414.
Southern plains, Moab, Dead Sea. Area, Mediterranean region, Canaries, Abyssinia, North-west India, China, Japan.

89. Ditto, var. *Lappacea.* Maritime plains. Area, Southern Europe.

90. Ditto, var. *Pentacycla* Judæa, Moab.

91. Ditto, var. *Apiculata.* Littoral district. Area, Central and Southern Europe.

92. *Medicago maculata.* Willd. Sp. iii., p. 1412.

The Ghor, north of Dead Sea. Area, Europe, North Africa.

93. *Medicago minima.* Lam. Dict. iii., p. 636.

General in sandy and sunny places east of Dead Sea. Area, Europe, North Africa, Abyssinia, Canaries.

94. *Medicago laciniata.* All. Ped. i., p. 316.

Near Gaza. Area, North Africa, Canaries.

95. *Medicago ciliaris.* Willd. Sp. iii., p. 1411.

Maritime and central plains. Area, Mediterranean region, Madeiras, Canaries.

96. *Medicago lupulina.* L. Sp. 1097. Black Medick.

Highest parts of Lebanon, mountains of Moab. Area, Europe, Siberia, North China, Azores, Canaries, North Africa, Abyssinia.

97. *Medicago sativa.* L. Sp. 1096. Lucerne, or Purple Medick.

Generally cultivated. Area, Central and Southern Europe, North Africa.

98. *Medicago orbicularis.* All. Ped., No. 1150.

Moab plains. Area, Mediterranean region, Canaries, Madeira, Abyssinia.

99. *Melilotus sulcata.* Desf. Atl. ii., p. 193.

Hill and coast regions. Area, Mediterranean region, Canaries.

100. *Melilotus infesta.* Guss. Prodr. ii., p. 486.

General. Area, Corsica, South Italy, Sicily.

101. *Melilotus messanensis.* (L. Mant. 275.)

Moist places. General. Area, Mediterranean region.

102. *Melilotus elegans.* Salz. De Cand. Prodr. ii., p. 188.

Mount Carmel. Area, Mediterranean region, Madeira, Abyssinia.

103. *Melilotus parviflora.* Desf. Atl. ii., p. 192.
General. Area, Mediterranean region.

104. *Melilotus italica.* (L. Sp. 1088.)
Area, South France and Italy, North Africa.

105. *Melilotus officinalis.* Lam. Dict. iv., p. 63. Melilot.
In the north. Area, Europe, Siberia.

106. *Trifolium cassium.* Boiss. Diagn. Ser. i., ix., p. 23.
Woods in the north.

107. *Trifolium hirtum.* All. Auch. 20.
Lebanon valleys. Area, Mediterranean region.

108. *Trifolium cherleri.* L. Sp. 1081.
Near Beyrout. Area, Mediterranean region, Canaries.

109. *Trifolium arvense.* L. Sp. 1083. Hare's-foot, Trefoil.
Lebanon. Area, Europe, North Africa, Siberia.
Ditto, var. *Longisetum.* Lebanon.

110. *Trifolium stellatum.* L. Sp. 1083.
General. Area, Mediterranean region, Madeira, Canaries.

111. *Trifolium lappaceum.* L. Sp. 1082.
General in meadows. Area, Mediterranean region, Canaries, Azores, Madeira.

112. *Trifolium incarnatum.* L. Sp. 1083. Crimson Clover.
Probably introduced. Area, Central and Southern Europe.

113. *Trifolium augustifolium.* L. Sp. 1083.
On the coast. Area, Mediterranean region, Azores, Canaries, Madeira.

114. *Trifolium purpureum.* Lois. Gall. ii., p. 125, pl. 14.
General in Central and Eastern Palestine. Area, South France, Tuscany.

115. *Trifolium desvauxii.* Boiss. Diagn. Ser. ii., ii., p. 12.
On the coast. P.
Ditto, var. *Laxinsculum.* Beyrout.

116. *Trifolium blancheanum.* Boiss. Diagn. Ser. ii., ii., 13.
On the waysides, on the coast. P.

117. *Trifolium palestinum.* Boiss. Diagn. Ser. i., ix., p. 21.
In the south, hills and coast. P.

118. *Trifolium dichroathum.* Boiss. Diagn. Ser., i., ix., p. 20.
Hills and plains near the coast.

119. *Trifolium formosum.* Urv. Enum., p. 94. General.

120. *Trifolium supinum.* Savi. Trif., p. 46, f. 2.
Central districts. Area, Portugal, Italy, Dalmatia, the Danube.

121. Ditto, var. *Tuberculatum.* Phœnician plain.

122. *Trifolium carmeli.* Boiss. Diagn. Ser. ii., ii., p. 16.
Mount Carmel, etc. P.

123. *Trifolium alexandrinum.* L. Sp. 1085. Arab. برسين, *Bersyn.*
General. Area, Thrace.

124. *Trifolium panormitanum.* Presl. Fl. Sic. i., p. 20. = *T. squarrosum.* L. Syn., ii., 293.
Moist places on the coast. Area, Mediterranean region, Canaries.

125. *Trifolium maritimum.* Huds. Angl., p. 284.
Northern coast, in moist places. Area, Mediterranean region, West France, Dalmatia, Canaries.

126. *Trifolium plebicum.* Boiss. Diagn. Ser. i., ix., p. 23.
Anti-Lebanon. P.

127. *Trifolium scutatum.* Boiss. Diagn. Ser. i., ii., p. 27.
Northern and central districts.

128. *Trifolium clypeatum.* L. Sp. 1084. Universal.

129. *Trifolium scabrum.* L. Sp. 1084.
Moab, Judæa. Area, Central and Southern Europe, North Africa, Canaries, Azores.

130. *Trifolium globosum.* L. Sp. 1081.
Northern districts.

131. *Trifolium eriosphærum.* Boiss. Diagn. Ser. i., ix., p. 25.
Central hills, Jerusalem, Hebron, Gilead.

132. *Trifolium uniflorum.* L. Amœn. iv., p. 285.
Locality not given.

133. *Trifolium meduceum.* Boiss. Fl. Or. ii., p. 134.
Barren rocky and sandy spots. Lebanon and Anti-Lebanon. P.

134. *Trifolium pilulare.* Boiss. Diagn. Ser. i., ii., p. 29.
General.

135. *Trifolium physodes.* Stev. M. B. Taur. Cauc. ii., p. 217.
Lebanon. Area, Sicily.

136. Ditto, var. *Psilocalyx.* (= *T. sclerorhizum*, Boiss.)
Lower slopes of Lebanon and Anti-Lebanon.

137. *Trifolium modestum.* Boiss. Diagn. Ser. i., ix., p. 27.
Lebanon and Anti-Lebanon, 8,500 feet. P.

138. *Trifolium resupinatum.* L. Sp. 1086, and vars.
In cultivated and grassy places. General. Area, Mediterranean region, Azores, Canaries.

139. *Trifolium tomentosum.* L. Sp. 1086.
General in grassy places. Area, Mediterranean region, Canaries.

140. *Trifolium spumosum.* L. Sp. 1085.
In the plains. Area, Mediterranean region.

141. *Trifolium xerocephalum.* Fenz. Pug., No. 10.
Northern district.

142. Ditto, var. *Minus* (*T. moriferum*, Boiss.)
Higher parts of Lebanon.

143. *Trifolium nervulosum.* Boiss. Diagn. Ser. i., ix., p. 25.
Maritime plains. Eastern Gilead. Ditto var. *Galliæum.* Galilee.

144. *Trifolium petrisavii.* Clem. Sert. Or., p. 32, pl. vii., f. 2.
Near Beyrout.

145. *Trifolium nigrescens.* Viv. Fl. Ital. Fasc. i., p. 12, pl. 13. Area, Southern Europe.

146. *Trifolium meneghinianum.* Clem. Sert. Or., p. 31, pl. vii., f. 1.
Maritime and central plains. Area, South Russia.

147. *Trifolium repens.* L. Sp. 1080. Dutch Clover.
General. Area, Europe, Siberia, North America.

148. *Trifolium hybridum.* L. Fl. Succ., ed. ii., p. 258. Alsike Clover. Area, Europe, North Africa.

149. *Trifolium comosum.* Labill. Déc. v., p. 15. Near Beyrout.

150. *Trifolium stenophyllum.* Boiss. Diagn. Ser. i., ix., p. 30.
Sandy deserts, Gaza. P.

151. *Trifolium speciosum.* Willd. Sp. v., p. 1382.
Northern and central districts. Area, Sicily.

152. *Trifolium erubescens.* Fenz. Pug., No. 11.
Universal in rocky places.

153. *Trifolium boissieri.* Gress. Syn. ii., Add., p. 858.
Central hills.

154. *Trifolium agrarium.* L. Sp. 1087. =*T. procumbens,* De Cand, nec L.

General. Area, Europe, North Africa, Abyssinia.

155. *Trifolium velivolum.* Paine. Pal. Expl. Soc., No. 3, p. 103.

Hills of Gilead. P.

156. *Physanthyllis tetraphylla.* (L. Sp. 1011.)

General in the north, and Gilead. Area, Southern Europe.

157. *Hymenocarpus circinnatus.* (L. Sp. 1096.)

General from Carmel to Moab, and by Dead Sea. Area, Mediterranean region.

158. *Cytisopsis dorycniifolia.* Jaub. ed. Sp. Ill. Or., p. 154, pl. 84.

Middle region of Lebanon. Mount Carmel. Tabor.

159. *Dorycnium hirsutum.* (L. Sp. 1091.) And var. *Syriaca,* Boiss.

Lebanon and Anti-Lebanon. Area, Southern Europe.

160. *Dorycnium rectum.* (L. Sp. 1092.)

Northern districts to 4,000 feet. Area, Mediterranean region.

161. *Dorycnium haussknechtii.* Boiss. Fl. Or. ii., p. 163. Var. *libanoticum.* Boiss.

Dry districts of sub-alpine Lebanon.

162. *Lotus creticus.* L. Sp. 1091.

Sandy coasts. Moab. Area, Mediterranean region.

163. Ditto, var. *Cytisoides.*

Northern coasts, Sidon.

164. Ditto, var. *Collinus.* (=*L. judaicus.* Boiss.)

Lebanon and Hermon up to 6,000 feet, and southern hills.

165. *Lotus tenuifolius.* Rchb. Fl. Exc. 506.

Lebanon, and coast. Area, Europe, North Africa, Songraria.

166. *Lotus lamprocarpus.* Boiss. Diagn. Ser. i., ix., p. 33.
General. Abounds in the Jordan valley.

167. *Lotus corniculatus.* L. Sp. 1092. Var. *Alpinus.* Bird's-foot Trefoil.
Lebanon and Anti-Lebanon. Area, Europe, Asia, Abyssinia, Australia.

168. *Lotus gebelia.* Vent. Hort. Cels., pl. 57. Cœle Syria.

169. Ditto, var. *Villosus.*

170. Ditto, var. *Libanoticus.* Boiss. Diagn. Ser. i., ix., p. 32.
Lebanon and Anti-Lebanon.

171. *Lotus lanuginosus.* Venten. Malm., p. 92.
The southern desert.

172. *Lotus arabicus.* L. Mant., 104.
South end of the Dead Sea. Area, Senegambia, Nubia, Abyssinia, Canaries.

173. *Lotus angustissimus.* L. Sp. 1090.
Northern coast. Area, Central and Southern Europe, North Africa, Siberia, Azores, Madeira, Canaries.

174. *Lotus peregrinus.* L. Sp. 1090.
On the coast in sandy places. Area, Dalmatia.

175. *Lotus carmeli.* Boiss. Diagn. Ser. i., ix., p. 34.
Mount Carmel. P.

176. *Lotus pusillus.* Viv. Fl. Lib., p. 47. And varieties.
South desert. Maritime plains in sand. Pine-woods at Beyrout. Area, Sicily.

177. *Lotus ornithopodioïdes.* L. Sp. 1091.
Coast and central districts. Moab. Area, Mediterranean region.

178. *Lotus edulis.* L. Sp. 1090.

Maritime districts, general. Area, Mediterranean region.

179. *Tetragonolobus purpureus.* Mœnch. Meth. 164.

Highlands of Moab, Jordan valley. Area, Mediterranean region.

180. *Tetragonolobus palæstinus.* Boiss. Diagn. Ser. ii., ii., p. 20.

Universally distributed.

181. *Securigera coronilla.* De Cand. Fl. Fr. iv., p. 609.

On the coast and inland plains. Area, Mediterranean region.

182. *Ornithopus compressus.* L. Sp. 1049.

Coast and hill-districts. Area, Mediterranean region, West France, Canaries.

183. *Scorpiurus muricata.* L. Sp. 1050.

Var. *Lævigata.* Sibth. Ruins in Moab. Area, Mediterranean region.

184. *Scorpiurus sulcata.* L. Sp. 1050.

Central and maritime plains. Area, Canaries, Spain, North Africa, Abyssinia.

185. *Scorpiurus subvillosus.* L. Sp. 1050.

Plains of Moab. Area, Mediterranean region, Madeira, Canaries.

186. *Coronilla emeroides.* Boiss. Diagn. Ser. i., ii., p. 100.

Galilee and Lebanon.

187. *Coronilla varia.* L. Sp. 1048.

Ditto, var. *Pauciflora.* (= *C. libanotica.* Boiss.)
Lebanon. Area, Central and Southern Europe.

188. *Coronilla cretica.* L. Sp. 1048.

Coast and north. Area, Italy, Istria, Dalmatia.

189. *Coronilla parviflora.* Willd. Sp. iii., p. 1155.

Coast and central district.

190. *Coronilla scorpioides.* (L. Sp. 1049.)

Mountains, central, east regions, Moab. Area, Mediterranean region.

191. *Hippocrepis unisiliquosa.* L. Sp. 1050.

Lebanon region, Moab, and Gilead. Area, Mediterranean region.

192. *Hippocrepis biflora.* Spreng. Pug. ii., p. 73.

Central and southern district.

193. *Hippocrepis multisiliquosa.* L. Sp. 1050.

Moab, Mount Carmel, Tabor. Area, Mediterranean region.

194. *Hippocrepis ciliata.* Willd. Mag. N. Ges. Berol., 1806, p. 173.

East side of Dead Sea. Area, Mediterranean region.

195. *Hippocrepis cornigera.* Boiss. Diagn. Ser. i., ii., p. 102.

Desert south of Dead Sea. Area, North-east Africa.

196. *Psoralea bituminosa.* L. Sp. 1075 (= *P. palæstina.* Jacq.).

By the Dead Sea, Galilee, Esdraelon. Area, Mediterranean region, Canaries.

197. *Indigofera argentea.* L. Mant. 27 = *I. tinctoria.* Forsk. Fl. Eg. Arab., p. 138. Arab., نيله, *Neyleh.*

Cultivated on east side of Dead Sea, in Ghor Safieh, and other places.

198. *Galega officinalis.* L. Sp. 1063.

The Bukâà. Area, Southern Europe.

199. *Colutea arborescens.* L. Sp. 1045.

In the north. Area, Central and Southern Europe, North Africa.

200. *Glycyrrhiza glandulifera.* Reg. et Herd. Pl. Sem. p. 37.

General by the rivers, not Jordan valley. Area, Mediterranean and Danubian region, Central and Southern Russia.

201. *Glycyrrhiza echinata.* L. Sp. 1046.

Maritime plains. Area, East Italy, Danubian region.

202. *Glycyrrhizopsis flavescens.* Boiss. Diagn. Ser. i., vi., p. 33.

Northern mountains.

203. *Biserrula pelecinus.* L. Sp. 1073.

Coast, central, and eastern districts. Area, Mediterranean region, Canaries, Madeira.

204. *Astragalus epiglottis.* L. Mant. 274.

Coast region. Area, Mediterranean region.

205. *Astragalus tribuloides.* Del. Ill. Eg., p. 22.

The Southern Desert. Area, North-west India.

206. *Astragalus cruciatus.* Link. Enum. ii., p. 256.

In barren places, East and West Palestine.

207. *Astragalus radiatus.* Bge. Astr., p. 8.

Southern Desert.

208. *Astragalus eremophilus.* Boiss. Diagn. Ser. i., ii., p. 54.

South of Dead Sea. Area, Arabia.

209. *Astragalus tenuirugis.* Boiss. Diagn. Ser. i., ix., p. 61.

Southern Desert. Area, Sahara.

210. *Astragalus damascenus.* Boiss. Diag. Ser. ii., vi., p. 56.

Near Damascus.

211. *Astragalus conduplicatus.* Bertol. Nov. Comm. Bonon. vi., p. 231.

Near Damascus.

212. *Astragalus gyzensis.* Del. Fl. Eg. Suppl., p. 64.

South of Dead Sea.

213. *Astragalus hispidulus.* De Cand. Astr., p. 105, pl. 13.

Southern desert.

214. *Astragalus callichrous.* Boiss. Diagn. Ser. i., ix., p. 62.

Wilderness of Judæa. Gaza. P.

215. *Astragalus annularis.* Forsk. Eg. Arab. 139.

Southern desert. Philistine plain.

216. *Astragalus bœticus.* L. Sp. 1068.

Maritime plains. Jordan valley. Area, Mediterranean region, Canaries.

217. *Astragalus aulacolobus.* Boiss. Diagn. Ser. i., ix., p. 64.
Fields and plains.

218. *Astragalus hamosus.* L. Sp. 1067. Arab. الكورن, *el Koren.*
Coast district. Plains of Moab. Jordan valley. Area, Mediterranean region, Canaries.

219. *Astragalus oxytropifolius.* Boiss. Diagn. Ser. i., ix., p. 37.
Mount Gilead. Mount Nebo.

220. *Astragalus tuberculosus.* De Cand. Astr., p. 33, pl. 22.
Generally distributed.

221. *Astragalus bombycinus.* Boiss. Diagn. Ser. i., ii., p. 50.
Near Gaza.

222. *Astragalus berytheus.* Boiss. Diagn. Ser. ii., p. 33.
Northern maritime plains. P.

223. *Astragalus peregrinus.* Vahl. Symb. i., p. 57.
Near Gaza. Area, North Africa.

224. *Astragalus hirsutissimus.* De Cand. Astr. 119, pl. 19.
Lebanon, 8,000 feet.

225. *Astragalus lanatus.* Labill. Déc. Syr. i., p. 21, pl. 10.
Top of Hermon. Higher parts of Lebanon.

226. *Astragalus chrysophyllus.* Boiss. Diagn. Ser. ii., ix., p. 38.
In the north.

227. *Astragalus cedreti.* Boiss. Diagn. Ser. i., ix., p. 39.
Cedars of Lebanon. P.

228. *Astragalus emarginatus.* Labill. Déc. Syr. i., p. 19, pl. 9.
Lebanon and Anti-Lebanon. P.

229. *Astragalus cretaceus.* Boiss. Diagn. Ser. ii., v., p. 84.
Near Hebron. Bethlehem.

230. *Astragalus nanus.* De Cand. Astr., p. 114, pl. 17.
Rocky places, North Lebanon.

231. *Astragalus antilibani.* Bge. Astr. p. 90.
Anti-Lebanon, near Bludan, 6,000 feet. P.

232. *Astragalus stramineus.* Boiss. Diagn. Ser. ii., v., p. 85.
Anti-Lebanon. P.

233. *Astragalus neurocarpus.* Boiss. Diagn. Ser. i., ii., p. 59.
Fields and plains.

234. *Astragalus macrocarpus.* De Cand. Astr., p. 143, pl. 28.
Generally distributed.

235. *Astragalus christianus.* L. Sp. 1064.
Plain of Gennesaret.

236. *Astragalus græcus.* Boiss. Diagn. Ser. i., ii., p. 57.
Gadara. Bashan.

237. *Astragalus aleppicus.* Boiss. Diagn. Ser. i., ii., p. 58.
The Bukâà. Gilead.

238. *Astragalus alexandrinus.* Boiss. Diagn. Ser. i., ix., p. 74.
Jordan valley.

239. *Astragalus trichopterus.* Boiss. Fl. Or. ii., p. 292.
Alpine region of Lebanon. P.

240. *Astragalus pinetorum.* Boiss. Diagn. Ser. i., ii., p. 77.
Lebanon and Anti-Lebanon, 6,000—9,000 feet.

241. *Astragalus sparsus.* Decaisn. Fl. Sin., p. 43.
Southern desert.

242. *Astragalus trigonus.* De Cand. Astr., No. 81.
Desert south-west of Dead Sea.

243. *Astragalus gummifer.* Labill. Journ. Phys., 1790, p. 46, Ic.
Lebanon and Hermon.

244. *Astragalus roussæanus.* Boiss. Diagn. Ser. i., ii., p. 61.
Barren plains; the Bukâà.

245. *Astragalus echinus.* De Cand. Astr., p. 197, pl. 34.
Alpine region of Lebanon and Anti-Lebanon.

246. Ditto, var. *virens* = *A. libanoticus.* Boiss.
Sub-alpine region of Lebanon and Anti-Lebanon.

247. *Astragalus argyrothamnus.* Boiss. Diagn. Ser. i., ix., p. 89.
Lebanon and Anti-Lebanon, 6,000 feet. P.

248. *Astragalus psilodontius.* Boiss. Diagn. Ser. i., ix., p. 86.
The lower region of Anti-Lebanon. P.

249. *Astragalus zachlensis.* Bge. Astr., p. 154.
In the north, and on Lebanon and Anti-Lebanon. P.

250. *Astragalus compactus.* Willd. Act. Berol, 1794, p. 29, pl. 1.
Lebanon, 8,000 feet.

251. *Astragalus cruentiflorus.* Boiss. Diagn. Ser. i., ix., p. 82.
Hermon and Lebanon, 7,000—9,000 feet. P.

252. *Astragalus lepidanthus.* Boiss. Diagn. Ser. i., ix., p. 85.
Northern plains.

253. *Astragalus bethlehemiticus.* Boiss. Diagn. Ser. i., ix., p. 85.
Hill country from South Judæa to Lebanon and Hermon. P.

254. *Astragalus argyrophyllus.* Boiss. Fl. Or. ii., p. 358.
Anti-Lebanon.

255. *Astragalus andrachnefolius.* Fenz. Pug., No. 6.
In the north.

256. *Astragalus andrachne.* Bge. Astr., p. 158.
Barren plains, Bukâà.

257. *Astragalus drusorum.* Boiss. Diagn. Ser. i., ix., p. 78.
Galilee and Lebanon above Deir el Kammar. P. Ditto var. *Maroniticus.* Boiss. Northern Lebanon.

258. *Astragalus hasbeyanus.* Boiss. Diagn. Ser. i., ix., p. 77.
Above Hasbeya, Anti-Lebanon. P.

259. *Astragalus deinacanthus.* Boiss. Diagn. Ser. i., ix., p. 76.
The Bukâà.

260. *Astragalus longifolius.* Lam. Encycl. i., p. 319.
Hermon, 4,000 feet.

261. *Astragalus forskahlei.* Boiss. Diagn. Ser. i., ix., p. 101.
= *A. tumidus.* Willd. Round the Dead Sea. Moab.

262. *Astralagus russellii.* Boiss. Diagn. Ser. i., ix., p. 102.
Northern plains.

263. *Astralagus coluteoides.* Willd. Act. Berol., 1794, p. 27.
Lebanon and Anti-Lebanon, 6,000—9,000 feet. P.

264. *Astragalus macrocephalus.* Willd. Sp. iii., p. 1260.
Anti-Lebanon.

265. *Astragalus ehrenbergii.* Bge. Astr., p. 102.
Anti-Lebanon and Lebanon. P.

266. *Astragalus echinops.* Boiss. Diagn. Ser. i., ii., p. 57.
Anti-Lebanon and Bukâà.

267. *Astragalus kahiricus.* De Cand. Prodr. ii., p. 292.
Southern desert. Area, the Cyrenaica.

268. *Astragalus kotschyanus.* Boiss. Diagn. Ser. i., ii., p. 44.
Chalky hills near Rascheya, Anti-Lebanon, etc.

269. *Astragalus trifoliolatus.* Boiss. Diagn. Ser. i., ii., p. 48.
Base of Lebanon.

270. *Astragalus angulosus.* De Cand. Astr., No. 122, pl. 45.
Throughout the country, from Lebanon to Dead Sea. P.

271. *Astragalus dictyocarpus.* Boiss. Diagn. Ser. i., ii., p. 84.
Alpine region, Lebanon and Anti-Lebanon. P.

272. *Astragalus gladiatus.* Boiss. Diagn. Ser. i., ii., p. 45.
Lebanon alpine region.

273. *Astragalus sanctus.* Boiss. Diagn. Ser. i., ix., p. 47.
Hill districts of the south, Judæan wilderness, Southern Desert, Sahara of Damascus. P.

274. *Astragalus amalecitanus.* Boiss. Diagn. Ser. i., ix., p. 46.
Downs near Gaza. P.

275. *Astragalus angustifolius.* Lam. Enc. Méth. i., p. 321.
Lebanon and Anti-Lebanon, 6,000 feet.

276. *Astragalus hermoneus.* Boiss. Diagn. Ser. i., ix., p. 94.
Higher parts of Lebanon and Anti-Lebanon. P.

277. *Astragalus leucophæus.* De Cand. Prodr. ii., p. 293.
Thickets on the plains of Moab.

278. *Hedysarum spinosissimum.* Sibth. Fl. Gr. viii., p. 16, pl. 721.
Maritime hills. Area, Spain.

279. *Hedysarum atomarium.* Boiss. Diagn. Ser. i., ix., p. 3. Var. *suffrutescens.* Boiss.
Northern district, on dry hills.

280. *Hedysarum pannosum.* (Boiss. Diagn. Ser. i., ix., p. 106.)
Eastern desert of Moab.

281. *Onobrychis crista-galli.* (L. Syst. 563.)
Southern wilderness, Eastern Gilead. Area, North Africa.

282. *Onobrychis gærtneriana.* Boiss. Diagn. Ser. i., ix., p. 108.
Throughout Central Palestine north to south.

283. *Onobrychis æquidentata.* (Sibth. Prodr. ii., p. 84.)
Galilee. Area, South Italy, Sicily, Dalmatia.

284. *Onobrychis megataphros.* Boiss. Diagn. Ser. ii., ii., p. 97.
Northern plains.

285. *Onobrychis galegifolia.* Boiss. Diagn. Ser. i., ii., p. 91.
In the north.

286. *Onobrychis caput-galli.* (L. Sp. 1059.)
Maritime region, north. Area, Mediterranean region.

287. *Onobrychis sativa.* Lam. Fl. Fr. ii., p. 652. Sainfoin.
Cultivated. Jordan valley, naturalized? Area, Central Europe, Siberia.

288. *Onobrychis gracilis.* Bess. En., p. 74.

Coast region. Area, Southern Russia.

289. *Onobrychis kotschyana.* Fenz. Pug., No. 2.

Under Hermon. Nazareth.

290. *Onobrychis cornuta.* (L. Sp. 1060.) = *O. tragacanthoïdes.*

Sub-alpine and alpine Lebanon, and Anti-Lebanon.

291. *Onobrychis ptolemaica.* Del. Fl. Eg., p. 328, pl. 39 fr.

South end of Dead Sea.

292. *Onobrychis gaillardoti.* Boiss. Fl. Or. ii., p. 548.

Sahara, west of Damascus.

293. *Onobrychis aurantiaca.* Boiss. Diagn. Ser. i., ix., p. 105.

The Bukâà.

294. *Alhagi maurorum.* De Cand. Prodr. ii., p. 352.

Shores of the Dead Sea.

295. *Alhagi camelorum.* Fisch. Ind. Gor. 1812, p. 72. Var. *turcorum.* Boiss.

Northern district. Area, South-east Russia, Scinde.

296. *Cicer pinnatifidum.* Jaub. et Sp. Ann. Sc. N. xviii., 1842, p. 227. = *C. judaicum.* Boiss.

Universal.

297. *Cicer arietinum.* L. Sp. 1040.

Cultivated everywhere. Area, Southern Europe.

298. *Cicer ervoides.* (Sieb. Reis. ii., p. 325, pl. 11.) Var. *Libanoticum.*

Lebanon and Anti-Lebanon.

299. *Cicer minutum.* Boiss. Diagn. Ser. i., ix., p. 130.

300. *Vicia hybrida.* L. Sp. 1037.

General. Area, Central and Southern Europe.

301. *Vicia lutea.* L. Sp. 1037. Var. *hirta.* Yellow Vetch.

General, Moab, Dead Sea. Area, Central and Southern Europe, North Africa.

302. *Vicia sericocarpa.* Fenz. Pug., No. 4.

General in fields.

303. *Vicia galeata.* Boiss. Diagn. Ser. i., ii., p. 103.

Common in maritime and central districts.

304. *Vicia sativa.* L. Sp. 1037, var. L. Common Vetch, Tare.

General in fields. Area, Europe, North Africa, North America.

305. *Vicia angustifolia.* Roth. Teut. i., p. 310. Var. *Amphycarpa.*

Philistine plain. Area, Europe, North Africa.

306. *Vicia lathyroides.* L. Sp. 1037. Spring Vetch.

Hills and bare places. Area, Central and Southern Europe, North Africa.

307. *Vicia cuspidata.* Boiss. Diagn. Ser. i., ii., p. 104.

Central Palestine.

308. *Vicia peregrina.* L. Sp. 1038.

From Lebanon southwards general, Mount Nebo, etc. Area, Mediterranean region.

309. *Vicia michauxii.* Spreng in Willd. En., p. 764.

Galilee.

310. *Vicia aintabensis.* Boiss. Fl. Or. ii., 577.

Jacob's Well, Nablus, Nazareth.

311. *Vicia narbonensis.* L. Sp. 1038.

By the Dead Sea and Moab. Area, Mediterranean region.

312. *Vicia canescens.* Labill. Pl. Syr. Dec. i., p. 17, pl. 7.

Lebanon, 8,000—9,000 feet. P.

313. *Vicia tenuifolia.* Roth. Germ. i., p. 309.

Lebanon, sub-alpine. Area, North Europe, Siberia.

Ditto, var. *Stenophylla.* Lebanon and Anti-Lebanon.

314. *Vicia cassia.* Boiss. Diagn. Ser. i., ix., p. 119.
Northern woods.

315. *Vicia calcarata.* Desf. Ath. ii., p. 160. Var. *Cinerea.*
Fields in the plains. Area, South Europe.

316. *Vicia villosa.* Roth. Germ. ii., p. 182.
Maritime plains. Area, Europe.

317. *Vicia palæstina.* Boiss. Diagn. Ser. i., ix., p. 117.
Generally distributed, in rocky, shady places, east and west.

318. *Vicia ervilia.* (L. Sp. 1040.)
General, east and west. Area, South Europe.

319. *Ervum lens.* L. Sp. 1039. Lentiles. Heb., צְרָשִׁים.
General, probably from cultivation; certainly wild in Moab.

320. *Ervum orientale.* Boiss. Diagn. Ser. i., ix., p. 115.
Northern region, Gilead, Moab.

321. *Ervum lenticulum.* Schreb. in Sturm. D. Fl. i., fasc. 32.
The hill-districts. Area, South Europe.

322. *Lathyrus ochrus.* (L. Sp. 1027.)
Maritime and central plains. Area, Mediterranean region.

323. *Lathyrus aphaca.* L. Sp. 1029. Yellow Pea.
General. Area, Central and Southern Europe, North Africa.

324. *Lathyrus polyanthus.* Boiss. Diagn. Ser. ii., ii., p. 43.
Maritime and central plains.

325. *Lathyrus stenolobus.* Boiss. Diagn. Ser. i., ix., p. 124.
Woods in the north.

326. *Lathyrus nissolia.* L. Sp. 1029. Grass Pea.
Lower slopes, Lebanon and Anti-Lebanon. Area, Central and Southern Europe, Mount Atlas.

327. *Lathyrus annuus.* L. Sp. 1032.
Mountain plains, Jordan valley. Area, Mediterranean region, Canaries, Madeira.

328. *Lathyrus hierosolymitanus.* Boiss. Diagn. Ser. i., ix., p. 127.
Hill districts.

329. *Lathyrus cassius.* Boiss. Diagn. Ser. i., ix., p. 128.
Lebanon.

330. *Lathyrus cicera.* L. Sp. 1030.
General, in fields. Area, Central and Southern Europe, North Africa, Canaries.

331. *Lathyrus amœnus.* Fenz. in Russ. Reis. i., p. 890, pl. 2.
Maritime plains.

332. *Lathyrus stenophyllus.* Boiss. Diagn. Ser. i., ix., p. 126.
Maritime districts.

333. *Lathyrus marmoratus.* Boiss. Fl. Or. ii., p. 606.
Lebanon, and maritime desert plains.

334. *Lathyrus blepharicarpus.* Boiss. Diagn. Ser. i., ix., p. 126.
Hilly districts everywhere, Gilead.

335. *Lathyrus chrysanthus.* Boiss. Diagn. Ser. i., vi., p. 46.
The Bukâà.

336. *Lathyrus erectus.* Lag., p. 22.
Near Jerusalem. Area, Spain, South France, Istria, Dalmatia.

337. *Lathyrus pratensis.* L. Sp. 1033. Meadow Pea.
Northern hills and mountains. Area, Europe, Siberia, West Himalayas, Abyssinia.

338. *Lathyrus sativus.* L. Sp. 1030.
Thickets and wâdies, Moab, common. Area, South Europe.

339. *Orobus sessilifolius.* Sibth. Prodr. ii., p. 64.
Lebanon, Anti-Lebanon, Galilee, Moab.

340. Ditto, var. *Ovalifolius.* Lebanon at Afka.

341. *Orobus hirsutus.* L. Sp. 1027. Rough Pea.
Hilly districts near the coast. Area, Central and Southern Europe.

342. *Orobus grandiflorus.* Boiss. Fl. Or. ii., p. 622.

Lebanon woods, 6,000 feet.

343. *Pisum sativum.* L. Sp. 1026.

On the coast. Cultivated in Europe.

344. *Pisum arvense.* L. Sp. 1027.

On the coast and central districts, hills of Moab. Area, North Europe.

345. *Pisum elatius.* M. B. Taur. Cauc. ii., p. 151.

Carmel, Galilee, Lebanon, Anti-Lebanon. Area, South Europe.

346. *Pisum fulvum.* Sibth. Fl. Gr. vii., p. 79, pl. 688.

Jerusalem, Mount Tabor, Sidon, etc.

347. *Vigna nilotica.* (Del. Fl. Eg. 323, pl. 38, f. 1.)

Marshes near the coast, plain of Sharon. Area, Nubia.

348. *Dolichos lablab.* L. Sp. 1019.

El Huleh. Area, Egypt.

349. *Cassia obovata.* Collad. Mon., p. 92.

Wâdys east and south opening on the Dead Sea. Area, Senegal, Nubia, Abyssinia, Arabia, Scinde.

350. *Ceratonia siliqua.* L. Sp. 1513. Locust Tree or Carob Tree. Arab. خروب, *Kharoob.*

Everywhere except on the higher and colder hills. Area, Mediterranean region.

This is the tree alluded to in Luke xv. 16, where its fruit is mentioned under the name 'husks' (κεράτια).

351. *Cercis siliquastrum.* L. Sp. 534. Judas Tree.

Carmel, Tabor, Gerizim, and all the lower wooded hills. Area, Southern Europe.

352. *Prosopis spicigera.* L. Mant. 68.

Jordan valley. Area, Scinde, Tropical India.

353. *Prosopis stephaniana.* (Willd. Sp. iv. 1034.) Arab. الخارمبة, *el Charembâ.* Coast and Jordan valley.

354. *Acacia nilotica.* Del. Ill. Eg., p. 3. Hebr. סְנֶה, Ex. iii. 2, etc., A.V. 'Bush.'

Shores of the Dead Sea.

355. *Acacia tortilis.* Hayne. Ex. Schweinf. in Linn. Jour., 1867, p. 327.

Southern desert. Area, Nubia, Tropical Arabia, Senegal.

356. *Acacia seyal.* Del. Fl. Eg., p. 384, pl. 52, f. 2. Shittim Tree, Heb. שִׁטָּה.

Round the Dead Sea. Area, Nubia, Abyssinia, Senegal, Tropical Arabia.

357. *Acacia farnesiana.* Benth. On the coast.

358. *Acacia albida.* Del. Fl. Eg. 385, pl. 52, f. 3.

Plain of Phœnicia. Area, Nubia, Abyssinia, Senegal.

ORDER XLI., ROSACEÆ.

1. *Amygdalus spartioides.* Spach. Ann. Sc. N. Ser. ii., xix., p. 108.

Rocky slopes of Anti-Lebanon.

2. *Amygdalus agrestis.* Boiss. Diagn. Ser. i., x., p. 1.

The Bukâà. P.

3. *Amygdalus communis.* L. Sp. 677. Almond Tree. Heb. שָׁקֵד.

Hermon and Lebanon to 5,000 feet, Moab, Gilead, Bashan (Hauran). Area, South Europe.

4. *Amygdalus orientalis.* Ait. Kew. i., p. 161.

Anti-Lebanon.

5. *Persica vulgaris.* Mill. Dict. iii., p. 465. Peach.

Probably introduced from Persia.

6. *Cerasus microcarpa.* (C. A. Mey. En. Casp., p. 167.) = *C. orientalis.* Spach. Anti-Lebanon, 4,500 feet.

7. *Cerasus prostrata.* (Labill. Dec. Syr. iv., pl. 6.)

Galilæan hills, Lebanon, Anti-Lebanon. Area, mountains of South Spain, Sardinia, North Africa, and Dalmatia.

8. *Prunus ursina.* Ky. V. Zool. Bot. Ges. Vindob., 1864.

Lebanon and Anti-Lebanon, 5,000 feet. P.

9. *Armeniaca vulgaris.* Lam. Dict. i., p. 2. (=*Prunus armeniaca.* L. Sp. 679.) Apricot. Hebr. תַּפּוּחַ, probably.

Not truly indigenous, but cultivated and naturalized from very early times. Originally from Armenia.

10. *Pyrus syriaca.* Boiss. Diagn. Ser. i., x., p. 1. Arab. انجاص, *Endschas.*

In the littoral and northern hill districts, and Gilead.

11. *Pyrus boveana.* Decaisn. Jard. Fr. Mus., pl. 10.

Lower parts of Anti-Lebanon. P.

12. *Sorbus aucuparia.* L. Sp. 683. Mountain Ash, Rowan.

Northern hills. Area, Europe, Siberia, Dahuria, Madeira.

13. *Sorbus trilobata.* (Labill. Dec. iv., p. 15.)

Lebanon, 3,000—5,000 feet.

14. *Sorbus aria.* Crantz. Austr., f. 2, p. 46. Var. *Græca.* White Beam Tree.

Lebanon and Anti-Lebanon. Area, Europe, North Africa, Siberia.

15. *Mespilus germanica.* L. Sp. 684. Medlar.

About Nablus and Samaria. Probably introduced.

16. *Cratægus azarolus.* L. Sp. 683.

Universal. Area, South Europe.

17. *Cratægus monogyna.* (Willd. En. 524.) Hawthorn.

Coast and mountain districts, Gilead. Area, Europe, North-west India.

18. *Cratægus orientalis.* M. B. Taur. Cauc. i., p. 387.

Mountain thickets, Gilead and Moab.

19. *Cotoneaster nummularia.* F. and M. Ind. Petrop. ii., p. 34.
Area, North Africa, North-west India.

20. *Cotoneaster pyracantha.* (L. Sp. 685.)
Perhaps introduced. Area, South France, Italy.

21. *Rosa glutinosa.* Sibth. Prodr. i., p. 348 = *R. libanotica.* Boiss.
Hills and mountains north. Area, South Spain, Corsica, Sicily.

22. *Rosa canina.* L. Sp. 704.
Galilee northwards. Area, Europe, North Africa, Siberia.

23. Ditto, var. *Collina.* Jacq. General.

24. *Rosa schergiana.* Boiss. Fl. Or. ii., p. 686.
Anti-Lebanon, 5,000 feet. P.

25. *Rosa phœnicia.* Boiss. Diagn. Ser. i., x., p. 4.
Common in the cultivated parts of the country.

26. *Rosa tomentosa.* Smith. Brit. ii., p. 539.
Highest mountains. Area, Europe, Siberia, North Africa.

N.B.—There are several other species of rose in my herbarium which I am unable to identify.

27. *Rubus tomentosus.* Borckh. in Rœm. Bot. Mag. 1.
Northern Lebanon.

28. *Rubus collinus.* De Cand. Fl. Fr. v., p. 545.
Phœnicia, Galilee, and Lebanon. Area, Spain, South France, Italy.

29. *Rubus discolor.* W. and Nees. Rub. Germ., p. 46, pl. 20.
Common everywhere. Area, Central and Southern Europe, North Africa, Canaries.

30. *Geum heterocarpum.* Boiss. Voy. Esp., p. 201, pl. 63.
Galilee, Lebanon, and Anti-Lebanon. Area, mountains of South Spain and East Algeria.

31. *Geum urbanum.* L. Sp. 716.
Woods in northern districts. Area, Europe, Siberia, North Africa, West Himalayas.

32. *Potentilla libanotica.* Boiss. Diagn. Ser. i., iii., p. 4.

Rocks in Lebanon, up to 8,000 feet. P.

33. *Potentilla anserina.* L. Sp. 710.

Waysides, Lebanon. Area, Europe, Siberia, North America.

34. *Potentilla hirta.* L. Sp. 712.

Hilly districts, especially in the north. Area, Mediterranean region.

35. *Potentilla supina.* L. Sp. 711.

Coast and plains. Area, Central and Southern Europe, Siberia, India, Africa (tropical).

36. *Potentilla argentea.* L. Sp. 712.

Northern mountains. Area, Europe, Siberia.

37. *Potentilla granioïdes.* Willd. Sp. ii., p. 1101.

Dry plains in sub-alpine Lebanon and Anti-Lebanon.

38. *Potentilla reptans.* L. Sp. 714.

General. Area, Europe, West Siberia, North Africa, Abyssinia.

39. *Agrimonia eupatoria.* L. Sp. 643.

Coast and mountain-districts. Area, Europe, Siberia, North Africa, Canaries, Madeira.

40. *Alchunilla arvensis.* (L. Sp. 179.)

Northern coast. Area, Central and Southern Europe, North America.

41. *Poterium villosum.* Sibth. Prodr. Fl. Gr. ii., p. 238.

Eastern deserts, east of Gilead, Hauran.

42. *Poterium compactum.* Boiss. Diagn. Ser. i., x., p. 7.

Sub-alpine Lebanon. P.

43. *Poterium gaillardoti.* Boiss. Diagn. Ser. ii., ii., p. 52.

Southern slopes of Lebanon, etc. ; forest of Gilead.

44. *Poterium muricatum.* Spach. Ann. Sc. N.; 1846, p. 36.

Lebanon and Galilee. Area, Central and Southern Europe.

45. *Poterium verrucosum.* Ehr. Ind. Hort. Berol., 1829.

Coast and Jordan valley. Area, Canaries, Madeira, Spain, North Africa.

Ditto, var. *Microcarpum.* Beyrout and thence inland.

46. *Poterium spinosum.* L. Sp. 1411.

Universal. Area, Sardinia, Sicily, Calabria, Dalmatia.

47. *Neurada procumbens.* L. Sp. 631.

Wilderness of Judæa and round the Dead Sea. Area, Sahara, Scinde.

ORDER XLII., MYRTACEÆ.

1. *Myrtus communis.* L. Sp. 673. Myrtle. Heb., הֲדַס.

General, chiefly in maritime and northern districts. Area, Mediterranean region.

ORDER XLIII., GRANATEÆ.

1. *Punica granatum.* L. Sp. 676. Pomegranate. Heb., רמון.

Cultivated everywhere. Apparently indigenous in Gilead. Area, North-west India. Cultivated in South Europe, and subspontaneous.

ORDER XLIV., LYTHRARIEÆ.

1. *Lythrum salicaria.* L. Sp. 640. Var. *Syriacum.* Boiss. Loosestrife.

Northern valleys. Area, Temperate Northern Regions, Australia.

2. Ditto, var. *Tomentosum.* De Cand.

General. Area, Europe, Siberia, North Africa, North America, Australia.

3. *Lythrum hyssopifolium.* L. Sp. 642.

Wet and saline spots, littoral districts. Area, Central and Southern Europe, Siberia, Africa, North and South America.

4. *Lythrum græfferi.* Ten. Fl. Nap. Prodr. Sup. ii., p. 27 = *L. flexuosum.* Lag.

Central districts, Gilead, and Jordan valley. Area, Mediterranean region, Azores, Canaries, Madeira.

5. *Lawsonia alba.* Lam. Dict. i., p. 106. Henna. Heb., כֹּפֶר. Arab., حنع., *Henna.*

Engedi. Area, India.

ORDER XLV., ONAGRARIÆ.

1. *Epilobium angustifolium.* L. Sp. 493. Rose Bay, or Willow-herb.

Lebanon. Area, Europe, Siberia, North America.

2. *Epilobium hirsutum.* L. Sp. 494. Codlins and Cream. And var. *Tomentosum.* Vent.

Coast and mountain regions. Area, Europe, Siberia, North Africa, Abyssinia, the Cape.

3. *Epilobium parviflorum.* Schreb. Spic., p. 146.

Coast and mountain regions. Area, Europe, North Africa, Madeira, Canaries.

4. *Epilobium tetragonum.* L. Sp. 494.

Littoral districts. Area, Europe, Siberia, North America, North Africa, the Cape, Canaries.

5. *Epilobium tournefortii.* Michal. Bull. Soc. Bot. Fr., 1855, p. 731.

Moist places, Northern Palestine. Area, Mediterranean region.

6. *Epilobium roseum.* Schreb. Spic. Lips. 147. Var. *Subsessile.*

Lebanon. Area, Northern and Central Europe, Siberia.

7. *Jussiæa repens.* L. Mant. 381.

Upper Jordan valley, Huleh, and Tiberias. Area, Algeria, Senegal, Nubia, the Cape, Tropical Asia, Tropical America.

8. *Jussiæa angustifolia.* Lam. Dict. iii., p. 331.

Lake Huleh.

ORDER XLVI., HALORAGEÆ.

1. *Myriophyllum spicatum.* L. Sp. 1410. Water Milfoil.

Upper Jordan valley and central districts. Area, Europe, Siberia, North Africa, North America, the Cape.

2. *Callitriche vernalis.* Kütz. Linn. vii., p. 175. Water Startwort.

In standing water, general. Area, Europe, North Africa, Siberia, North America.

ORDER XLVII., CUCURBITACEÆ.

1. *Momordica balsamina.* L. Sp. 1453.

Nooks by the east side of Dead Sea. Area, Arabia, India, tropical Africa.

2. *Cucumis prophetarum.* L. Sp. 1436.

Round the Dead Sea.

3. *Cucumis trigonus.* Roxb. Fl. Ind. iii., p. 722.

Plain of Huleh. Area, India.

4. *Cucumis sativus.* L. Sp. 1437. Heb., קִשֻּׁאִים. Cucumber.

5. *Cucumis melo.* L. Sp. 1436. Heb., אֲבַטִּחִים. Melon.

6. *Cucumis dudain.* L. Sp. 1437.

7. *Cucumis chate.* L. Sp. 1437. Hairy Cucumber.

8. *Cucurbita citrullus.* L. Sp. 1435. Water Melon.

All five cultivated, but often found growing spontaneously.

9. *Citrullus colocynthis.* (L. Sp. 1433.) Colocynth.

Maritime plains and Jordan valley, round the Dead Sea. Area, North Africa, South Spain, Nubia, Arabia, the Cape, India.

10. *Ecballium elaterium.* (L. Sp. 1434.)

Very common in sandy plains. Area, Mediterranean region.

11. *Bryonia multiflora.* Boiss. Diagn. Ser. i., x., p. 8. General.

12. *Bryonia syriaca.* Boiss. Diagn. Ser. ii., ii., p. 59. Everywhere in thickets and brushwood.

13. *Bryonia dioica.* Jacq. Austr. ii., p. 59. Bryony. Area, Central and Southern Europe, North Africa.

ORDER XLVIII., DATISCEÆ.

1. *Datisca cannabina.* L. Sp. 1460. Sides of mountain streams. Area, the Himalayas.

ORDER XLIX., FICOIDEÆ.

1. *Mesembryanthemum crystallinum.* L. Sp. 688. Near Askalon. Area, Mediterranean shores, Canaries.

2. *Mesembryanthemum nodiflorum.* L. Sp. 687. Mount Carmel. Area, Mediterranean coasts, South Africa, Canaries.

3. *Mesembryanthemum forskahlei.* Hochst. in Schimp. Pl. Ar. Exs., ed. 2. South end of Dead Sea.

4. *Aizoon hispanicum.* L. Sp. 700. Judæan wilderness, towards Dead Sea. Area, Mediterranean region.

5. *Aizoon canariense.* L. Sp. 700. Wâdys south-west and south-east of Dead Sea. Area, deserts, North Africa, Canaries, Madeira, North-west India.

ORDER L., CRASSULACEÆ.

1. *Tillæa trichopoda.* Fenz. in Ky. Pl. Pers. Austr. Maritime districts.

2. *Umbilicus erectus.* De Cand. Fl. Fr. iv., p. 384. Under Lebanon and Hermon, south.

3. *Umbilicus pendulinus.* De Cand. Pl. Grass., pl. 156. Penny-wort, Navelwort.

Hill-districts, on rocks. Area, Central and Southern Europe, Algeria.

4. *Umbilicus intermedius.* Boiss. Fl. Or. ii., p. 769.

Mount Tabor, Hebron, and intermediate hill-region.

5. *Umbilicus libanoticus.* (Labill. De Cand. iii., p. 3, pl. 1.)

General, on rocks, in hilly districts, Lebanon, etc. P.

Ditto, var. *Glaber.* Alpine Lebanon.

6. *Umbilicus globulariæfolius.* Fenz. Pug., No. 52.

Rocks near the coast.

7. *Umbilicus lineatus.* Boiss. Diagn. Ser. i., x., p. 14.

Cliffs overhanging Lake of Galilee. P.

8. *Umbilicus horizontalis.* (Guss. Syn. i , p. 513.)

On rocks.

9. *Sedum album.* L. Sp. 619.

Lebanon, etc. Area, Europe, Siberia, North Africa.

10. *Sedum laconicum.* Boiss. Diagn. Ser. i., vi., p. 55.

Hills about Nazareth and Samaria, Banias, etc.

11. *Sedum altissimum.* Poir. Dict. iv., p. 634.

Hills about Jerusalem. Area, Mediterranean region.

12. *Sedum amplexicaule.* De Cand. Rapp. ii., p. 80.

Lebanon and Anti-Lebanon. Area, Mediterranean region on mountains.

13. *Sedum glaucum.* W. K. Pl. Rar. Hung., p. 198, pl. 181. Var. *Eriocarpum.* Sibth.

Everywhere in mountain, rocky districts. Area, central and eastern mountain ranges of Europe.

14. Ditto, var. *Polypetalum.* Boiss.

On the coast.

15. *Sedum littoreum.* Guss. Pl. Rar., p. 185, pl. 37.

Maritime districts. Area, Sardinia, South Italy, Sicily.

16. *Sedum palæstinum.* Boiss. Diagn. Ser. i., x., p. 18.

Mount Carmel, Lebanon above the Cedars, Gilead, Moab. P.

17. *Sedum cespitosum.* (Cav. Icon., pl. 69, f. 2.)

Sandy and rocky districts, Northern Palestine. Area, Mediterranean coast region.

18. *Sedum rubens.* L. Sp. 619.

Lebanon. Area, Central and Southern Europe, North Africa, Canaries.

19. *Telmissa microcarpa.* (Sm. Prodr. Fl. Gr. i., p. 217.)

Hills about Jerusalem, South Judæa, Moab.

ORDER LI.,* CACTACEÆ.

1. *Opuntia vulgaris.* Mill. Dict., ed. 8, No. 1. (*Cactus Opuntia.* L. Sp. 669.) Prickly Pear.

Overruns the whole lower country in sandy places, being used as a hedge for enclosures, and is one of the most conspicuous features in the flora of modern Palestine. It is of American origin, but it is now naturalized in all the Mediterranean region.

2. *Opuntia cochinellifera.* (L. Sp. 670.)

This, the Cochineal Cactus, on which lives the female cochineal insect, which, when dried, forms the cochineal dye, has also become naturalized, especially on Mount Ebal, and in the neighbourhood of Samaria. It is a native of tropical America.

ORDER LI., DROSERACEÆ.

1. *Drosera rotundifolia.* L. Sp. 402. Sundew.

Area, Northern and Central Europe.

ORDER LII., SAXIFRAGACEÆ.

1. *Saxifraga tridactylites.* L. Sp. 578.

Galilee, Lebanon, Anti-Lebanon. Area, Europe, North Africa, Siberia.

2. *Saxifraga hederacea.* L. Sp. 579.
Moist rocks in the hill district, chiefly in the north. Area, Sicily.

3. *Ribes orientale.* Poir. Encycl. Suppl. ii., p. 856.
Lebanon and Hermon. Area, Western Himalayas.

ORDER LIII., HAMAMELIDEÆ.

1. *Liquidambar orientalis.* Mill. Dict., No. 2.
On warm hill-sides. Probably introduced. Area, India.

ORDER LIV., UMBELLIFERÆ.

1. *Hydrocotyle natans.* Cyr. Pl. Rar. Neap. i., p. 20, pl. 6, f. B.
Lake Huleh. Area, South Italy, Sardinia, Sicily, Abyssinia, America.

2. *Eryngium barrelieri.* Boiss. Ann. Sc. Nat. 1844, p. 125.
Marshes near the coast. Area, South Italy, Sardinia, Sicily.

3. *Eryngium glomeratum.* Lam. Dict. iv., p. 755.
Generally distributed.

4. *Eryngium billardieri.* Laroch. Er., p. 25, pl. 2.
Sub-alpine and alpine Lebanon and Anti-Lebanon.

5. *Eryngium heildreichii.* Boiss. Diagn. Ser. i., x., p. 20.
Alpine Lebanon and Hermon, 7,000—9,000 feet.

6. *Eryngium tricuspidatum.* L. Sp. 337.
Locality uncertain. Area, South Spain, Sardinia, Sicily, North Africa.

7. *Eryngium falcatum.* Laroch. Eryng., p. 40, pl. 13.
Dry hills, Lebanon to Carmel.

8. *Eryngium creticum.* Lam. Dict. iv., p. 754.
Coast, plains, and Jordan valley, universal. Area, Dalmatia.

9. *Actinolema eryngioides.* Fenz. Pug. Pl. Taur., No. 53.
Barren plains in the north.

10. *Lagœcia cuminoides.* L. Sp. 294.

General in the plains, Mount Gilead. Area, South Spain.

11. *Bupleurum protractum.* Link. Fl. Portug. ii., p. 387. And var. *Heterophyllum.* Link.

General, east and west. Area, South Europe, North Africa, Madeira.

12. *Bupleurum odontites.* De Cand. Prodr. iv., p. 129.

General, plains and hills. Area, Mediterranean region.

13. *Bupleurum croceum.* Fenz. Pug., No. 55. In fields.

14. *Bupleurum dichotomum.* Boiss. Diagn. Ser. i., x., p. 28.

15. *Bupleurum nodiflorum.* Sibth. Fl. Gr., pl. 260.

Universal and common.

16. *Bupleurum brevicaule.* Schlecht. in Linn. xvii., pl. 124.

Galilee, Lebanon, Anti-Lebanon.

17. *Bupleurum glaucum.* De Cand. Suppl. Fl. Fr., p. 515.

The Bukâà. Area, Italy, North Africa, Canaries.

18. *Bupleurum gerardi.* Jacq. Austr. iii., p. 31, pl. 256.

Northern plains. Area, South Europe.

19. *Bupleurum libanoticum.* Boiss. Diagn. Ser. ii., ii., p. 82.

Lebanon. P.

20. *Bupleurum irregulare.* Boiss. et Ky. in Pl. Exs., 1855.

Alpine and sub-alpine Lebanon. P.

21. *Bupleurum junceum.* L. Sp. 342.

Northern hills. Area, Central Europe, Siberia.

22. *Bupleurum trichopodum.* Boiss. Ann. Sci. Nat., 1844, p. 145.

Northern Lebanon.

23. *Bupleurum fruticosum.* L. Sp. 343.

Lower slopes of Lebanon. Area, Portugal and Mediterranean region.

24. *Apium graveolens.* L. Sp. 379. Celery.

Marshy places, maritime. Area, Europe, North Africa, Canaries, the Cape, America.

25. *Helosciadium nodiflorum.* (L. Sp. 362.)

Jordan valley, Fountain of Moses, under Nebo. Area, Europe, North Africa, Abyssinia.

26. *Petroselinum sativum.* Hoffm. Umb. Gen. i., p. 78. Parsley.

Lebanon. Area, South Europe, from cultivation.

27. *Ridolfia segetum.* Moris. Fl. Sardin. ii., p. 212.

Maritime plains. Area, Portugal and Mediterranean region.

28. *Deverra tortuosa.* (Desf. Abl. i., p. 357, pl. 73.)

Southern Desert.

29. *Pimpinella anisum.* L. Sp. 399. Aniseed.

Common. Probably introduced. Cultivated in South Europe.

30. *Pimpinella cretica.* Poir. Suppl. i., p. 684. General.

31. Ditto, var. *Arabica.* Southern Desert.

32. *Pimpinella eriocarpa.* Russ. Aless. ii., p. 249.

Judæan wilderness, towards Dead Sea.

33. *Pimpinella peregrina.* L. Mant. 357.

General. Area, South France, Italy, Dalmatia.

34. *Pimpinella corymbosa.* Boiss. Ann. Sc. Nat., 1844, p. 131.

Hills in the north.

35. *Pimpinella tragium.* Vill. Dauph. ii., p. 606.

Lebanon, 6,000 feet.

36. Ditto, var. *Depauperata* = *P. triradiata.* Boiss.

Lebanon, 7,000—9,000 feet.

37. *Pimpinella saxifraga.* L. Sp. 378. Burnet Saxifrage.

Area, Northern and Central Europe, Siberia to Dahuria.

38. *Scaligeria cretica.* (Urv. Enum., p. 31.)

Carmel, Tabor, Lower Lebanon. Area, Dalmatia.

39. *Scaligeria meifolia.* (Fenz. Fl., 1843, p. 458.) = *Butinia libanotica.* Boiss.

Lebanon, Anti-Lebanon, Bukâå.

40. *Carum elegans.* Fenz. Pug., No. 54.

Northern plains and mountains.

41. *Carum ferulæfolium.* (Desf. Coroll., p. 55, pl. 43.)

Coast and northern plains. Area, Istria, Dalmatia.

42. *Carum pestalozzæ.* Boiss. Diagn. Ser. i., x., 24. And vars.

Alpine Lebanon and Anti-Lebanon.

43. *Berula angustifolia.* (L. Sp. 1672.) Water Parsnip.

Huleh and elsewhere, in water. Area, Europe, North America, Australia.

44. *Ammi copticum.* L. Mant. 56.

Plains near Jordan.

45. *Ammi majus.* L. Gen. 334.

General. Area, Mediterranean region, West France, Canaries, Abyssinia.

46. *Ammi visnaga.* (L. Sp. 348.)

Fields and plains, and Jordan valley. Area, Mediterranean region.

47. *Falcaria rivini.* Host. Austr. i., 381.

Wayside on hills. Area, Central Europe, Siberia.

48. *Sison exaltatum.* Boiss. Diagn. Ser. i., x., p. 21.

Lower parts of Lebanon. P.

49. *Chærophyllum syriacum.* Hemp. and Ehr. Boiss. Fl. Or. ii., 904.

Lebanon. P.

50. *Chærophyllum libanoticum.* Boiss. Diagn. Ser. ii., vi., p. 89.

Lebanon, 4,600—6,000 feet.

51. *Physocaulos nodosus.* (L. Sp. 369.)

Mount Carmel, Samaria. Area, Southern Europe, North Africa.

52. *Anthriscus nemorosa.* M. B. Taur. Cauc. iii., p. 237.
Wooded districts, Lebanon, to 8,000 feet. Area, Silesia, Croatia, Servia.

53. *Anthriscus lamprocarpa.* Boiss. Ann. Sc. Nat., 1844, p. 59.
Shady rocky places in hill districts from Jerusalem northwards.

54. *Anthriscus.* Sp.? Jordan valley.

55. *Anthriscus sylvestris.* (L. Sp. 369.)
Area, Europe (except extreme south), Siberia.

56. *Scandix pecten-veneris.* L. Sp. 368. Shepherd's Needle.
In fields, Northern Palestine. Area, Central and Southern Europe, North Africa.

57. *Scandix iberica.* M. B. Taur. Cauc. i., p. 425, iii., p. 236.
Cornfields, coast and north, and Gilead.

58. *Scandix pinnatifida.* Vent. Hort. Cels., pl. 14.
Hills of Central Palestine, Lebanon and Anti-Lebanon, to 6,500 feet. Area, South-east Spain, North Africa, South of the Atlas.

59. *Scandix grandiflora.* L. Sp. 369.
Base of Lebanon. Area, Dalmatia.

60. *Scandix palæstina.* Boiss. Fl. Or. ii., p. 918.
Alluvial plains of Esdraelon and Hattin. P.

61. *Astoma seselifolium.* De Cand. Mém. v., p. 71, pl. 17.
Bare hills, Southern Palestine and Lebanon.

62. *Coriandrum sativum* L. Sp. 567. Coriander. Heb. גַּד.
In all cultivated ground and Jordan valley. Area, Southern Europe, probably introduced, Africa, Arabia.

63. *Coriandram tordylioides.* Boiss. Fl. Or. ii., p. 921.
Uncultivated plains in the north, the Bukâà.

64. *Bifora testiculata.* (L. Sp. 367.) Var. *Microcarpa.*
Near Jerusalem, Moab. Area, Mediterranean region.

65. *Conium maculatum.* L. Sp. 349. Hemlock.
Gilead, etc. Area, Europe, North Africa, Abyssinia, Siberia.

66. *Physospermum aquilegifolium.* (All. Ped. No. 1392, pl. 63.)
Lebanon. Area, Portugal, Spain, North and Central Italy.

67. *Smyrnium connatum.* Boiss. and Ky. Ins. Cyp., p. 309. =*S. ægyptiacum.*
Lower and sub-alpine Lebanon.

68. *Smyrnium olusatrum.* L. Sp. 376. Alexanders.
Maritime, mountain, and northern districts. Area, Central and Southern Europe, North Africa, Canaries.

69. *Smyrnium perfoliatum.* L. Sp. 376.
Area, South Europe, Danubian region.

70. *Smyrniopsis syriaca.* Boiss. Diagn. Ser. i., x., p. 50.
Lower valleys of the Lebanon and Anti-Lebanon. P.

71. *Lecokia cretica.* (Lam. Dict. i., p. 259.) Sub-alpine Lebanon.

72. *Hippomarathrum boissieri.* Reut. Apud. Boiss. Fl. Or. ii., p. 933.
Everywhere except the Ghor.

73. *Cachrys goniocarpa.* Boiss. Diagn. Ser. i., x., p. 53.
Plain of Philistia.

74. *Prangos asperula.* Boiss. Diagn. Ser. i., x., p. 54.
Galilee, Phœnicia, Lebanon. P.

75. *Prangos hermonis.* Boiss. Fl. Or. ii., p. 943.
Ravines on east of Hermon, 5,000 feet. P.

76. *Colladonia microcarpa.* Boiss. Fl. Or. ii., p. 946.

77. *Colladonia anisoptera.* Boiss. Fl. Or. ii., 946.
All the hills of Central Palestine, Anti-Lebanon. P.

78. *Colladonia crenata.* (Fenz. Flor. 1843, p. 460.)
Vineyards and thickets of Lebanon and Anti-Lebanon to 4,500 feet.

79. *Athamantha cretensis.* L. Sp. 352.
Area, alpine and sub-alpine, Central Europe.

80. *Pycnocycla tomentosa.* Decaisn. Ann. Sc. Nat. ii., 719.
Wâdy Ghurundel, south-east of Dead Sea.

81. *Œnanthe media.* Griesb. Spic. i., p. 352.

Lake Huleh, marshes under Lebanon. Area, Hungary, Servia, Dalmatia.

82. *Œnanthe prolifera.* L. Sp. 365.

General in moist or grassy plains.

83. *Fœniculum officinale.* All. Ped. ii., p. 25. Fennel.

Plain above Lake Huleh. Area, Central and Southern Europe, North Africa.

84. *Fœniculum piperitum.* De Cand. Prodr. iv., p. 142.

Lebanon and hills of Galilee. Area, Mediterranean region.

85. *Kundmannia syriaca.* Boiss. Diagn. Ser. i., x., p. 31.

In fissures of rocks.

86. *Siler trilobum.* (L. Sp. 357.)

Wooded parts of Lebanon. Area, Central Europe, Danube region.

87. *Ferula hermonis.* Boiss. Fl. Or. ii., p. 985. Arab., ثلاوة, *Selloa.*

Hermon, 8,000 feet. P.

88. *Ferula communis.* (L. Sp. 355.) = *F. nodiflora.* Sibth.

Mount Tabor and northern plains. Area, South Europe, North Africa, Canaries.

89. *Ferula tingitana.* L. Sp. 355. = *F. sancta.* Boiss.

Plain of Esdraelon. Area, North Africa.

90. *Ferrulago cassia.* Boiss. Diagn. Ser. i., x., p. 38.

Base of the northern hills.

91. *Ferrulago syriaca.* Boiss. Diagn. Ser. i., x., p. 38.

Judæan hills, Lebanon, Anti-Lebanon.

92. *Ferrulago frigida.* Boiss. Diagn. Ser. i., x., p. 39. = *F. trachycarpa.* Var. *Libanotica.* Boiss.

Lebanon and Hermon, 6,500—9,000 feet. P.

93. *Johrenia dichotoma.* De Cand. Prodr. iv., p. 196.

Hills and mountain region, Northern Palestine.

94. *Johrenia juncea.* Boiss. Diagn. Ser. ii., x., p. 33.
Barren ground below Hermon. P.

95. *Peucedanum depauperatum.* Boiss. Diagn. Ser. ii., v., p. 99.
District round Lebanon and Hermon.

96. Ditto, var. *Alpinum.* Alpine region of Lebanon and Hermon.

97. *Anethum graveolens.* L. Sp. 377. Dill.
Plain of Sharon. Area, South Europe.

98. *Krubera peregrina.* (L. Mant. 55.)
Central Palestine, Samaria, etc. Area, Mediterranean region, Portugal, Madeira.

99. *Tordylium lanatum.* (Boiss. Ann. Sc. Nat., 1844, p. 347.)
Mount Tabor.

100. *Tordylium ægyptiacum.* (L. Amœn. iv., p. 270.)
Universally distributed.

101. *Tordylium syriacum.* L. Sp. 345.
Maritime and northern districts.

102. *Tordylium apulum.* L. Sp. 345.
Wâdys in Gilead. Area, Mediterranean region.

103. *Ainsworthia cordata.* (L. Pl. Suppl. 179.)
Kanobin valley, Lebanon.

104. *Ainsworthia trachycarpa.* Boiss. Diagn. Ser. i., x., p. 43.
The hill-country, north and south, Moab.

105. *Ainsworthia carmeli.* Boiss. Diagn. Ser. i., x., p. 44.
Mount Carmel, Moab. P.

106. *Zozimia absinthifolia.* (Vent. Choix., pl. 22.)
Near Gaza.

107. *Heracleum humile.* Sibth. Prodr. i., p. 193.
Lebanon and Hermon, 7,000—9,000 feet.

108. *Synelcosciadium carmeli.* (Labill. Déc. Syr. v., p. 3, pl. 1.)

Among bushes in Central and Northern Palestine, Shechem, Carmel, Tiberias, Sidon, Hasbeya, etc. P.

109. *Malabaila sekakul.* Russ. Alepp. ii., p. 249. Arab., سكاكول, *Sekakul.*

Shrubby hills and fields through the whole country, including Jordan valley.

110. Ditto, var. *Platycarpa.* Lebanon.

111. *Pastinaca teretiuscula.* Boiss. Fl. Or. ii., p. 1060.

Lebanon, near Hazroun.

112. *Opoponax orientale.* Boiss. Ann. Sc. Nat., 1844, p. 330.

On the plains.

113. *Opoponax chironium.* Koch. Umb. 96.

Area, South Europe.

114. *Laserpitium glabrum.* Crantz. Austr. iii., p. 54.

115. *Exoacantha heterophylla.* Labill. Déc. i., p. 10.

In fertile ground throughout the country.

116. *Artedia squamata.* L. Sp. 347.

Hills and fields in sunny places, general, Jordan valley.

117. *Orlaya grandiflora.* (L. Sp. 346.)

Plains of Moab. Area, Central and Southern Europe, North Africa,

118. *Orlaya platycarpus.* (L. Sp. 347.)

Maritime and mountain regions. Area, Mediterranean region.

119. *Orlaya maritima.* Gou. Hort. Monsp. 135.

Southern deserts, near Gaza. Area, Mediterranean region.

120. *Daucus broteri.* Ten. Syll., p. 591.

Maritime districts. Area, South Italy.

Ditto, var. *Bicolor.* Mount Carmel.

121. *Daucus littoralis.* Sibth. Fl. Gr. iii., p. 65, sub. 272.

Phœnician plains.

122. Ditto, var. *Forskahlii.* Maritime and inland districts alike.

123. *Daucus setulosus.* Guss. in De Cand. Prodr. iv., p. 211. Var. *Brachylænus.* (=*D. guttatus.* Sibth.)

Northern maritime plains. Area, South Italy, Dalmatia, Servia.

124. *Daucus aureus.* Desf. Abl. i., p. 242, pl. 61.

Coast and inland fertile districts. Area, Portugal, South Spain, South Italy, North Africa.

125. *Daucus carota.* L. Sp. 348. Carrot.

General. Area, Europe, Siberia, North Africa, Abyssinia, Madeira, Canaries.

126. *Daucus maximus.* Desf. Atl. i., p. 241.

Galilee and Anti-Lebanon. Area, Mediterranean region.

127. *Daucus blanchei.* Reut. Hort. Genev., 1867.

Lower Lebanon. P.

128. *Chætosciadium trichospermum.* (L. Mant. 57.)

Rocky shady places, Jerusalem, Carmel, Sidon, etc. Ravines of Callirrhoe, Moab.

129. *Cuminum cyminum.* L. Sp. 365. Cummin. Heb. כַּמֹּן.

Generally cultivated.

130. *Turgeniopsis fœniculacea.* (Fenz. Pug., No. 51.)

Lebanon and Anti-Lebanon.

131. *Torilis anthriscus.* (L. Sp. 346.) Hedge Parsley.

Hills of Gilead and Moab. Area, Europe, North Africa.

132. *Torilis nodosa.* L. Sp. 346.

Area, Central and Southern Europe, North Africa.

133. *Torilis fallax.* Boiss. Fl. Or. ii., p. 1086. =*T. chrysocarpa.*

134. *Torilis infesta.* (L. Syst. ii., 732.)

General in hedges, brushwood, and waste land. Area, Central and Southern Europe, North Africa.

135. *Torilis heterophylla.* Guss. Prodr. i., p. 326.

Round Lebanon and Anti Lebanon. Area, Mediterranean region.

136. *Torilis neglecta.* Rœm. et Schult. Syst. vi., p. 484. = *T. syriaca.* Boiss.

Galilee, Lebanon, and Hermon. Area, Spain, Canaries, Madeira, Austria, North Italy, Danube region.

137. *Caucalis tenella.* Del. Fl. Eg., p. 58, pl. 21, f. 3.

Jerusalem, and central and maritime region to the north.

138. *Caucalis leptophylla.* L. Sp. 347.

Universal. Area, Central and Southern Europe, North Africa, Canaries.

139. *Caucalis fallax.* Boiss. Fl. Or. ii., p. 1086.

Valleys and fields, Lebanon, and Phœnicia.

140. *Turgenia latifolia.* (L. Sp. 545). And varieties.

Universal in cultivated ground. Area, Central and Southern Europe, Siberia, North Africa.

141. *Lisæa syriaca.* Boiss. Ann. Sc. Nat., 1844, p. 55.

Central districts, Samaria, Nablus, Bukâà, etc.

ORDER LV., ARALIACEÆ.

1. *Hedera helix.* L. Sp. 292. Ivy.

Lebanon. Area, Central and Southern Europe, North India, Japan.

ORDER LVI., CORNACEÆ.

1. *Cornus australis.* C. A. Mey. Bull. Acad. Petersb., 1844, No. 72.

Lebanon, 4,500 feet.

2. *Cornus mas.* (L. Sp. 171.)

Bludan; probably introduced. Area, Central Europe.

SUB-CLASS, CALYCIFLORÆ, GAMOPETALÆ.

ORDER LVII., CAPRIFOLIACEÆ.

1. *Sambucus ebulus.* L. Sp. 385. Dwarf Elder.

Waste places on and under Lebanon. Area, Europe, North Africa, Madeira.

2. *Sambucus racemosa.* L. Sp. 386. *Non vidi.*

Area, South Europe.

3. *Viburnum tinus.* L. Sp. 383. Lauristinus.

The valleys under Lebanon, Mount Carmel, Tabor. Area, Mediterranean region.

4. *Lonicera etrusca.* Santi. Viag. i., p. 113, pl. 1. Italian Honeysuckle.

Through the whole of the hill districts east and west of Jordan, where there is brushwood, from Hebron to Lebanon.

Area, Mediterranean region, Portugal, Madeira.

5. Ditto, var. *Viscidula.* Boiss.

Anti-Lebanon and Bukâà.

6. *Lonicera nummularifolia.* Jaub. and Sp. Ill. Or. i., p. 133.

Lebanon, Anti-Lebanon, Hauran.

7. *Lonicera implexa.* Ait. Kew, i., p. 131.

Banias, south of Hermon. Area, Southern Europe.

8. *Lonicera caprifolium.* L. Sp. 246.

Lebanon. Area, Central and Southern Europe, Danubian region.

ORDER LVIII., RUBIACEÆ.

1. *Oldenlandia capensis.* Thaub. Fl. Cap. i., p. 507.

Mouth of Nahr el Kelb, Beyrout (Boissier). Area, South and Tropical Africa.

2. *Putoria calabrica.* (L. Fil. Suppl. 120).

Rocky places, Jerusalem to Lebanon. Area, South Italy, Sicily, North Africa.

3. *Rubia aucheri.* Boiss. Diagn. Ser. i., iii., p. 54.

Wooded and higher parts of Lebanon, Mount Carmel.

4. *Rubia tinctorum.* L. Sp. 158. Madder.

Galilee, Lebanon, Anti-Lebanon. Area, South Europe.

5. *Rubia olivieri.* A. Rich. Mém. Soc. Nat. Par. v., p. 132.

6. Var. *Elliptica=R. brachypoda.* Boiss.

Hill districts east and west of Jordan, from Jerusalem to Lebanon and Anti-Lebanon.

7. Var. *Stenophylla=R. doniettii.* Griseb.

Anti-Lebanon and Bukâà.

8. *Rubia peregrina.* L. Sp. 158.

Jordan valley, upper part. Area, West and South Europe.

9. *Sherardia arvensis.* L. Sp. 149. Field Madder.

Coast, Gennesaret, Upper Jordan valley. Area, Europe, North Africa, Canaries.

10. *Crucianella latiflora.* L. Sp. 158.

Coast, Carmel, Lebanon. Area, Mediterranean region.

11. *Crucianella imbricata.* Boiss. Diagn. Ser. i., x., p. 59.

12. *Crucianella macrostachya.* Boiss. Diagn. Ser. i., iii., p. 27.

Coast, hills and north. Ditto, var. *hispidula.* Boiss. Lebanon.

13. *Crucianella maritima.* L. Sp. 158.

On loose sands on the coast. Area, Mediterranean region.

14. *Asperula glomerata.* (M. B. Taur. Cauc. i., p. 107.) Var. *Capitata.* Labill. Alpine, Lebanon, and Anti-Lebanon.

15. *Asperula arvensis.* L. Sp. 150.

General. Area, Central and Southern Europe, North Africa.

16. *Asperula orientalis.* Boiss. in Ky. Pl. Exs. Jan. 1843.
Northern plains.

17. *Asperula setosa.* Jaubr. et Sp. Ill. Or. i., p. 152, pl. 82 B.
Lebanon, Anti-Lebanon, 6,000 feet.

18. *Asperula stricta.* Boiss. Diagn. Ser. i., iii., p. 33. Var. *Longifolia*=*A. fasciculata.* Boiss.
Lebanon, sub-alpine.
Ditto, var. *Alpina.* Alpine, Lebanon and Anti-Lebanon.

19. *Asperula libanotica.* Boiss. Diagn. Ser. i., iii., p. 59.
Alpine and sub-alpine, Lebanon and Anti-Lebanon.

20. *Asperula aparine.* M. B. Taur. Cauc., 102.
Banks of streams, Northern Palestine, Lebanon and Anti-Lebanon, Gilead. Area, Danubian region, South Russia.

21. *Asperula humifusa.* M. B. Taur. Cauc. Suppl., p. 105.
Plains in the north. Area, South Russia.
Ditto, var. *Pycnantha.* Boiss. Plains in the north.

22. *Asperula breviflora.* Boiss. Diagn. Ser. i., x., p. 63.
Rocky dry places, Lebanon, Anti-Lebanon, Bukâà. P.

23. *Asperula mollis.* Boiss. Diagn. Ser. i., iii., p. 31. *Non vidi.*

24. *Galium bocconi.* All. De. Cand. Prodr. iv., p. 594.
Pine forests of Gilead.

25. *Galium schlumbergeri.* Boiss. Fl. Or. iii., p. 51.
Mount Hermon. P.

26. *Galium pestalozzæ.* Boiss. Diagn. Ser. i., x., p. 65.
South side of Lebanon. P.

27. *Galium ehrenbergii.* Boiss. Fl. Or. iii., p. 53.
Jisr el Hajar, Lebanon. P.

28. *Galium orientale.* Boiss. Diagn. Ser. i., iii., p. 38.
Lebanon and Hermon to 9,000 feet.

29. Ditto, var. *Latifolium.* Boiss.
Hills about Nazareth.

30. *Galium aureum.* Vis. Ind. Ort. Bot. Padov., 1842, p. 134.

31. Var. *Scabrifolium.* Boiss. = *G. corrudæfolium.* Kosh.
Lebanon and Hermon to 7,000 feet, general.

32. Ditto, var. *Incurvum.* Sibth. = *G. melanantherum.* Boiss.
Eastern highlands, Moab.

33. *Galium verum.* L. Sp. 155. Lady's bedstraw.
General. Area, Europe, Siberia, North Africa, North-west India.

34. *Galium canum.* De Cand. Prodr. iv., p. 602.
General; on rocks in sunny places.

35. Ditto, var. *Musciforme.* Boiss. Hazrun, Lebanon.

36. *Galium jungermannioides.* Boiss. Diagn. Ser. i., iii., p. 43.
Alpine Lebanon and Hermon. P.

37. *Galium saccharatum.* All. Ped. i., p. 9.
Southern hills, Jerusalem, etc. Area, Central and Southern Europe, North Africa, Madeira, Canaries.

38. *Galium tricorne.* With. Brit. Ed. ii., p. 153.
Common; cultivated, and waste places. Area, Central and Southern Europe, North Africa, India.

39. *Galium aparine.* L. Sp. 157. Goosegrass, Cleavers.
Universal; fields and thickets. Area, Europe, North Africa, Siberia, North America.

40. *Galium spurium.* L. Sp. 154.
Northern plains. Area, Europe, Siberia, North Africa.

41. *Galium pisiferum.* Boiss. Diagn. Ser. i., x., p. 67.
Shady rocky places and thickets everywhere except Jordan valley.

42. *Galium adhærens.* Boiss. Diagn. Ser. ii., v., p. 106.
Lebanon valleys.

43. *Galium peplidifolium.* Boiss. Diagn. Ser. i., iii., p. 46.

Forest region of Lebanon.

44. *Galium divaricatum.* Lam. Dict. ii., p. 580.

Lebanon. Area, Southern Europe, Danubian region, Azores.

45. *Galium tenuissimum.* M. B. Taur. Cauc. i., p. 104.

Coast, and Lower Lebanon. Area, Danubian region.

46. *Galium nigricans.* Boiss. Diagn. Ser. i., iii., p. 48.

Vineyards and fields in the north.

47. *Galium judaicum.* Boiss. Diagn. Ser. i., x., p. 60.

Rocky places, Jerusalem to Lebanon and Anti-Lebanon. P.

48. *Galium hierosolymitanum.* L. Sp. 156. = *G. trachyanthum.* Boiss.

Every part of the country.

49. *Galium cassium.* Boiss. Diagn. Ser. i., x., p. 68.

Plains and Lower Lebanon.

50. *Galium murale.* (L. Sp. 1490.)

Coast and base of Lebanon and Hermon. Area, Mediterranean region, Madeira, Canaries.

51. *Galium coronatum.* Sibth. Prodr., p. 90, Fl. Gr., pl. 125.

Along the central range of Palestine from north to south, to 4,000 feet, hills of Gilead and Moab.

52. Ditto, var. *Stenophyllum.* Boiss. (= *G. persicum.* De Cand.) Under Anti-Lebanon.

53. *Galium articulatum.* (L. Hort. Upsal. 303.)

Coast and plains.

54. *Galium cruciatum.* (L. Sp. 1491.) Crosswort.

Hermon, 7,000 feet. Area, Northern and Central Europe, Siberia.

55. *Galium verticillatum.* Danth. Lam. Enc. Méth. ii., p. 585.

Mountain districts. Area, Mediterranean region.

56. *Vaillantia muralis.* L. Sp. 1490.

Lower Lebanon, Moab and Gilead. Area, Mediterranean region.

57. *Vaillantia hispida.* L. Sp. 1490.

Carmel, Gerizim, plain of Sharon, etc. Area, Mediterranean region, Nubia, Canaries.

58. *Mericarpæa vaillantioides.* Boiss. Diagn. Ser. i., iii., p. 52.

Jerusalem and central range of hills. Hills of Gilead and Moab.

59. *Callipeltis cucularia.* (L. Am. Acad. iv., p. 295.)

Universal on hills east and west of Jordan. Area, South Spain, North Africa.

ORDER LIX., VALERIANEÆ.

1. *Valeriana dioscoridis.* Sibth. Fl. Gr. i., p. 24, pl. 33.

Maritime and northern mountain region.

2. *Valeriana sisymbriifolia.* Desf. Choix. Tourn., p. 53, pl. 41.

Crest of mountains of Gilead and Moab.

3. *Centranthus ruber.* (L. Sp. 44.) Red Valerian.

On rocks, Lebanon. Area, Southern Europe.

4. *Centranthus longiflorus.* Stev. Obs. Pl. Ross. 76. Var. *Latifolius.* =*C. elatus.* Boiss.

Rocky places, Lebanon and Anti-Lebanon.

5. *Valerianella tuberculata.* Boiss. Diagn. Ser. i., iii., p. 59.

In bare places, alpine and sub-alpine of Anti-Lebanon. Area, Persia.

6. *Valerianella dactylophylla.* Boiss. Diagn. Ser. i., x., p. 75.

Round the base of Lebanon and Anti-Lebanon, Galilee.

7. *Valerianella diodon.* Boiss. Diagn. Ser. i., iii., p. 57.

Fields in the north.

8. *Valerianella szovitsiana.* F. and M. Ind. iii., Petrop., p. 48. = *Vaucheri.* Boiss.

Desert south-west of Dead Sea.

9. *Valerianella echinata.* (L. Sp. 47.)

Cultivated ground about Lebanon, Anti-Lebanon, and Galilee. Area, Mediterranean region.

10. *Valerianella orientalis.* (Sch echt. Linn. xvii., p. 126.)

Northern plain of the Bukâà.

11. *Valerianella truncata.* Rchb. Pl. Crit. ii., p. 7, pl. 115, f. 225.

Littoral and hill districts. Area, Spain, Sicily.

12. *Valerianella carinata.* Loisel. Not. 149.

Littoral districts and Lebanon. Area, Central and Southern Europe, North Africa.

13. *Valerianella morisoni.* (Spreng. Pug. i., p. 4.)

Higher parts of Mount Gilead. Area, Europe, North Africa, Canaries, Azores.

14. *Valerianella tridentata.* (Stev. Obs. in Mém. Morg., 1817, p. 346.)

The Hauran. Area, Mediterranean region.

15. *Valerianella coronata.* (Willd. Sp. i., p. 184.)

Lebanon, lower region, highland, and plains, Gilead and Moab. Area, Central and Southern Europe and North Africa.

16. *Valerianella kotschyi.* Boiss. Diagn. Ser. i., iii., p. 60.

Under Anti-Lebanon.

17. *Valerianella obtusiloba.* (Boiss. Diagn. Ser. i., iii., p. 59.)

Southern slopes of Lebanon.

18. *Valerianella vesicaria.* (Willd. Sp. i., p. 183.)

General in waste land, hills and plains, east and west. Area, Sicily.

19. *Valerianella soyeri.* Boiss. Diagn. Ser. i., x., p. 74.

Slopes of Mount Nebo.

ORDER LX., DIPSACEÆ.

1. *Morina persica.* L. Sp. 39.
Lower and sub-alpine Lebanon and Anti-Lebanon.

2. *Dipsacus sylvestris.* L. Sp. 141. Var. *Cornosus.* Wild Teasel.
Lebanon, near Dimas. Area, Central and Southern Europe, North Africa, Canaries.

3. *Cephalaria setosa.* Boiss. Diagn. Ser. i., ii., p. 107.
Lower slopes of Anti-Lebanon, Mount Gilead.

4. *Cephalaria joppensis.* (Sprengl. Syst. i., p. 1378.)
Littoral of Palestine.

5. *Cephalaria syriaca.* (L. Sp. 141.) And var. *Boissieri.* Reut.
General. Area, South Spain, South France, North Africa.

6. *Cephalaria ambrosioides.* (Sibth. Fl. Gr. ii., p. 5, pl. 103.)
Near Nazareth.

7. *Cephalaria tenella.* Paine. Pal. Expl. Soc., pt. iii., p. 108.
Summit of Mount Gilead. P.

8. *Cephalaria stellipilis.* Boiss. Diagn. Ser. i., x., p. 76.
Lower slopes, Lebanon and Hermon.

9. *Knautia hybrida.* (All. Auct., p. 9.) Field Scabious.
General. Area, Central and Southern Europe.

10. *Knautia arvensis.* L. Sp. 143.
Non vidi. Area, Europe, Siberia.

11. *Scabiosa ochroleuca.* L. Sp. 146.
Higher parts of Lebanon. Area, Central Europe, Siberia.

12. *Scabiosa arenaria.* Forsk. Fl. Eg., p. 61. = *S. argentea.* L.
Southern Desert.

13. *Scabiosa ucranica.* L. Sp. 144.
Lower regions of Lebanon and Anti-Lebanon. Area, South Europe.

14. *Scabiosa prolifera.* L. Sp. 144.

Maritime districts, inland plains and hills. Extremely abundant.

15. *Scabiosa sicula.* L. Mant. ii, p. 196.

The Bukâà. Area, Spain.

16. *Scabiosa rotata.* M. B. Taur. Cauc. iii., p. 102.

17. *Scabiosa palæstina.* L. Mant. 37. And varieties.

Universal in grassy places.

18. Ditto, var. *Microcephala.* = *S. phrygia.* Boiss.

Lebanon and northern plains.

19. *Scabiosa aucheri.* Boiss. Diagn. Ser. i., ii., p. 111.

Jerusalem, northern plains and hills.

20. *Pterocephalus plumosus.* (L. Mant. 147.)

Universal in hilly districts.

21. *Pterocephalus involucratus.* Sibth. Fl. Gr. ii., p. 11, pl. 112.

Everywhere in dry, hilly places.

ORDER LXI., COMPOSITÆ.

SUB-ORDER, TUBULIFLORÆ.

1. *Eupatorium cannabinum.* L. Sp. 1173. Hemp Agrimony. Var. *Syriacum.* Jacq.

General in moist and shrubby places east and west of Jordan. Area, Europe, Siberia to Japan, North Africa.

2. *Solidago virga-aurea.* L. Sp. 1250. Golden Rod.

Lebanon and Anti-Lebanon. Area, Arctic and Temperate Europe and Asia, Himalayas, America.

3. *Erigeron alpinum.* L. Sp. 1211.

Alpine Lebanon. Area, mountains of Europe, Siberia and Arctic America, South Chili, Fuegia.

4. *Erigeron trilobum.* (Decaisn. Fl. Sin., p. 23.)

Arabah, south of Dead Sea.

5. *Erigeron linifolium.* Willd. Spec. iii., p. 1955. =*Conyza ambigua.* De Cand.

Maritime plains. Area, Mediterranean region, Madeira, Canaries, Cape Verde.

6. *Erigeron ægyptiacum.* (L. Mant., 112.)

The lower Jordan valley.

7. *Bellis perennis.* L. Sp. 1248. Daisy.

Lower slopes of Lebanon. Area, Central and Southern Europe, Madeira.

8. *Bellis sylvestris.* Cyrill. Pl. Rar. ii., p. 12, pl. 4.

Maritime region. Area, Portugal, Mediterranean region.

9. *Bellis annua.* L. Sp. 1249.

Phœnician plain. Area, Mediterranean region, Madeira.

10. *Asteriscus aquaticus.* (L. Sp. 1274.)

Ditches, maritime plains. Area, Mediterranean region, Canaries.

11. *Asteriscus pygmæus.* Coss. et Dur. Sert. Tunet, p. 26.

The Ghor, near Jericho. Area, Algerian and Tunisian Sahara.

12. *Asteriscus graveolens.* (Forsk. Descr., p. 151.)

Wâdy Ghurundel, south-east end of Dead Sea. Area, Sahara.

13. *Anvillæa garcini.* Burm. Fl. Ind., pl. 60, f. 1.

Wâdy Zuweirah, south-west of Dead Sea, Callirrhoe, Moab.

14. *Pallenis spinosa.* (L. Sp. 1274.)

Cultivated land and waysides, general. Area, Mediterranean region, Canaries.

15. *Chrysophthalmum montanum.* (De Cand. Prodr. vii., p. 287.)

16. *Postia lanuginosa.* (De Cand. Prodr. vii., p. 287.)

Rocks, el Bukâà, northern plains.

17. *Inula crithmoides.* L. Sp. 1240. Golden Samphire.

Maritime rocks and marshes, and all round the Dead Sea. Area, Mediterranean and West European coasts, from Britain southwards.

41—2

18. *Inula heterolepis.* Boiss. Diagn. Ser. ii., iii., p. 12.

Southern Lebanon.

19. *Inula viscosa.* L. Sp. 1209.

Plains and central hill range, Jerusalem and Jordan valley, not mountain. Area, Mediterranean region, Madeira, Canaries.

20. *Inula graveolens.* (L. Sp. 1210.)

On the coast. Area, Mediterranean region.

21. *Pulicaria sicula.* (L. Sp. 1210.) On the coast. Area, Mediterranean region.

22. *Pulicaria odora.* (L. Sp. 1236.)

23. *Pulicaria dysenterica.* (L. Sp. 1237.) Fleabane. Var. *Microcephala = I. dentata.* Sibth. = *P. uliginosa.* De Cand.

Moist places, coast, Esdraelon, Moab. Area, Europe, North Africa.

24. *Pulicaria undulata.* (L. Mant., 115).

Shores of the Dead Sea.

25. *Pulicaria arabica.* Cass. Dict. 44, p. 94.

Ravines, Moab, Galilee.

26. *Francoeuria crispa.* (Forsk. Eg. Arab. Descr., p. 150.)

Wâdys south of Dead Sea. Area, Sahara, Tropical Arabia, North India, Canaries.

27. *Iphiona scabra.* De Cand. Ann. Nat. 1834, p. 263. Arab. دفري, *Dafra.*

Wâdys south of Dead Sea.

28. *Varthemia iphionoides.* Boiss. Diagn. Ser. ii., iii., p. 9.

Central hill-range throughout Palestine. P.

29. *Conyza dioscoridis.* Rauw. It., tab. 54.

Ravines down to the Dead Sea, maritime coast. Area, Nubia, Abyssinia, Tropical Arabia.

30. *Phagnalon rupestre.* (L. Mant., 113.)

Jerusalem, central hills, Tabor, hills round Dead Sea. Area, Mediterranean region Portugal, Canaries.

31. *Phagnalon kotschyi.* Schultz. Bip. in Ky. Pl. Exs., 1843.

Clefts in higher parts of Lebanon.

32. *Lasiopogon muscoides.* Desf. Atl. ii., p. 267, pl. 231. Arab. *Krescht igetti.*

Cliffs south of Dead Sea. Area, South-east Spain, North Africa, the Cape.

33. *Gnaphalium luteo-album.* L. Sp. 1196.

Universal in sandy places. Area, temperate and warmer regions of the world.

34. *Helichrysum siculum.* (Spreng. Syst. iii., p. 476.)

Littoral region. Area, Sicily, South Italy.

Ditto, var. *Brachyphyllum.* Boiss.

Rocks near Beyrout.

35. *Helichrysum pallasii.* (Spreng. Syst. iii., p. 470.)

Lebanon, 4,000 feet.

36. *Helichrysum plicatum.* De Cand. Prodr. vi., p. 183. = *H. anatolicum.* Boiss.

Alpine and sub-alpine Lebanon.

37. *Helichrysum sanguineum.* (L. Amœn. Acad. iv., p. 78.) Everlasting. Universal.

38. *Helichrysum billardieri.* Boiss. Diagn. Ser. ii., v., p. 111.

Rocky places, Lebanon.

39. *Leyssera capillifolia.* (Willd. Mag. Nat. Ges., 1811, p. 160.)

Wâdy Zuweirah, south-west end of Dead Sea. Area, South-east Spain, North Africa, Nubian coast.

40. *Gymnarrhena micrantha.* Desf. Mém. Mus. iv., p. 1, pl. 1.

Wilderness of Judæa, Moab. Area, North Sahara.

41. *Micropus erectus.* L. Sp. 1313.

Lebanon and Anti-Lebanon. Area, Central and Southern Europe, North Africa.

42. *Micropus longifolius.* Boiss. Fl. Or. iii., p. 242.

Under Hermon.

43. *Micropus supinus.* L. Sp. 1311.

The Ghor, Mount Tabor. Area, Portugal, Spain, North Africa, Dalmatia.

44. *Evax pygmæa.* (L. Sp. 1311.)

On the coast near Beyrout. Area, Mediterranean region, Canaries.

45. *Evax contracta.* Boiss. Diagn. Ser. i., ii., p. 3.

Wilderness of Judæa, Philistine plain.

46. *Evax anatolica.* Boiss. Diagn. Ser. i., ii., p. 2. = *E. palæstina.* Boiss.

Central hill range from Jordan northwards, Lebanon and Anti-Lebanon.

47. *Evax eriosphæra.* Boiss. Diagn. Ser. i., ii., p. 3.

Maritime districts.

48. *Filago spathulata.* Presl. Del. Prag., p. 93. Var. *Prostrata.*

Southern desert, maritime sandy plains. Area, Mediterranean region, Canaries, North-west India.

49. *Filago germanica.* L. Sp. 1311. Cudweed.

Littoral region, Lebanon, Moab. Area, Central and Southern Europe, Siberia, Canaries.

50. *Filago arvensis.* L. Sp. 1312. Var. *Lagopus,* De Cand.

On bare dry places, northern hills. Area, Central and Southern Europe, Siberia, Canaries.

51. *Ifloga spicata.* (Forsk. Cat. Eg., 483.)

Littoral sands. Ghor, near Jericho. Area, South-east Spain, Canaries, North Africa, North-west India.

52. *Xanthium strumarium.* L. Sp. 1400. And var. *Antiquorum* = *X. anatolica.* Boiss.

General. Area, Central and Southern Europe, North Africa, Siberia, Abyssinia.

53. *Ambrosia maritima.* L. Sp. 1481.

Maritime plains. Area, Spain, Italy, Dalmatia, Nubia, Senegal.

54. *Diotis maritima.* Sm. Engl. Fl. iii., p. 403. Cotton weed.

Maritime sands. Area, shores of Europe from Ireland, southward and eastward, North Africa, Canaries.

55. *Achillea odorata.* Koch. Syn., p. 412. = *A. kotschyi.* Boiss.

Lebanon, alpine and sub-alpine. Area, mountains of Spain, Italy and Illyria.

56. *Achillea ligustica.* All. Ped. i., p. 181, pl. 53.

Lebanon, 6,000 feet. Area, Spain, Italy.

57. *Achillea micrantha.* M. B. Taur. Cauc. ii., p. 336.

Bare hills, Northern Palestine, and by Lake Phiala. Area, South Russia.

58. *Achillea santolina.* L. Sp. 1264.

Hills round the Dead Sea. Area, Algerian Sahara.

59. *Achillea falcata.* L. Sp. 1264. = *A. sulphurea.* Boiss.

Northern hills, Lebanon and Anti-Lebanon.

60. *Achillea tomentosa.* L. Sp. 1264.

Mount Gilead. Area, South Europe, West Siberia.

61. *Achillea aleppica.* De Cand. Prodr. vii., p. 296.

Waste places, round Dead Sea, Lebanon, Bukâà.

62. *Achillea fragrantissima.* (Forsk. Descr. 147.) Arab. كيسوم, *Keysoum.*

Southern desert, Bukâà, northern plains.

63. *Achillea spinulifolia.* Fenz. in Ky. P. C. Cil. Exs.

64. *Achillea membranacea.* (Labill. Syr. Dec. iii., p. 14, pl. 9.)

Barren hills in the north.

65. *Achillea oligocephala.* De Cand. Prodr. vi., p. 32.

Eastern barren plains.

66. *Achillea teretifolia.* Willd. Sp. iii., p. 2198.

Hills above Ain Fidjeh.

67. *Achillea nobilis.* L. Sp. 1268.

Summits of Gilead, Hauran. Area, Central Europe and Siberia.

68. *Anthemis tinctoria.* L. Sp. 1263. Yellow Camomile. Var. *Discoidea.* Sibth.

Dry hilly and mountain districts. General. Area, Europe, Siberia.

69. *Anthemis palæstina.* Boiss. Fl. Or. iii., p. 283.

General. Hills and plains.

70. *Anthemis altissima.* L. Sp. 1259.

Plain of Hattin, the Bukâà. Area, South France, North Spain, Italy, Dalmatia, South Russia.

71. *Anthemis cassia.* Boiss. Diagn. Ser. i., ii., p. 10.

Northern mountains.

72. *Anthemis lyonnetioides.* Boiss. Fl. Or. iii., p. 286.

Anti-Lebanon. P.

73. *Anthemis blancheana.* Boiss. Fl. Or. iii., p. 292.

Lebanon, Hazrun, and Kanobin.

74. *Anthemis pauciloba.* Boiss. Diagn. Ser. i., vi., p. 83.

Lebanon and Anti-Lebanon, alpine and sub-alpine districts.

75. *Anthemis philistea.* Boiss. Plant. Palest. Exs. 1846.

Sands near Gaza. P.

76. *Anthemis leucanthemifolia.* Boiss. Diagn. Ser. ii., iii., p. 20.

Maritime plains in sand by the shore. P.

77. *Anthemis deserti.* Boiss. Fl. Or. iii., p. 305. = *A. peregrina.* Decaisn.

Southern desert.

78. *Anthemis rascheyana.* Boiss. Diagn. Ser. i., ii., p. 8.

Dry hills under Hermon. P.

79. *Anthemis hyalina.* De Cand. Prodr. vi., p. 4.

Galilee, Lebanon, and Anti-Lebanon regions.

80. *Anthemis melampodina.* Del. Fl. Eg., p. 351, pl. 45, f. 1.

Southern desert.

81. *Anthemis chia.* L. Sp. 1260.

Waste and cultivated ground, Lower Lebanon, and maritime districts. Area, South Italy, Dalmatia.

82. *Anthemis scariosa.* De Cand. Prodr. vi., p. 4.

Anti-Lebanon.

83. *Anthemis cornucopiæ.* Boiss. Diagn. Ser. i., ii., p. 8.

Plains of Esdraelon. P.

84. *Anthemis hebronica.* Boiss. Diagn. Ser. ii., v., p. 108.

Judæa and Ghor. P.

85. *Anthemis cotula.* L. Sp. 1261.

General through the country, except Jordan valley, to the top of Lebanon, Moab. Area, Europe, Siberia, North Africa, Canaries, Madeira.

86. *Anthemis pseudocotula.* Boiss. Diagn. Ser. i., vi., p. 86.

Among crops on the maritime plains.

87. *Anthemis tripolitana.* Boiss. Diagn. Ser. ii., iii., p. 22.

Sandy fields on the coast.

88. *Ormenis mixta.* (L. Sp. 1260.)

Fields on all the plains. Area, South-west Europe, Canaries.

89. *Anacyclus radiatus.* Loisel. Gall. ed. 1, p. 583.

Waste lands on the coast. Area, Mediterranean region.

90. *Anacyclus membranacea.* Labill. Pl. Syr. Dec. iii. pl. 9.

East plains of Moab.

91. *Anacyclus nigellæfolius.* Boiss. Diagn. Ser. i., ii., p. 13.

General in cultivated land and brushwood.

92. *Matricaria chamomilla.* L. Sp. 1256. Wild Camomile.

Near Banias. Area, Europe, Siberia, Canaries.

93. *Matricaria aurea.* (L. Sp. 1267.)

Universal, Lebanon to Dead Sea basin and Gaza, in barren places. Area, Spain, North Sahara.

94. *Chamæmelum præcox.* (M. B. Taur. Cauc. ii., p. 324.)

Cultivated fields.

95. *Chamæmelum auriculatum.* Boiss. Diagn. Ser. i., ii., p. 23.

Wilderness of Judæa. Dead Sea border. Ravines of Moab. P.

96. *Chamæmelum oreades.* Boiss. Diagn. Ser. i., ii., p. 21.

Rocky places. Mountain and alpine regions of Lebanon and Anti-lebanon.

97. *Chamæmelum inodorum.* L. Sp. 1253.

Gennesaret. Area, Northern and Cen tral Europe.

98. *Chrysanthemum myconis.* L. Sp. 1254.

Maritime and central districts. Area, Portugal and Mediterranean region.

99. *Chrysanthemum segetum.* L. Sp. 1254. Corn Marigold.

Universal in cultivated land. Area, Central and Southern Europe, North Africa.

100. *Chrysanthemum viscosum.* Boiss. Cat. Par. 1821.

Plain of Sharon. Area, South Spain, North Africa.

101. *Chrysanthemum coronarium.* L. Sp. 1254.

Coast and southern districts, Plain of Philistia, Gilead. Area, South Europe, Azores, Canaries, Madeira.

102. *Pyrethrum cilicicum.* Boiss. Diagn. Ser. ii., v., p. 29.

Woods on Lebanon.

103. *Pyrethrum densum.* Labill. Dec. iii., p. 12, pl. 8 = *P. syriacum.* Boiss.

Alpine districts of Lebanon and Anti-Lebanon to 9,500 feet.

104. *Pyrethrum tenuilobum.* Boiss. Fl. Or. iii., p. 352.

Galilee, Lebanon and Anti-Lebanon districts.

105. *Pyrethrum argenteum.* (Willd. Ach. 51, pl. 2, f. 4.) Var. *tenuisectum.* Boiss.

Alpine Lebanon.

106. *Pyrethrum achilleæfolium.* M. B. Taur. Cauc. ii., p. 327.
Sub-alpine Lebanon. Area, South-east Russia.

107. *Pyrethrum myriophyllum.* C. A. May. Enum., p. 74.

108. *Pyrethrum cassium.* Boiss. Diagn. Ser. i., ii., p. 23.
Woods in the north.

109. *Pyrethrum myconis.* Moench.

110. *Brocchia cinerea.* (Del. Fl. Eg., p. 131, pl. 41.)
Southern desert towards Petra. Area, Sahara, Nubia.

ARTEMISIA GENUS, WORMWOOD. Hebr. לַעֲנָה.

111. *Artemisia monosperma.* Del. Fl. Eg. Descr., p. 120, pl. 43, f. 1. Arab. عدة, *Adeh.*
Along the coast on barren sands and in Southern Desert, Gaza, etc.

112. *Artemisia herba-alba.* Ass. Fl. Arag., p. 117, pl. 8. And many varieties.
In bare, dry places through all the warmer parts of Palestine. Area, North Africa, Canaries.

113. *Artemisia judaica.* L. Mant. 281. Arab., باعتران, *Baatheran.*
Southern Desert.

114. *Artemisia arborescens.* L. Sp. 1188.
Mount Carmel. Area, Mediterranean region.

115. *Artemisia crithmifolia.* L. Sp. 1186.
Northern coast, on sands. Area, Spain.

116. *Artemisia maritima.* L. Sp. 1186.
On the coast. Area, Northern and Central Europe, Siberia.

117. *Artemisia campestris.* L. Sp. 1185.
Area, Central and Southern Europe, North Africa.

118. *Tussilago farfara.* L. Sp. 1214. Colt's-foot.
Lebanon. Area, Europe, Siberia, Western Himalayas, North Africa.

119. *Senecio nebrodensis.* L. Sp. 1217.

Mountain region. Area, mountains of Central and Southern Europe and North Africa.

120. *Senecio ægyptius.* L. Sp. 1216. Jordan valley.

121. *Senecio decaisnei.* De Cand. Prodr. vi., p. 342.

Ravines east and west of Dead Sea. Area, Sahara, Arabia, Canaries, the Cape.

122. *Senecio vulgaris.* L. Sp. 1216. Groundsel.

On cultivated land, littoral and inland plains. Area, Europe, North Africa, Northern and Central Asia.

123. *Senecio leucanthemifolius.* Poir. Voy. ii., p. 283. = *S. glaucus.* De Cand.

Maritime plains. Area, South Italy, Sardinia, Sicily, North Africa.

124. *Senecio vernalis.* W. K. Pl. Rar. Hung. i., p. 23, pl. 24.

Littoral and mountain districts, east and west. Area, South-east Europe, North Africa.

125. *Senecio doriæ.* Koch. Linn. xvii., p. 48.

126. *Senecio doronicum.* Roch.

127. *Senecio coronopifolius.* Desf. Atl. ii., p. 273.

Common in all sandy and barren places, except the mountains. Area, Sahara, Thibetian desert, Canaries.

128. *Senecio erraticus.* Bertol. Amœn., p. 92.

Meadows, Galilee and north. Area, Mediterranean and Danubian regions.

129. *Senecio doriæformis.* De Cand. Prodr. vi., p. 352.

Lebanon and Anti-Lebanon, moist alpine opens.

130. *Calendula sinuata.* Boiss. Diagn. Ser. ii., vi., p. 109.

In cultivated land, Plain of Kedes.

131. *Calendula palæstina.* Boiss. Diagn. Ser. i., x., p. 83.

Carmel, Gerizim, about Jerusalem, Judæan wilderness, Moab.

132. *Calendula arvensis.* L. Sp. 1303.

Cultivated land everywhere, Moab. Grows to a gigantic size in the Jordan valley. Area, Central and Southern Europe, North Africa, Canaries.

133. *Calendula ægyptiaca.* Desf. Cat. Hort. Par., 1804, p. 100. = *C. parviflora.* De Cand.

Abundant in the Ghor, Sharon, Philistine plains, and Southern Desert. Area, Tunis.

134. *Gundelia tournefortii.* L. Sp. 1315.

Hills, and waste places under Lebanon, Nazareth, Jerusalem, Gaza.

135. *Echinops glaberrimus.* De Cand. Ann. Sc. Nat., 1834, p. 260.

Desert south of Beersheba.

136. *Echinops spinosus.* L. Mant. 119.

Near Gaza.

137. *Echinops viscosus.* De Cand. Prodr. vi., p. 525.

General in the mountain and hill-districts and Lebanon. Area, Sicily.

138. *Echinops gaillardoti.* Boiss. Diagn. Ser. ii., iii., p. 38.

Phœnician plain and Lebanon. P.

139. *Echinops blancheanus.* Boiss. Fl. Or. iii., p. 430.

At the base of Hermon, the Bukâà.

140. *Echinops sphærocephalus.* L. Sp. 1314. Var. *Taygeteus.* Boiss.

Gennesaret. Area, Central and Southern Europe, North Africa, South Siberia.

141. *Cardopatium corymbosum.* (L. Sp. 1164.) = *C. orientale.* Spach.

Maritime district near Beyrout. Area, South Italy.

142. *Xeranthemum longepapposum.* F. et Mey. N. Mem. Nat. Mosc. iv., p. 337.

Gilead, Galilee.

143. *Acantholepis orientalis.* Less. Linn., 1831, p. 88.
Plains of Hauran.

144. *Xeranthemum inapertum.* Willd. Sp. iii., p. 1902.
Lebanon. Area, Portugal, Mediterranean and Danubian regions.

145. *Xeranthemum cylindraceum.* Sibth. Prodr. ii., p. 172.
Lebanon. Area, South Europe, Danubian region.

146. *Chardinia xeranthemoides.* Desf. Mém. Mus. Par. iii., p. 455, pl. 21.
Lebanon and Anti-Lebanon, Gilead.

147. *Siebera pungens.* (Lam. Dict., p. 236.)
Lebanon and Anti-Lebanon.

148. *Carlina corymbosa.* L. Sp. 1160. Var. *Libanotica.* Boiss.
Country round Lebanon and Anti-Lebanon.
Ditto, var. *Involucrata.* Coast region under Lebanon. Area, Mediterranean region.

149. *Carlina lanata.* L. Sp. 1160.
Maritime districts. Area, South Europe.

150. *Atractylis flava.* Desf. Atl. ii., p. 254.
Wâdy Ghurundel, south of Dead Sea. Area, North Africa.

151. *Atractylis prolifera.* Boiss. Diagn. Ser. i., x., p. 96.
Deserts near Gaza, Eastern Ghor. Area, Sahara.

152. *Atractylis cancellata.* L. Sp. 1162.
Cultivated land on the coast, waste places, Moab. Area, Mediterranean region, Canaries, Arabia.

153. *Atractylis comosa.* Sieb. ex Cass. Dict. 50, p. 58.
General, except Jordan valley.

154. *Stæhelina apiculata.* Labill. Syr. Dec. 4, p. 3, pl. 1.
Cliffs in higher Lebanon.

155. *Lappa major.* Gærtn. Fruct. ii., p. 379. Burdock.
Galilee, Lebanon. Area, Europe, Siberia, Japan, Himalayas.

156. *Cousinia pestalozzæ.* Boiss. Fl. Or. iii., p. 471.
Under Anti-Lebanon.

157. *Cousinia ramosissima.* De Cand. Prodr. vi., p. 552.
Eastern slopes of Lebanon, Bukâà, under Hermon.

158. *Cousinia libanotica.* De Cand. Prodr. vi., p. 556.
Alpine Lebanon and Hermon. P.

159. *Cousinia hermonis.* Boiss. Diagn. Ser. i., x., p. 102.
Round the base of Hermon and Lebanon. P.

160. *Cousinia aleppica.* Boiss. Diagn. Ser. i., x., p. 101.
Northern hills.

161. *Carduus nutans.* L. Sp. 1150.
Hills, coast. Area, Central and Southern Europe, North Africa.

162. *Carduus pycnocephalus.* Jacq. Hort. Vind. i., p. 17, pl. 44.
In cultivated land, universal. Area, Central and Southern Europe, North Africa, Canaries.

163. Ditto, var. *Arabicus.* Jacq.
Desert districts, north and south.

164. *Carduus argentatus.* L. Mant. 280.
General, littoral and central hills, from Hebron northwards, Lebanon.

165. Ditto, var. *Esdraëlonicus.* Boiss.
Moist places, Esdraelon and Huleh.

166. *Cirsium lappaceum.* (M. B. Taur. Cauc. ii., p. 277.) Var. *Hermonis.* Boiss.
Lebanon and Hermon.

167. *Cirsium phyllocephalum.* Boiss. Diagn. Ser. ii., iii., p. 39.
Sub-alpine Lebanon. P.

168. *Cirsium lanceolatum.* (L. Sp. 1149.)
Northern districts and Lebanon. Area, Europe, Siberia, North Africa.

169. *Cirsium gaillardoti.* Boiss. Diagn. Ser. ii., iii., p. 42.

On Lebanon, rocky places ; local. P.

170. *Cirsium siculum.* Spreng. Neu. Ent., p. 36.

Northern maritime district and Lebanon. Area, Corsica, Italy, Sicily, Danubian region.

171. *Cirsium libanoticum.* De Cand. Prodr. vi., p. 647.

Alpine and sub-alpine Lebanon and Anti-Lebanon.

172. *Cirsium acarna.* (L. Sp. 820.)

Northern districts, waste and cultivated lands. Area, Mediterranean region, Portugal.

173. *Cirsium diacantha.* (Labill. Syr. Déc. ii., p. 7, pl. 3.)

Sub-alpine districts of Lebanon and Anti-Lebanon.

174. *Cirsium arvense.* (L. Sp. 1149.)

Mount Carmel. Area, Europe, Siberia, North-west India.

175. *Notobasis syriaca.* (L. Sp. 1133.) Probably the 'Thistle' (Heb. חוֹח) of Scripture. A most powerful and noxious thistle.

General, except in the mountains. Area, Mediterranean region, Portugal, Canaries, Madeira.

176. *Chamæpeuce alpina.* Jaub. et Spach. Ill. Or. v., pl. 425.

177. Ditto, Var. *Camptolepis* = *C. polycephala.* De Cand.

General in all Northern and Central Palestine.

Ditto, var. *Mutica.* Coast, rocks.

178. *Tyrimnus leucographus.* (L. Sp. 1149.)

Waste ground from the coast to the east of Hermon. Area, Southern Europe.

179. *Silybum marianum.* (L. Sp. 1153.) Milk Thistle.

Universal. Mount Tabor, Gerizim, etc. Area, Central and Southern Europe, North Africa, Canaries, Madeira.

180. *Cynara syriaca.* Boiss. Diagn. Ser. i., x., p. 94.

Fields on the coast.

181. *Galactites tomentosa.* Mœnch. Méth. 558.
Area, Mediterranean region, Canaries, Madeira.

182. *Onopordon illyricum.* L. Sp. 1158. Var. *Libanoticum.* Boiss.
Littoral, Lebanon, Anti-Lebanon, Moab. Area, Southern Europe.

183. *Onopordon sibthorpianum.* Boiss. et Heldr. in Pl. Gr. Exs.
Banias.

184. Ditto, var. *Alexandrinum.* Boiss.
Philistine plain. Area, interior Algeria.

185. *Onopordon ambiguum.* Fres. Mus. Senck., p. 85.
Southern Desert and Northern Desert south of Damascus.

186. *Onopordon cynarocephalum.* Boiss. Diagn. Ser. ii., iii., p. 48.
District round southern base of Lebanon. P.

187. *Onopordon anisacanthum.* Boiss. Diagn. Ser. i., x., p. 93.
The Bukâà.

188. *Jurinea stæhelinæ.* (De Cand. Prodr. vi., p. 544.)
Alpine and sub-alpine Lebanon and Anti-Lebanon. P.

189. *Serratula cerinthefolia.* Sibth. Prodr. ii., p. 567.
Dry hills and fields, Northern Palestine.

190. *Rhaponticum pusillum.* (Labill. Déc. iii., p. 2, pl. 7.) = *R. pygmæum.* De Cand.
General on hills of Benjamin, Samaria, Galilee, Lebanon, and Anti-Lebanon.

191. *Phæopappus libanoticus.* (Boiss. Diagn. Ser. i., x., p. 107.)
Alpine Lebanon and Hermon. P.

192. *Amberboa lippii.* (L. Sp. 1286.)
Desert south-west of Dead Sea. Area, North Africa, South-east Spain, Canaries.

193. *Amberboa crupinoïdes.* (Desf. Atl. ii., p. 293.)
Judæan wilderness, and Southern Desert, Gaza. Area, Nubia, Sahara.

194. *Amberboa moschata.* L. Sp. 1286.
Probably introduced.

195. *Ptosimopappus bractoatus.* Boiss. Diagn. Ser. i., x., p. 104.
Galilean hills.

196. *Centaurea cyanoides.* Berg. et Wahl. *Isis*, 1828, v., 21.
Universal. P.

197. *Centaurea axillaris.* Willd. Sp. iii., p. 2290. Var. *Cana* = *C. acmophylla.* Boiss.

Alpine Lebanon and Anti-Lebanon. Area, mountains of Central and Southern Europe.

198. *Centaurea ammocyana.* Boiss. Diagn. Ser. i., x., p. 110.
Sandy deserts near Beersheba and Gaza, plain of Moab. P.

199. *Centaurea virgata.* Lam. Dict. i., p. 670. Var. *Squarrosa.* Willd.

The Bukâà. Area, South Siberia.

200. *Centaurea leptocephala.* Boiss. Diagn. Ser. i., x., p. 110.
Bare hills, Anti-Lebanon, and Bukâà. P.

201. *Centaurea dumulosa.* Boiss. Diagn. Ser. i., x., p. 111.
Dry places under Lebanon and Anti-Lebanon. P.

202. *Centaurea depressa.* M. B. Taur. Cauc. ii., p. 346.
Northern Galilee, Mount Tabor, Gilead, Moab.

203. *Centaurea damascena.* Boiss. Diagn. Ser. i., x., p. 112.
Base of Hermon, on barren ground. P.

204. *Centaurea antiochia.* Boiss. Diagn. Ser. i., x., p. 115.
North Lebanon.

205. *Centaurea speciosa.* Boiss. Diagn. Ser. i., x., p. 116.
Galilee, Lebanon, Anti-Lebanon. P.

206. *Centaurea eryngioides.* Lam. Dict. i., p. 675.
Northern Palestine, Lebanon, Anti-Lebanon, Eastern Ghor.

207. Ditto, var. *Brachyantha.* Boiss. Judæan wilderness, Anti-Lebanon.

208. Ditto, var. *Ainetensis.* Boiss. Eastern Lebanon, Bukâà.

209. *Centaurea arifolia.* Boiss. Diagn. Ser. i., x., p. 112.
Northern Lebanon, in woods.

210. *Centaurea onopordifolia.* Boiss. Diagn. Ser. i., x., p. 114.
Waste land, Northern Palestine.

211. *Centaurea crocodylium.* L. Sp. 1299.
Plains of Esdraelon, Galilee, Banias.
Ditto, var. *Crocodyloides.* Boiss. Same localities.

212. *Centaurea heterocarpa.* Boiss. Fl. Or. iii., p. 680.
Below Anti-Lebanon. P.

213. *Centaurea babylonica.* L. Mant. 460.
General.

214. *Centaurea myriocephala.* Sch. Bip. in Ky. Pl. Alep. Kurd., 1843.
Fields and inland plains, mountains of Moab.

215. *Centaurea behen.* L. Sp. 1292. = *C. alata.* Lam. = *C. acuta.* Vahl.
Lower parts of Lebanon, Galilee.

216. *Centaurea verutum.* L. Amœn. iv., p. 292.
Cultivated land on all the plains, Esdraelon, Gennesaret, etc.

217. *Centaurea solstitialis.* L. Sp. 1297.
Universal. Area, Southern Europe and Danubian region.

218. *Centaurea sinaïca.* De Cand. Prodr. vi., p. 592.
Hills overhanging Dead Sea.

219. *Centaurea egyptiaca.* L. Mant., p. 118.
Southern Desert.

220. *Centaurea postii.* Boiss. Fl. Or. iii., p. 688.
Eastern Lebanon. P.

221. *Centaurea procurrens.* Sieb. Pl. Exs.

Southern Desert, plains of Sharon and Philistia, in sandy places.

222. *Centaurea cassia.* Boiss. Diagn. Ser. i., x., p. 108.

Hills and woods.

223. *Centaurea calcitrapa.* L. Sp. 1297.

Upper plains of Moab. Area, Central and Southern Europe, North Africa, Canaries.

224. *Centaurea iberica.* Trev. in Spreng. Sept. iii., p. 406.

General in waste and bushy ground.

225. Ditto, var. *Meryonis.* De Cand. Littoral, Lebanon, Anti-Lebanon.

226. Ditto, var. *Hermonis.* Boiss. Lebanon and Anti-Lebanon.

227. *Centaurea pallescens.* Del. Fl. Eg., p. 370, pl. 49, f. 1. Var. *Hyalolepis.* Boiss.

Universal, except northern mountains.

228. *Centaurea araneonosa.* Boiss. Diagn. Ser. i., xi., p. 121.

Maritime plains. P.

229. *Centaurea cheiracantha.* Fenz. in Ky. Exs., 1836.

Sub-alpine Lebanon.

230. *Centaurea hololeuca.* Boiss. Fl. Or. iii., p. 694.

Alpine Lebanon, near the Cedars. P.

231. *Centaurea diffusa.* Lam. Dict. i., p. 675.

The Bukâà. Area, South Russia.

232. *Centaurea jacea.* L. Sp. 1293.

Area, Europe, West Siberia, North Africa.

233. *Centaurea spicata.* Boiss. Diagn. Ser. i., x., p. 113.

Plains and northern lower hills.

234. *Zoegia purpurea.* Fresen. Beitr., p. 86.

Rocky places south-east and south-west of Dead Sea.

235. *Zoegia.* Sp. *nov.?*
The Bukâà.

236. *Crupina vulgaris.* Cass. Dict. xlv., p. 39.
Mount Carmel, Mount Gilead.

237. *Crupina crupinastrum.* (Mor. Sem. Taur., 1842, Fl. Sard. ii., p. 443.)
Littoral and Central Palestine. Area, Mediterranean and Danubian regions.

238. *Cnicus benedictus.* L. Sp. ed. i., p. 626.
In cultivated land.
Ditto, var. *Kotschyi.* C. Bip. in Ky. Lebanon.

239. *Carthamus dentatus.* Vahl. Synt. i., p. 69, pl. 17.
Mountains of Moab.

240. *Carthamus lanatus.* L. Sp. 1163.
Waste and cultivated land. General. Area, Mediterranean region, Abyssinia, Canaries.

241. *Carthamus glaucus.* M. B. Taur. Cauc. ii., p. 284. Var. *Syriacus.* Boiss.
General.
Ditto, var. *Tenuis.* Boiss. Maritime and sandy places.

242. *Carthamus nitidus.* Boiss. Fl. Or. iii., p. 708.
Jordan valley. P.

243. *Carthamus flavescens.* Willd. Spec. iii., p. 1706.
The Bukâà.

244. *Carthamus cæruleus.* L. Sp. 1163.
Maritime districts. Area, Mediterranean region, Canaries.

245. *Carthamus oxyacantha.* M. B. Taur. Cauc. ii., p. 283.
Ain Fidjeh, Lebanon.

246. *Carduncellus eriocephalus.* Boiss. Diagn. Ser. i., x., p. 100.
Southern Desert, Gaza. Area, North Sahara.

SUB-ORDER, LIGULIFLORÆ.

247. *Scolymus maculatus.* L. Sp. 1143.

Jordan valley. Area, Mediterranean region, Nubia, Canaries.

248. *Scolymus hispanicus.* L. Sp. 1143.

Waste lands. General. By Lake Huleh. Area, Mediterranean region, Canaries, Madeira.

249. *Catananche lutea.* L. Sp. 1142.

Universal. Area, Mediterranean region.

250. *Cichorium intybus.* L. Sp. 1142. Chicory.

Waysides. General.

251. *Cichorium divaricatum.* Schoësb. Mar., p. 197.

General. Area, Southern Europe.

252. *Hyoseris scabra.* L. Sp. 1138.

Littoral districts. Area, Mediterranean region.

253. *Hedypnois cretica.* (L. Sp. 1139.) = *H. polymorpha.* De Cand.

Maritime and central regions, Jordan valley. General. Area, Mediterranean region, Madeira, Canaries.

254. *Lapsana peduncularis.* Boiss. Fl. Or. iii., p. 720.

Alpine and sub-alpine Lebanon and Anti-Lebanon.

255. *Kœlpinia linearis.* Pall. Itin. iii., App., p. 755.

Southern Desert. Area, North Africa, South Russia, South Siberia, Western Himalayas.

256. *Rhagadiolus stellatus.* De Cand. Prodr. vii., p. 77.

Universal. Jordan valley, as well as hills and fields. Area, Mediterranean region, Canaries, Madeira.

Ditto, var. *Leiolænus.* Boiss. Near Jerusalem.

257. *Garhadiolus hedypnois.* (F. and M. Ind. iv., Petrop., p. 48.)

Waste lands. General.

258. *Tolpis altissima.* Pers. Ench. ii., p. 377.
Waste plains. Universal. Area, Mediterranean region.

259. *Thrincia tuberosa.* (L. Sp. 1123.)
Maritime and central districts. General. Area, Southern Europe.

260. *Leontodon hispidulum.* (Del. Fl. Eg., p. 117, pl. 42, f. 1.)
Southern Desert.

261. *Leontodon arabicum.* Boiss. Fl. Or. iii., p. 728.
Southern Desert. P.

262. *Leontodon libanoticum.* Boiss. Diagn. Ser. i, ii., p. 40.
Sub-alpine Lebanon. P.

263. *Leontodon asperrimum.* (Willd. Sp. 111, p. 1507.)
Lower and sub-alpine Lebanon, Galilee.

264. *Leontodon oxylepis.* Boiss. Diagn. Ser. i., ii., p. 40.

265. *Picris stricta.* Jord. Cat. Dij., 1848, p. 19.
Base of Lebanon. Area, Southern Europe.

266. *Picris strigosa.* M. B. Taur. Cauc. ii., p. 250.
Northern plains, Anti-Lebanon.

267. *Picris sprengeriana.* (L. Sp. 1130.)
Maritime districts, Jordan valley. Area, South France, Dalmatia.

268. *Picris pauciflora.* Willd. Sp. iii., 1557.
Judæan wilderness. Area, Southern Europe.

269. *Picris radicata.* (Forsk. Fl. Eg., p. 145.)
Desert near Gaza. Area, Tunisian desert.

270. *Picris damascena.* Boiss. Fl. Or. iii., p. 740.
Top of Jebel Antar. P.

271. *Picris sulphurea.* Del. Fl. Eg., p. 114, pl. 40, f. 2.
Southern Desert.

272. *Hagioseris galilæa.* Boiss. Diagn. Ser. i., ii,, p. 36.
Grassy places, Galilee, maritime plains. P.

273. Ditto, var. *Diffusa.* = *H. amalecitana.* Boiss.
Sandy places, maritime plains, Beyrout to Gaza.

274. *Helminthia echioides.* (L. Sp. 1114.)
Cultivated ground, general. Area, North Africa, Canaries, Madeira.

275. *Urospermum picroides.* (L. Sp. 1111.)
Universal in cultivated and bushy land. Area, Mediterranean region, Canaries, Madeira.

276. *Urospermum dalechampii.* (L. Sp. 1110.)
Plain of Huleh. Area, Mediterranean region.

277. *Geropogon glabrum.* L. Sp. 1109.
Coast, Carmel, Galilee. Area, Mediterranean region, Canaries, Madeira.

278. *Tragopogon longirostre.* Webb. Phyt. Can. ii., p. 469. = *T. cœlesyriacum.* Boiss.
Universal.

279. *Tragopogon plantagineum.* Boiss. Diagn. Ser. ii., iii., p. 91.
Around and upon Lebanon and Anti-Lebanon.

280. *Tragopogon persicum.* Boiss. Diagn. Ser. i., vii., p. 4.
Stony places on and round Lebanon and Anti-Lebanon to 4,000 feet.

281. *Tragopogon palæstinum.* Boiss. Diagn. Ser. i., ii., pp. 46, 47.
On the highest part of Lebanon.
N.B.—These three have since been united by Boissier under *T. buphtalmoides.*

282. *Tragopogon porrifolium.* L. Sp. 1110.
Glens of Moab. Area, Mediterranean region.

283. *Scorzonera jacquiniana.* (Koch. Syn., p. 489.)
Lower and sub-alpine Lebanon. Area, Danubian region.

284. Ditto, var. *Subintegra.* Boiss.
Summits of Lebanon and Hermon.

285. *Scorzonera mollis.* M. B. Taur. Cauc. iii., p. 522.
Lebanon, Anti-Lebanon, Galilæan hills.

286. *Scorzonera syriaca.* Boiss. Diagn. Ser. ii., iii., p. 93.
On rocks, maritime districts.

287. *Scorzonera phæopappa.* Boiss. Fl. Or. iii., p. 764.
The Bukâà.

288. *Scorzonera multiscapa.* Boiss. Diagn. Ser. i., ii., p. 41. = *S. phæopappa.* Var. *Minor.* Fl. Or. Lebanon and Anti-Lebanon.

289. *Scorzonera humilis.* L. Sp. 1112.
Mount Nebo (*Fide* Paine). Area, Central Europe, West Siberia.

290. *Scorzonera papposa.* De Cand. Prodr. vii., p. 119.
Universal in every kind of country.

291. *Scorzonera libanotica.* Boiss. Diagn. Ser. i., ii., p. 43.
Sub-alpine Lebanon. P.

292. *Scorzonera mackmeliana.* Boiss. Diagn. Ser. i., ii., p. 44.
Lebanon and Anti-Lebanon districts. P.

293. *Scorzonera lanata.* M. B. Taur. Cauc. ii., p. 237.
Northern district.

294. *Hypochæris glabra.* L. Sp. 1141. Littoral district.

295. *Taraxacum serotinum.* (W. K. Pl. Hung. ii., p. 119, pl. 114.) = *T. syriacum.* Boiss. = *T. libanoticum.* De Cand.

Sub-alpine and Lower Lebanon and Anti-Lebanon. Area, Danubian region, South Siberia.

296. *Taraxicum montanum.* C. A. Mey. Enum., p. 58.

297. *Taraxicum officinale.* Wigg. Prim. Fl. Hols., p. 56. = *T. dens-leonis.* Desf. Atl. 2, p. 228. Dandelion.

General, mountains and coast. Area, Europe, Siberia, North America, North Africa.

Ditto, var. *Lævigatum.* Bisch. Northern district.

298. *Taraxicum gymnanthum.* (Link. Linn., 1834, p. 582.) Littoral district. Area, South France, Italy.

299. *Taraxicum assemani.* Boiss. Fl. Or. iii., p. 791. = *T. bithynicum.* Boiss. By streams, Lebanon.

300. *Chondrilla juncea.* L. Sp. 1120. Maritime hill and mountain regions. Area, Europe, West Siberia.

301. *Sonchus oleraceus.* L. Sp. 1116. Sow Thistle. Universal. Area, world-wide.

302. *Sonchus glaucescens.* Jord. Obs. v., p. 75, pl. 5. Maritime plains.

303. *Sonchus tenerrimus.* L. Sp. 1117. Universal. Area, Portugal, Mediterranean region, Abyssinia.

304. *Mulgedium plumieri.* De Cand.

305. *Lactuca cretica.* Desf. Cor. Tourn. 44, pl. 34. All the hill-region of Southern, Central, and Eastern Palestine.

306. *Lactuca tuberosa.* (L. Fil. Suppl., p. 346.) Sands on the coast.

307. *Lactuca scariola.* L. Sp. 1119. Prickly Lettuce. Cultivated and waste land, general. Area, Central and Southern Europe, North Africa, Siberia, Abyssinia, Madeira.

308. *Lactuca saligna.* L. Sp. 1119. Willow Lettuce. Cultivated land, general. Area, Central and Southern Europe, North Africa.

309. *Lactuca viminea.* (L. Sp. 1120.) Lebanon. Area, South Europe, Atlas mountains.

310. *Lactuca orientalis.* Boiss. Fl. Or. iii., p. 819. Rocky and dry places, Hebron up to Lebanon and Anti-Lebanon.

311. *Lactuca triquatra.* (Labill. Syr. Dec. iii., p. 4, pl. 2.) In rocks, from the coast up to Lebanon.

312. *Cephalorrhynchus candolleanus.* Boiss. Fl. Or. iii., p. 820. = *L. hispida.* De Cand.

Lebanon and Anti-Lebanon, alpine and sub-alpine.

313. *Zollikoferia nudicaulis.* L. Mant. 273.

Wâdy Zuweirah, south-west of Dead Sea. Area, South-east Spain, Morocco, Canaries.

314. *Zollikoferia mucronata.* (Forsk. Desc. 144.)

Southern desert, near Gaza. Area, Sahara.

315. *Zollikoferia tenuiloba.* Boiss. Diagn. Ser. i., ii., p. 50.

Desert near Gaza; all the maritime plains. P.

316. *Zollikoferia arabica.* Boiss. Diagn. Ser. i., vii., p. 12.

Wâdys south of Dead Sea.

317. *Zollikoferia spinosa.* (Forsk. Desc. 144.)

Wâdys south-west of Dead Sea. Area, South-east Spain, Morocco, Canaries.

318. *Picridium tingitanum.* (L. Sp. 1114.) = *P. orientale.* De Cand.

Wâdy Zuweirah, south-west of Dead Sea. Area, North Africa, Abyssinia, Arabia, North-west India.

319. *Picridium intermedium.* Sch. Bip. in Phyt. Can. ii., p. 451.

Maritime plains. Area, South Spain, Portugal, Canaries, Sicily.

320. *Picridium vulgare.* Desf. Atl. ii., p. 221.

Waste places on the coast. Area, Mediterranean region.

321. *Picridium dichotomum.* (M. B. Taur. Cauc. ii., p. 240.)

Sub-alpine Lebanon.

322. *Zacyntha verrucosa.* Gærtn. Ic. Fl. Gr., pl. 820.

Maritime districts, Moab. Area, Mediterranean region.

323. *Cymboseris palæstina.* Boiss. Diagn. Ser. i., ii., p. 50.

Carmel, Tabor, Lebanon, and coast.

324. *Crepis bulbosa.* (L. Sp. 1122.)

Maritime districts. Area, Mediterranean region.

325. *Crepis robertioides.* Boiss. Diagn. Ser. i., ii., p. 59.
Alpine Lebanon and Anti-Lebanon. P.

326. *Crepis sieberi.* Boiss. Diagn. Ser. i., ii., p. 53.
Mount Gilead. (*Fide* Paine.)

327. *Crepis hierosolymitana.* Boiss. Diagn. Ser. i., ii., p. 54.
Shady rocks, Hebron and Lebanon. P.

328. *Crepis reuteriana.* Boiss. Diagn. Ser. i., ii., p. 55.
Galilee, Phœnicia, Lebanon, Anti-Lebanon.
Ditto, var. *Alpina.* Alpine Lebanon.

329. *Crepis pterothecoides.* Boiss. Fl. Or. iii., p. 850.
Anti-Lebanon, 5,000 feet. P.

330. *Crepis fœtida.* L. Sp. 1133. Var. *rheadifolia.*
Lower and sub-alpine Lebanon.

331. *Crepis arabica.* Boiss. Fl. Or. iii., p. 853.
Southern Desert. P.

332. *Crepis neglecta.* L. Mant., p. 107. Moab. (*Fide* Paine.)

333. *Crepis alpina.* L. Sp. 1134.
Lower and sub-alpine northern districts. Area, South Siberia.

334. *Crepis aculeata.* (De Cand. Prodr. vii., p. 159.)
Sands on the Phœnician plain, Lake of Galilee.

335. *Crepis aspera.* L. Sp. 1138.
Wilderness of Judæa, Ghor, bare hills near Jerusalem, coast.

336. *Hieracium prœaltum.* Vill. Voy. 62, pl. 11, f. 1. Var. *Hispidissimum.* Fries.
Northern Palestine. Central and Southern Europe.

337. *Hieracium balansæ.* Boiss. Diagn. Ser. ii., vi., p. 119.
Lebanon, 5,500 feet.

338. *Hieracium libanoticum.* Boiss. Fl. Or. iii., p. 870.
Rocks in alpine Lebanon. P.

339. *Audryala dentata.* Sibth. Fl. Græc. ix., p. 7, pl. 811.
Sands on the northern coast. Area, South Italy, Sicily, North Africa.

340. *Lagoseris bifida.* Vis. Stirp. Dalmat., p. 19, pl. 7.
Maritime region, Lebanon, Southern Desert, barren northern plains.

ORDER LXII., LOBELIACEÆ.

1. *Laurentia tenella.* (Biv. Cent. i., p. 53, pl. 2.)
Lower parts of Lebanon. Area, Corsica, Sardinia, Sicily.

ORDER LXIII., CAMPANULACEÆ.

1. *Michauxia campanuloides.* L'Her. Diss. Lam. Ill. ii., pl. 295.
Litany valley, Kulah esh Sherkif.

2. *Campanula ephesia.* Boiss. Fl. Or. iii., p. 898.
Gerash, Gilead.

3. *Campanula euclasta.* Boiss. Diagn. Ser. i., xi., p. 70.
Deserts in the north, near Damascus.

4. *Campanula damascena.* Labill. Dec. Syr. v., p. 7, pl. 5.
Mount Hermon, *not* near Damascus. P.

5. *Campanula trichopoda.* Boiss. Diagn. Ser. i., ii., p. 68.
Fissures of rocks, Lebanon. P.

6. *Campanula cymbalaria.* Sibth. Prodr. Fl. Gr. i., p. 139. = *C. billardieri.* De Cand.
Sub-alpine Lebanon.

7. *Campanula trachelium.* L. Sp. 235. Nettle-leaved Bell Flower.
Lebanon. Area, Europe, Western Siberia, North Africa.

8. *Campanula stricta.* L. Sp. 238. Var. *Libanotica.* Boiss.
Lebanon and Hermon.

9. *Campanula dulcis.* Decaisn. Ann. Sc. Nat. Ser. ii., ii., p. 258.

Bare rocky slopes, Ainat, Lebanon.

10. *Campanula stellaris.* Boiss. Diagn. Ser. i., ii., p. 63.

Mount Carmel, Phœnician plain, Moab, and by roadsides. P.

11. *Campanula strigosa.* Russ. Aleppo, ii., p. 246.

On the hills and in cultivated ground, Central Palestine from Hebron to Lebanon and Anti-Lebanon.

12. *Campanula sulphurea.* Boiss. Diagn. Ser. i., ii., p. 64.

On all the maritime plains. P.

13. *Campanula hierosolymitana.* Boiss. Diagn. Ser. i., ii., p. 62.

Jerusalem, Nablus, etc. P.

14. *Campanula camptoclada.* Boiss. Diagn. Ser. i., ii., p. 63.

Fissures of rocks round Hermon.

15. *Campanula erinus.* L. Sp. 240.

Hill-districts. Area, Mediterranean region, Canaries, Madeira.

16. *Campanula peregrina.* L. Syst. 601.

Maritime districts and Lebanon to 4,500 feet.

17. *Campanula rapunculus.* L. Sp. 232. Rampion.

Universal. Area, Europe, West Siberia, North Africa.

18. *Campanula retrorsa.* Labill. Syr. Dec. v., p. 5, pl. 3.

Galilee, Phœnicia, Lower Lebanon.

19. *Campanula sidoniensis.* Boiss. Diagn. Ser. ii., iii., p. 114.

Phœnician coast. P.

20. *Campanula dichotoma.* L. Amœn. iv., p. 306.

Moab only. Area, West Mediterranean region, Canaries.

21. *Campanula ramosissima.* Sibth. Fl. Gr. iii., p. 3, pl. 204.

Gilead, in open places. Area, North-east Italy, Dalmatia.

22. *Campanula primulæfolia.* Brot. Beyrout.

23. *Podanthum virgatum.* (Labill. Dec. ii., p. 11, pl. 6.)
Sub-alpine and alpine Lebanon.

24. *Podauthum lanceolatum.* Willd. Spec. i., p. 924. = *Camp. tauricola.* Boiss.
Sub-alpine Lebanon.

25. Ditto, var. *Alpinum.* = *Campanula leptopetala.* Ehr. = *Phyteuma sinaï.* De Cand.
Alpine Lebanon, Cedars, etc.

26. *Podanthum controversum.* (Boiss. Diagn. Ser. ii., iii., p. 115.)

27. *Podanthum cappadocicum.* (Boiss. Diagn. Ser. i., ii., p. 93.)
On Northern Lebanon.

28. *Specularia speculum.* (L. Sp. 238.)
Coast and plains. Area, Central and Southern Europe, North Africa.
Ditto, var. *Libanensis.* De Cand. Sands near Beyrout.

29. *Specularia pentagonia.* (L. Sp. 239.)
Hebron, Jerusalem, central hills, Lebanon, Anti-Lebanon, Gennesaret.

30. *Specularia falcata.* (Ten. Prodr., p. 16.)
Mount Carmel, base of Hermon. Area, Mediterranean region, Canaries.

31. *Trachelium tubulosum.* Boiss. Diagn. Ser. i., ii., p. 60.
On vertical rocks, Lebanon, from base to 5,000 feet.

ORDER LXIV., SPHENOCLEACEÆ.

Absent.

ORDER LXV., VACCINIEÆ.

Absent.

ORDER LXVI., ERICACEÆ.

1. *Arbutus unedo.* L. Sp. 566. Strawberry-tree.

Woods of Gilead. Area, Central and Southern Europe, North Africa.

2. *Arbutus andrachne.* L. Sp. 566.

In woods and among shrubs in hill-country, from Hebron to Lebanon.

3. *Erica verticillata.* Forsk. Fl. Eg. Arab., p. 25. Mediterranean Heath.

Lower Lebanon, near the coast. Area, Istria, Dalmatia.

4. *Rhododendron ponticum.* L. Sp. 562. Var. *Brachycarpum.* Boiss. Common Rhododendron.

Sub-alpine Lebanon.

ORDER LXVII., PYROLACEÆ.

Absent.

ORDER LXVIII., MONOTROPEÆ.

Absent.

PLANTÆ VASCULARES.

CLASS, DICOTYLEDOENÆ.

SUB-CLASS, COROLLIFLORÆ.

ORDER LXIX., LENTIBULARIEÆ.

1. *Utricularia vulgaris.* L. Sp. 26. Bladderwort.

Above Lake Huleh, in stagnant waters. Area, Europe, Siberia, North America, North Africa.

ORDER LXX., PRIMULACEÆ.

1. *Samolus valerandi.* L. Sp. 243. Brookweed.

Pools in the Ghor, Esdraelon, Ayun Mûsa, Moab, etc. Area, Europe, North Africa, Siberia, India, the Cape, North America, Australia.

2. *Anagallis arvensis.* L. Sp. 211. Pimpernel.

Vars. *Phœnicea* and *Cœrulea.* Both equally universal. Area, world-wide, except Arctic and Antarctic regions.

3. *Anagallis latifolia.* L. Sp. 212.

In corn-land, Southern Palestine. Area, Portugal, South Spain, North Africa.

4. *Lysimachia dubia.* Ait. Kew. i., p. 199.

Coast and northern plains, Huleh, etc.

5. *Asterolinum linum-stellatum.* (L. Sp. 211.)

General. Area, Mediterranean region, Portugal.

6. *Cyclamen coum.* Mill. Dict., No. 6.

Sub-alpine Lebanon.

7. *Cyclamen latifolium.* Sibth. Fl. Gr. ii., p. 71, pl. 185. = *C. persicum.* Mill.

Universal, and extremely abundant.

8. *Cyclamen repandum.* Sibth. Fl. Græc. ii., p. 72, pl. 186. = *C. hederæfolium.* Ait. Area, South Europe.

9. *Androsace villosa.* L. Sp. 203.

Top of Lebanon, snow-line. Area, mountains of Central Europe and Asia.

10. *Androsace multiscapa.* De Cand. Prodr. viii., p. 51.

Alpine Lebanon.

11. *Androsace maxima.* L. Sp. 203.

Hill districts, Jerusalem, etc. Area, Central and Southern Europe, Danubian region, Siberia, interior of North Africa.

12. *Primula acaulis.* Jacq. Misc. i., 158. Primrose.

Galilee, Lebanon. Area, Europe, North Africa.

ORDER LXXI., MYRSINEACEÆ.

Absent.

ORDER LXXII., EBENACEÆ.

1. *Diospyrus lotus.* L. Sp. 1510. Cultivated.

ORDER LXXIII., AQUIFOLIACEÆ.

ORDER LXXIV., STYRACACEÆ.

1. *Styrax officinale.* L. Sp. 535. Storax Tree.

Hill regions abundant, Gilead, Carmel, Tabor, Galilee, etc. Area, Dalmatia.

ORDER LXXV., OLEACEÆ.

1. *Olea europæa.* L. Sp. 11. The Olive. Heb. זַיִת, Arab. زيتون, *Zaytoun.*

Universally cultivated, except in Jordan valley and the mountain region. Area, Mediterranean region.

2. *Phillyrhæa media.* L. Sp. 10, and var. *Latifolia.* Sibth.

Carmel, Tabor, Lower Lebanon. Area, Mediterranean region.

3. *Phillyrhæa angustifolia.* L. *Non vidi.*

4. *Fontanesia phillyreoides.* Labill. Syr. Dec. i., p. 9, pl. 1.

North of Lebanon. Area, East Sicily.

5. *Fraxinus ornus.* L. Sp. 1510.

Lower Lebanon. Area, Central and Southern Europe.

6. *Fraxinus oxyphylla.* M. B. Taur. Cauc. ii., p. 450. Var. *Syriaca.* Boiss.

General by waterside, Northern Palestine, Phœnicia. Area, Danubian region, South Russia.

7. *Fraxinus parvifolia.* Lam. Dict. ii., p. 540.

Northern mountains. Area, South Italy, Sicily.

ORDER LXXVI., JASMINEÆ.

1. *Jasminum fruticans.* L. Sp. 9. = *J. syriacum.* Boiss. Jasmine.

Mount Hermon and surrounding district. Area, Mediterranean region.

2. *Jasminum officinale.* L. Sp. 9. White Jasmine.

Beyrout; perhaps introduced.

ORDER LXXVII., SALVADORACEÆ.

1. *Salvadora persica.* Garcin. Ex. L. Gm., ed. vi., p. 168. Arab. خاردل, *Khardal.*

Ghor, Safieh, south-east of Dead Sea; Engedi, west of Dead Sea; Seisaban, north-east of Dead Sea.

ORDER LXXVIII., APOCYNEÆ.

1. *Vinca herbacea.* W. K. Pl. rar. Hung., p. 8, pl. 9.

Plains of Sharon and Esdraelon, Moab, etc. Area, Danubian region.

2. *Vinca libanotica.* Zucc. Act. Acad. Monac. iii., p. 246, pl. 8.

Lebanon and Anti-Lebanon to 6,500 feet.

3. *Nerium oleander.* L. Sp. 305. Oleander.

Jordan valley, and the banks of all the streams flowing into it, both east and west; maritime plains; very abundant. Area, Mediterranean region.

ORDER LXXIX., ASCLEPIADEÆ.

1. *Periploca græca.* L. Sp. 309.

Galilee, Leontes valley. Area, Central Italy, Dalmatia.

2. *Periploca lævigata.* Ait. Kew, i., p. 301.

Northern coast. Area, South-east Spain, Sicily, North Africa, Canaries.

3. *Vincetoxicum canescens.* (Willd. Nov. Act. Nat. Cur. iii., p. 418.)

The Hauran.

4. *Solenostoma argel.* (Delil. Fl. Eg., p. 216, pl. 20.) Arab. ارغل, *Argel.* Wâdys south of Dead Sea.

5. *Calotropis procera.* (Willd. Sp. i., p. 1263.) Arab. عشر, *Osher.*

Engedi, Zuweirah, Safieh, Callirrhoe, Arnon, plains of Shittim, all round Dead Sea. Area, Senegal, Sahara, Nubia, Abyssinia, North India,

6. *Dæmia cordata.* R. Br. Wern. Soc. i., p. 50.

Wâdy Zuweirah, Gorge of the Callirhoe, round Dead Sea. Always on sulphur. Area, Sahara, Nubia, Abyssinia.

7. *Oxystelma alpini.* De Cand. Prodr. viii., p. 543.

Southern Ghor.

8. *Cynanchum acutum.* L. Sp. 310.

Maritime districts. Area, Mediterranean region, South-west Siberia.

9. *Glossonema boveanum.* Decaisn. Ann. Sc. Nat. ix., p. 335, pl. 12, f. *d.*

Gorges east of Dead Sea.

10. *Gomphocarpus fruticosus.* (L. Sp. 315.)

Wâdy Zuweirah and south-east of Dead Sea. Area, Dalmatia, Spain, Corsica, Sardinia, Canaries, North Africa.

11. *Gomphocarpus sinaicus.* Boiss. Diagn. Ser. i., ii., p. 80.

Wâdys south of Dead Sea.

12. *Leptadenia pyrotechnica.* (Forsk. Descr., p. 53.) Arab., مارة *Mareh.*

Zara, east of Dead Sea.

ORDER LXXX., GENTIANEÆ.

1. *Chlora perfoliata.* (L. Sp. 335.) Yellow Wort.

On bare limestone hills under Lebanon. Area, Central Europe, Danubian region, North Africa.

2. *Erythræa ramosissima.* Pers. Syn. i., p. 283.

Plains, general. Area, Europe, West Siberia, North Africa.

3. *Erythræa centaurium.* Pers. Syn. i., p. 283. Centaury.

Lebanon (lower region). Area, Europe, North Africa.

4. *Erythræa maritima.* (Willd. Sp. i., p. 1069.)

Northern coast. Area, Portugal, Mediterranean region, Canaries.

ORDER LXXXI., BIGNONIACEÆ.

1. *Tecoma undulata.* (Roxb. Fl. Ind. iii., p. 101.
Beyrout. Probably cultivated. Area, North-west India.)

ORDER LXXXII., SESAMEÆ.

1. *Sesamum indicum.* L. Sp. 884.
Ghor es Safieh, south-east of Dead Sea. Cultivated and semi-spontaneous. Area, India.

ORDER LXXXIII., CYRTANDRACEÆ.

Absent.

ORDER LXXXIV., POLEMONIACEÆ.

Absent.

ORDER LXXXV., CONVOLVULACEÆ.

1. *Convolvulus hystrix.* Vahl. Symb. i., p. 16. Arab., برحما, *Brehema.*
Southern Desert.

2. *Convolvulus lanatus.* Vahl. Symb. i., p. 16.
Southern Desert, Beersheba.

3. *Convolvulus dorycnium.* L. Sp. 224. Var. *Oxysepalus.*
Gennesaret, Huleh, northern valleys, Moab.

4. *Convolvulus cantabricus.* L. Sp. 225.
Littoral and mountain districts. Area, Central and Southern Europe, North Africa, Danubian region.

5. *Convolvulus lineatus.* L. Sp. 224.
Under Anti-Lebanon and Lebanon, Tabor, Carmel. Area, South Europe, West Siberia, North Africa.

6. *Convolvulus libanoticus.* Boiss. Diagn. Ser. i., ii., p. 82.
Summits of Lebanon and Anti-Lebanon. P.

7. *Convolvulus secundus.* Descr. Encycl. iii., p. 553.
Littoral and central districts, Mount Tabor, etc.; general.

8. *Convolvulus hirsutus.* M. B. Taur. Cauc. i., p. 422.
General.

9. Ditto, var. *Tomentosus.* = *C. peduncularis.* Boiss.
Cedars of Lebanon.

10. *Convolvulus stenophyllus.* Boiss. Fl. Or. iv., p. 106.
Lebanon district.

11. *Convolvulus althæoïdes.* L. Sp. 122.
Littoral plains and deserts, crest of Moab range. Area, Portugal, Mediterranean region, Canaries.

12. *Convolvulus tenuissimus.* Sibth. Fl. Gr. ii., p. 79, pl. 195.
Sub-alpine Lebanon. Area, South Italy, Sicily, North Africa.

13. *Convolvulus palæstinus.* Boiss. Diagn. Ser. i., ii., p. 84.
Central Palestine. P.

14. *Convolvulus scammonia.* L. Sp. 218.
Northern Palestine, universal.

15. *Convolvulus arvensis.* L. Sp. 218. Small Bindweed.
Lebanon and Moab districts. Area, almost world-wide.

16. *Convolvulus cœlesyriacus.* Boiss. Diagn. Ser. i., ii., p. 85.
Cultivated land, Phœnicia, Anti-Lebanon, Bukâà, etc. P.

17. *Convolvulus siculus.* L. Sp. 223.
Jordan valley, Jericho, Moab. Area, Mediterranean region, Nubia, Canaries.

18. *Convolvulus stachydifolius.* De Cand. Prodr. ix., p. 408. = *C. damascenus.* Boiss. Northern and eastern plains.

19. *Convolvulus pentapetaloides.* L. Syst. N. iii., p. 229.
Littoral districts and plains. Area, Mediterranean region.

20. *Convolvulus sepium.* L. Sp. 218. Greater Bindweed.
Northern plains. Area, Europe, North Africa, Siberia, Temperate North and South America, Australia, New Zealand.

21. *Ipomœa littoralis.* (L. Sp. 227.)

Maritime plains. Area, Naples, North Africa, Azores, North America.

22. *Ipomœa palmata.* Forsk. Descript., p. 43.

Plains of Gennesaret, Huleh, and Phœnicia. Area, Nubia, Abyssinia, Senegal.

23. *Cressa cretica.* L. Sp. 325.

Plain of Philistia. Area, Portugal, Mediterranean region, Nubia, Abyssinia, India, Australia.

24. *Cuscuta epithymum.* L. Syst. Ed. Murr., p. 140. Dodder.

Common on *Genista.* Area, Europe, West Siberia, North Africa.

25. *Cuscuta planiflora.* Ten. Syll., p. 128.

General in all hill-districts, on shrubs. Area, Mediterranean region, South Siberia, Western Himalayas.

26. *Cuscuta palæstina.* Boiss. Diagn. Ser. i., ii., p. 86.

On *Poterium* and other shrubs; on all the hills, especially Gilead.

27. *Cuscuta epilinum.* Weihr. Prodr. Mon. 75.

Sharon, on Flax; Moab and Gilead. Area, Europe, Canaries.

28. *Cuscuta arabica.* Fres. Pl. Eg., p. 95.

Ghor, east side of Jordan.

29. *Cuscuta monogyna.* Vahl. Symb. ii., p. 32.

Gennesaret, on Oleander. Area, Portugal, South France, North Italy, Servia.

ORDER LXXXVI., BORRAGINEÆ.

1. *Cordia myxa.* L. Sp. 273.

Ghor, north of Dead Sea; on both sides of Jordan. Area, India.

2. *Heliotropium supinum.* L. Sp. 187.

On the coast. Area, Portugal, Mediterranean region, Abyssinia, Senegal.

3. *Heliotropium aleppicum.* Boiss. Diagn. Ser. i., ii., p. 88.
Phœnicia.

4. *Heliotropium europæum.* L. Sp. 187.
On the coast and inland plains. Area, Mediterranean and Danubian regions.

5. *Heliotropium villosum.* Willd. Sp. i., p. 741.
General.

6. *Heliotropium bovei.* Boiss. Diagn. Ser. i., ii., p. 87.
Cultivated land, maritime and inland plains.

7. *Heliotropium luteum.* Poir. Suppl. iii., p. 22.
Wâdy Zuweirah, south-west end of Dead Sea.

8. *Heliotropium rotundifolium.* Boiss. Fl. Or. iv., p. 144.
Hill-district of Judæa and Moab. P.

9. *Heliotropium arbainense.* Fresen. Beitr., p. 168.
Desert south-west of Dead Sea. Area, Nubia, Tropical Arabia.

10. *Heliotropium undulatum.* Vahl. Symb. i., p. 13. = *Lithospermum hispidum.* Forsk. Shores of Dead Sea, Gaza.

11. *Heliotropium persicum.* Lam. Dict. iii., p. 94.
Southern Desert. Area, Tropical Arabia.

12. *Cerinthe minor.* L. Sp. 772.
Northern district. Area, Central and Southern Europe, Danubian region.

13. *Cerinthe major.* Lam. Dict. iv., p. 67.
Central Palestine, Bashan. Area, Switzerland.

14. *Borago officinalis.* L. Sp. 197. Borage.
Cultivated and waste land. Area, Central and Southern Europe, North Africa.

15. *Anchusa hybrida.* Ten. Fl. Nap. i., p. 45, pl. 11.
General. Area, Italy, Sicily.

16. *Anchusa undulata.* L. Sp. 181.
Littoral, central, and mountain districts. Area, South Europe.

17. *Anchusa strigosa.* Labill. Syr. Dec. iii., p. 7, pl. 4.
Universally common.

18. *Anchusa neglecta.* De Cand. Prodr. x., p. 49.
Alpine and sub-alpine Lebanon and Anti-Lebanon to 8,000 feet.

19. *Anchusa milleri.* Willd. Enum. i., p. 179.
Eastern range of Gilead.

20. *Anchusa aggregata.* Lehm. Asp., p. 219, pl. 47.
Maritime plains, Jordan valley. Area, Sicily, North-east Africa

21. *Anchusa hispida.* Forsk. Eg., p. 40.
Southern Desert.

22. *Anchusa ægyptiaca.* (L. Sp. 198.)
Very common everywhere.

23. *Anchusa italica.* Retz. Obs. i., p. 12.
Common to 5,800 feet, east and west. Area, Mediterranean region, West Siberia, Madeira.

24. *Anchusa officinalis.* L. Sp. 191. Common Alkanet.
Jerusalem, etc. Area, Northern and Central Europe.

25. *Anchusa aucheri.* De Cand. Prodr., x., p. 49.
Hermon, 8,000 feet.

26. *Anchusa orientalis.* (L. Sp. 199.)
Area, Spain, South Russia, Thibet.

27. *Nonnea obtusifolia.* Willd. Sp. i., p. 780. = *N. lamprocarpa.* Gris. Hill-districts.

28. *Nonnea melanocarpa.* Boiss. Diagn. Ser. i., ii., p. 96.
Near Jerusalem.

29. *Nonnea philistæa.* Boiss. Diagn. Ser. i., ii., p. 96.
Plain of Philistia, Beersheba. P.

30. *Nonnea ventricosa.* Sibth. Fl. Gr. ii., p. 58, pl. 168.

Littoral, inland, and Moabite plains. Area, Spain, South France, Dalmatia.

31. *Symphytum orientale.* L. Sp. 195.

Gilead and Bashan. Cultivated in Europe.

32. *Symphytum palæstinum.* Boiss. Diagn. Ser. i., ii., p. 94.

Jerusalem, Tabor, Galilee, Anti-Lebanon, Gilead.

33. *Podonosma syriacum.* (Labill. Syr. Dec. iii., p. 8, pl. 5.) = *Onosma syriacum.*

Upon walls and ruins, Jerusalem, rocks in Moab, etc. ; general.

34. *Onosma aleppicum.* Boiss. Diagn. Ser. i., ii., p. 107.

Wilderness of Judæa, northern desert plains.

35. Ditto, var. *Xanthotrichum.* Boiss. Anti-Lebanon.

36. *Onosma flavum.* (Lehm. Mag. Ges. Nat. Berl. viii., p. 92, pl. 4.)

Lebanon and Anti-Lebanon region.

37. *Onosma frutescens.* Lam. Ill., No. 1837.

On all the hills and mountains of Palestine, in rocky places.

38. *Onosma cassium.* Boiss. Diagn. Ser. i., ii., p. 102.

Northern Lebanon.

39. *Onosma cærulescens.* Boiss. Diagn. Ser. i., ii., p. 110.

Round Anti-Lebanon. P.

40. *Onosma rascheyanum.* Boiss. Diagn. Ser. i., ii., p. 110.

Slopes of Hermon, near Rascheya.

41. *Onosma roussæi.* De Cand. Prodr. x., p. 49.

Alpine Lebanon, above the Cedars.

42. *Onosma stellulatum.* W. K. Pl. Rar. Hung. ii., p. 189, pl. 173. Var. *Brevifolium.* De Cand.

Lebanon. Area, Switzerland, South Germany, Italy, Danubian region.

43. *Onosma giganteum.* Lam. Ill. 1840.
Waste lands and hills, Jerusalem, Esdraelon, Sharon, etc.

44. *Onosma auriculatum.* De Cand. Prodr. x., p. 61.
Plains of Philistia, Sharon, and Esdraelon.

45. *Onosma echioides.* L. Sp. 196.
Galilean and Moabite hills. Area, Central and Southern Europe, West Siberia.

46. *Onosma syriacum.* Willd. Sp. i., p. 774. Ainât, Lebanon.

47. *Onosma setosum.* Ledeb. Fl. Atl. i., p. 181.
Desert plain of Sahra, under Hermon.

48. *Echium italicum.* L. Sp. 139.
Maritime plains. Area, Central and Southern Europe, North Africa.

49. *Echium glomeratum.* Poir. Dict. viii., p. 670.
Coast, central plains, Mount Tabor, Jordan valley.

50. *Echium sericeum.* Vahl. Symb. ii., p. 35.
Southern Desert. Area, Tunis.

51. *Echium calycinum.* Viv. Ann. Bot. i., p. 164.
Common near the coast. Area, Mediterranean region.

52. *Echium hispidum.* Sibth. Fl. Gr., p. 68, pl. 181. = *E. sericeum.* Var. *hispidum.* Boiss.
Plains of Sharon and Philistia, hills from Jerusalem southwards.

53. *Echium rauwolfii.* Del. Fl. Eg., p. 212, pl. 19, f. 3.
South of Dead Sea. Area, Nubia.

54. *Echium plantagineum.* L. Mant. 202. = *E. violaceum.* Koch.
On the coast, plain of Gennesaret, plains of Moab. Area, Mediterranean region, Madeira, Canaries.

55. *Echium arenarium.* Guss. Ind. Hort. Bocc., 1825.
Phœnician plain. Area, Italy, Sicily, Corsica, North Africa, Canaries.

56. *Echiochilon fruticosum.* Desf. Atl. i., p. 167, pl. 47.
Littoral and inland plains. Area, North Africa, Nubian coast.

57. *Arnebia hispidissima.* (Spreng. Syst. i., p. 556.)
Southern Desert. Area, Nubia, Scinde.

58. *Arnebia cornuta.* Ledeb. Pl. Atl. i., p. 175.
Desert south of Dead Sea.

59. *Arnebia linearifolia.* De Cand. Prodr. x., p. 95.
Wâdys south-east of Dead Sea.

60. *Arnebia tinctoria.* Forsk. Eg., p. 63. Arab., سغارط ال ارنب *Sagaret el Arneb.* Southern Desert.

61. *Lithospermum arvense.* L. Sp. 190. Corn Gromwell.
Universal from coast to the Moab desert. Area, Europe, North Africa, Siberia, North-west India.

62. *Lithospermum incrassatum.* Guss. Prod. Fl. Sic. i., p. 211.
Sub-alpine and alpine Lebanon and Anti-Lebanon. Area, mountains of Mediterranean region.

63. *Lithospermum tenuiflorum.* L. Fil. Suppl. 130.
General, except mountains. Area, Dalmatia, North Africa.

64. *Lithospermum apulum.* (L. Sp. 189.)
Northern districts. Area, Mediterranean region.

65. *Lithospermum callosum.* Vahl. Symb. i., p. 14.
By Dead Sea, deserts round Gaza. Area, East Sahara.

66. *Moltkia cærulea.* (Willd. Sp. i., p. 775.)
Eastern plains and deserts.

67. *Alkanna strigosa.* Boiss. Diagn. Ser. i., iv., p. 46.
Waste places, Jerusalem, Hebron, Northern Palestine, Gilead, and Moab. P.

68. *Alkanna tinctoria.* (L. Sp. 192.)
Maritime plains, plain of Esdraelon. Area, Mediterranean region.

69. *Alkanna orientalis.* (L. Syst. 156.) Arab., لبهط, *Lebett.* General from Jordan valley to Cedars of Lebanon.

70. *Alkanna gallilæa.* Boiss. Diagn. Ser. i., ii., p. 118. Waste land, plain of Hattin. P.

71. *Alkanna microphylla.* Boiss. Diagn. Ser. i., ii., p. 120. Higher Lebanon.

72. *Alkanna kotscheyana.* De Cand. Prodr. x., p. 98. = *A. amplexicaulis.* De Cand.
Hermon, 6,000 feet. (Doubtfully identified.)

73. *Myosotis hispida.* Schlecht. Mag. Nat. Ber., viii., p. 229. Forget-me-not.
Galilee, Lebanon, Anti-Lebanon. Area, Europe, North Africa.

74. *Myosotis stricta.* List. Enum. Ber. i., p. 164.
Northern Palestine and mountains. Area, Europe, Western Siberia.

75. *Myosotis refracta.* Boiss. Voy. Esp., p. 443, pl. 125.
Sub-alpine and alpine Lebanon and Anti-Lebanon. Area, mountains of Southern Spain.

76. *Rochelia stellulata.* Rchb. Flora, 1824, p. 243.
Northern plains, Anti-Lebanon. Area, Central Spain, Danubian region, Turkomania, West Himalayas, Algerian Sahara.

77. *Echinospermum spinocarpos.* (Forsk. Descr. 41.)
Southern Desert.

78. *Echinospermum barbatum.* (M. B. Taur. Cauc. i., p. 421.)
Anti-Lebanon district. Area, South-east Europe, Western Siberia.

79. *Paracaryum myosotoides.* (Labill. Dec. ii., p. 6, pl. 2.)
Alpine-Lebanon and Hermon.

80. *Paracaryum lamprocarpum.* Boiss. Diagn. Ser. i., ii., p. 131.
Rocks at Zebdany, Bukâà. P.

81. *Cynoglossum nebrodense.* Guss. Prodr. i., p. 216.

Sub-alpine and alpine Lebanon and Anti-Lebanon. Area, Southern Spain, Sicily, North Africa.

82. *Cynoglossum pictum.* Ait. Kew. i., p. 179.

Maritime plains, Gennesaret, Moab, etc. Area, Mediterranean region, Canaries.

83. *Cynoglossum officinale.* L. Sp. 192. Hound's Tongue.

Area, Europe, North Africa, Siberia.

84. *Trachelanthus foliosa.* (Paine. Pal. Expl. Soc., pt. 3, p. 114.)

Highlands of Gilead. P.

85. *Solenanthus tournefortii.* De Cand. Prodr. x., p. 164.

Lebanon and Anti-Lebanon.

86. *Solenanthus amplifolius.* Boiss. Diagn. Ser. i., ii., p. 126.

Rocky places, Anti-Lebanon. P.

87. *Cyphomattia lanata.* (Lam. Ill., No. 1802.) Northern hills.

88. *Mattia schlumbergeri.* Boiss. Fl. Or. iv., p. 274. Lebanon. P.

89. *Asperugo procumbens.* L. Sp. 198. Madwort.

Moab and highlands generally. Area, Europe, North Africa, Western Siberia.

90. *Caccinia russelii.* Boiss. Diagn. Ser. i., ii., p. 134.

Plains in the north.

91. *Trichodesma africanum.* (L. Sp. 197.)

Wâdys east of Dead Sea. Area, Northern India, Cape Verde, Senegal, the Cape.

92. *Trichodesma ehrenbergii.* Schweinf. in Sched. Pl. Eg., 1857.

Wâdy Zuweirah, south-west of Dead Sea. Area, Arabia.

ORDER LXXXVII. SOLANACEÆ.

1. *Solanum nigrum.* L. Sp. 266. Nightshade.

Waste ground everywhere. Area, all temperate and tropical regions.

2. *Solanum miniatum.* Berh. in Willd. Enum. Ber. i., 236.

In waste ground, Central Palestine. Area, Central and Southern Europe, Nubia.

3. *Solanum villosum.* Lam. Dict. iv., p. 289.

General. Area, Central and Southern Europe, North Africa.

4. *Solanum dulcamara.* L. Sp. 264. Bitter-sweet.

Shady places, under Lebanon. Area, Europe, Siberia, Japan, China, North Africa.

5. *Solanum sanctum.* L. Sp. 269. =*S. coagulans.* Forsk. Fl. Eg. Arab., p. 47. Hebr., חֵדֶק; Arabic, حادك, *Khadak.*

Round the Dead Sea. Area, Nubia, Abyssinia, Tropical Arabia.

6. *Physalis alkekengi.* L. Sp. i., p. 262.

Non vidi. Area, Central and Southern Europe.

7. *Withania somnifera.* (L. Sp. 261.)

Waste places on the coast and Jordan valley. Area, Southern Spain, Sardinia, Sicily, North Africa, Abyssinia, India.

8. *Lycium europæum.* L. Syst. i., 228.

In warm plains and upper Jordan valley. Area, Mediterranean region, Canaries.

9. *Lycium arabicum.* Boiss. Fl. Or. iv., p. 289.

Round the Dead Sea shore. Area, Nubia, Tropical Arabia, Northern India.

10. *Mandragora officinarum.* L. Sp. 181. = *M. vernalis.* Mandrake. Arab., ربوه, *Rabouhe.* Hebr., דּוּדָאִים.

In all the plains, littoral and inland; in valleys and Jordan basin; and plains of Moab. Gilead. Area, Spain, Italy, Dalmatia.

11. *Mandragora autumnalis.* Spreng. Syst. i., p. 693.

Non vidi. Area, Mediterranean region.

12. *Datura stramonium.* L. Sp. 179. Thorn Apple.

Near towns and among ruins; general. Area, Southern Europe, Asia, Africa, North America.

13. *Hyoscyamus muticus.* L. Mant. 45.

Judæan wilderness.

14. *Hyoscyamus reticulatus.* L. Sp. 257. = *H. pinnatifidus.* Schlecht.

Galilee, Lebanon and Anti-Lebanon, Moab.

15. *Hyoscyamus albus.* L. Sp. 257.

General in plains and lower grounds. Area, Mediterranean region, Canaries.

16. *Hyoscyamus aureus.* L. Sp. 257.

Gerizim, Jerusalem, etc., in central and littoral country. Gilead and Moab. Area, Southern Europe.

17. *Hyoscyamus pusillus.* L. Sp. 258.

The Hauran. Area, Central Asia.

18. *Hyoscyamus niger.* L. Sp. 257. Henbane.

Northern districts. Area, Europe, North Africa, Siberia.

ORDER LXXXVIII., SCROPHULARIACEÆ.

1. *Verbascum antiochium.* Boiss. Diagn. Ser. i., xii., p. 9.

Anti-Lebanon.

2. *Verbascum simplex.* Labill. Déc. iv., p. 10, pl. 5.

On the plains and low hills round Lebanon and Anti-Lebanon.

3. *Verbascum ptychophyllum.* Boiss. Diagn. Ser. i., xii., p. 8. = *V. undulatum.* Lam. (pt.)

Base of Anti-Lebanon ; western face of hills of Moab.

4. *Verbascum cæsareum.* Boiss. Diagn. Ser. i., xii., p. 7.

Galilee, Banias. P.

5. *Verbascum galilæum.* Boiss. Diagn. Ser. i., xii., p. 8.

General. Hills, plains, and Jordan valley, not mountains.

6. *Verbascum syriacum.* Schrad. Mon. ii., p. 6, pl. 1, f. 1.

Northern plains and deserts.

7. *Verbascum sinaiticum.* De Cand. Prodr. x., p. 236. Arab. خرما, *Cherma.*

Wâdys by Dead Sea, Anti-Lebanon.

8. *Verbascum tripolitanum.* Boiss. Diagn. Ser. i., xii., p. 9.

Lebanon, near Broummana.

9. *Verbascum sinuatum.* L. Sp. 254.

Common on the plains and lower hills. Area, Mediterranean region, Canaries.

10. *Verbascum cedreti.* Boiss. Diagn. Ser. i., xii., p. 19.

Sub-alpine and alpine Lebanon and Anti-Lebanon. P.

11. *Verbascum damascenum.* Boiss. Diagn. Ser. i., xii., p. 19.

Barren hills of Salahiyeh.

12. *Verbascum tiberiadis.* Boiss. Diagn. Ser. i., xii., p. 26.

Round Lake of Galilee and Upper Jordan valley. P.

13. *Verbascum chryserium.* Schrad. Mon. ii., p. 3.

Plain of Sharon.

14. *Verbascum scaposum.* Boiss. Diagn Ser. i., xii., p. 26.

Gilead and the Hauran.

15. *Verbascum berytheum.* Boiss. Diagn. Ser. i., xii., p. 28.

In hedges, etc., near Beyrout. P.

16. *Verbascum blancheanum.* Boiss. Fl. Or. iv., p. 345.

Waste places on the coast.

17. *Verbascum boerrhavii.* L. Mant., p. 45.

(Coll., W. A. Haynes.) Area, Mediterranean region.

18. *Verbascum blattaria.* L. Sp. 254. Moth Mullein.

Under Anti-Lebanon. Area, Central and Southern Europe, South Siberia, North Africa.

19. *Celsia alpina.* Boiss. Diagn. Ser. ii., iii., p. 150.

Glens of Anti-Lebanon, 5,500 feet. P.

20. *Celsia pinetorum.* Boiss. Diagn. Ser. i., xii., p. 30.
Higher woods, Gilead.

21. *Celsia heterophylla.* Desf. in Pers. Syn. ii., p. 161.
Wet places, lower and sub-alpine districts.

22. *Celsia orientalis.* L. Sp. 866.
General. Area, Dalmatia.

23. *Anarrhinum orientale.* De Cand. Prodr. x., p. 289.
In all the lower dry districts and on stony hills, Lebanon to Hebron.

24. *Linaria floribunda.* Boiss. Diagn. Ser. i., xii., p. 40.
Southern Desert, Ghor, etc., north and south of Dead Sea. P.

25. *Linaria lanigera.* Desf. Atl. ii., xxxviii., pl. 130.
Maritime plains. Area, Iberian Peninsula, Canaries.

26. *Linaria elatine.* (L. Sp. 821.) Var. *Villosa.* = *L. bombycina.* Boiss. Fluellen.

Jordan valley and sandy plains. Area, Central and Southern Europe, North Africa, Abyssinia, Canaries, Madeira.

27. *Linaria damascena.* Boiss. Diagn. Ser. ii., vi., p. 130.
Jebel el Assouad. P.

28. *Linaria græca.* (Bory et Ch. Fl. Pelop., No. 796, pl. 21.)
Coast and hill districts. Area, Mediterranean region, Dalmatia, Canaries.

29. *Linaria ægyptiaca.* (L. Sp. 851.)
Judæan wilderness, Galilee, Moab. Area, North-east Africa.

30. *Linaria aucheri.* Boiss. Diagn. Ser. i., vii., p. 44.
Waste ground near Lebanon. P.

31. *Linaria pelisseriana.* (L. Sp. 855.)
Maritime plains. Area, Southern Europe.

32. *Linaria arvensis.* (L. Sp. 855.) = *L. simplex.* De Cand.
General. Area, Central and Southern Europe, North Africa.

33. *Linaria micrantha.* (Cavan. Sc. i., p. 51.)

Northern plains. Area, Southern Europe.

34. *Linaria spuria.* L. Sp. 851.

In fields. Area, Central and Southern Europe, North Africa, Canaries.

35. *Linaria chalepensis.* (L. Sp. 857.)

Plains, maritime and inland, Moab. Area, Southern Europe.

36. *Linaria hælava.* (Forsk. Eg. Arab., p. iii.)

Plain of Philistia, in sand.

37. *Linaria ascalonica.* Boiss. Diagn. Ser. ii., iii., p. 165.

Near Askalon. P.

38. *Linaria albifrons.* (Sibth. Prodr. i., p. 432.)

Judæa and Philistia. Area, Tunis.

39. *Linaria triphylla.* (L. Sp. 852.)

Moab. Area, Mediterranean region.

40. *Linaria supina.* (L. Sp. 856.) Variety? Differs from typical *L. supina.*

Plains of Moab. Area, West Mediterranean region.

41. *Linaria persica.* Chav. Mon., p. 174. In the north.

42. *Antirrhinum orontium.* L. Sp. 860. Lesser Snapdragon.

General. Area, Europe, Siberia, Himalayas, North Africa, Abyssinia, Canaries, Madeira.

43. *Antirrhinum majus.* L. Sp. 859. Great Snapdragon.

Northern Palestine. Area, Central and Southern Europe, North Africa.

44. *Scrophularia peregrina.* L. Sp. 866.

Plains. Area, Mediterranean region.

45. *Scrophularia alata.* Gilib. Fl. Lithu. ii., p. 127. Var. *cordata* = *S. pisidica.* Boiss.

Northern plains and mountains. Area, West Siberia.

46. *Scrophularia macrophylla.* Boiss. Diagn. Ser. i., xii., p. 32.
Universal in wet places on the plains and Jordan valley.

47. *Scrophularia michoniana.* Coll. et Kral. Cat. Pl. Palest., p. 13 = *S. rubricaulis.* Boiss. Universal.

Ditto, var. *tenuisecta* = *S. hierochuntina.* Boiss. In wet places, Jordan valley, and maritime plains.

48. *Scrophularia lucida.* L. Sp. 865.
Non vidi. Area, South Italy.

49. *Scrophularia scariosa.* Boiss. Diagn. Ser. i., iv., p. 67.
The Bukâà.

50. *Scrophularia sphærocarpa.* Boiss. Diagn. Ser. ii., iii., p. 158.
Western base of Lebanon.

51. *Scrophularia xylorrhiza.* Boiss. Fl. Or. iv., 406.
Hills of Central Palestine.

52. *Scrophularia xanthoglossa.* Boiss. Diagn. Ser. i., xii., p. 38.
Common in hill districts, Hebron, Jerusalem, Moab, etc., to Lebanon.

53. Ditto, var. *decipiens* = *Scr. decipiens.* Boiss.
Lebanon and Anti-Lebanon.

54. *Scrophularia deserti.* Del. Fl. Eg., p. 96, pl. 33, f. 1.
Southern Desert, Wâdy Akabah.

55. *Scrophularia variegata.* M. B. Taur. Cauc. ii., p. 78.
Wâdy Zuweirah, south-west of Dead Sea.

56. Ditto, var. *libanotica.* = *Scr. libanotica.* Boiss.
Lebanon, above the Cedars, alpine Anti-Lebanon.

57. *Scrophularia canina.* L. Sp. 865.
Plain of Esdraelon. Area, Central and Southern Europe, North Africa.

58. *Scrophularia syriaca.* De Cand. Prodr. x., p. 316.
Near Nazareth.

59. *Anticharis glandulosa.* Aschers. Ber. Ac. Wiss. Berl. 1866, p. 880.

Southern Desert. Area, Tropical Arabia, Scinde.

60. *Digitalis ferruginea.* L. Sp. 867.

Lebanon. Area, South Italy, Danubian region.

61. *Wulfenia orientalis.* Boiss. Diagn. Ser. i., iv., p. 75.

Northern Lebanon.

62. *Veronica anagallis.* L. Sp., p. 16. Water Speedwell.

Wet places, and by streams; universal. Area, Europe, Asia, North America, North Africa, Abyssinia, Canaries.

63. *Veronica anagalloïdes.* Guss. Pl. Rar., p. 5, pl. 3.

On the coast. Area, Southern Europe.

64. *Veronica beccabunga.* L. Sp. 16. Brooklime.

In stagnant water everywhere. Area, Europe, North Africa, Siberia, Japan, Himalayas, Abyssinia.

65. *Veronica multifida.* L. Sp. 17.

Lebanon. Area, South Russia, West Siberia.

66. *Veronica orientalis.* Ait. Kew. i., p. 23.

Common in the north, and in hill districts.

67. *Veronica polifolia.* De Cand. Prodr. x., p. 473.

Alpine and sub-alpine Lebanon and Anti-Lebanon, 6,500 feet.

68. *Veronica aleppica.* Boiss. Diagn. Ser. ii., iii., p. 169.

Northern plains. *Non vidi.*

69. *Veronica stenobotrys.* Boiss. Diagn. Ser. ii., iii., p. 166. = *V. leiocarpa.* Boiss. Wooded parts of Lebanon.

70. *Veronica bombycina.* Boiss. Diagn. Ser. ii., iii., p. 171.

Lebanon, 8,500 feet. P.

71. *Veronica arvensis.* L. Sp. 18. Wall Speedwell.

Cultivated and rocky places on the coast. Area, Europe, Siberia, North Africa, Canaries.

72. *Veronica viscosa.* Boiss. Diagn. Ser. i., xii., p. 47. Sandy places, alpine Lebanon.

73. *Veronica conferta.* Boiss. Fl. Or. iv. 459. = *V. glaberrima.* Boiss. Diagn. Ser. ii., iii., p. 172.
Higher parts of Lebanon.

74. *Veronica syriaca.* Rœm. et Sch. Syst. i., p. 116. = *V. pedunculata.* Labill. General on the plains, Moab.

75. *Veronica chamædrys.* L. Sp. 17. Germander Speedwell. The Hauran. Area, Central and Southern Europe, Siberia, Canaries.

76. *Veronica triphyllos.* L. Sp. 19. The Bukâà. Area, Central and Southern Europe, North Africa, West Siberia.

77. *Veronica biloba.* L. Mant. 172.
Alpine Lebanon. Area, Siberia, Himalayas.

78. *Veronica campylopoda.* Boiss. Diagn. Ser. i., iv., p. 80. Lebanon and Anti-Lebanon.

79. *Veronica buxbaumii.* Ten. Nap. i., p. 7, pl. 1.
General. Area, Central and Southern Europe, North Africa.

80. *Veronica didyma.* Ten. Nap. Prodr. vi.
Common on cultivated land. Area, Europe, North Africa.

81. *Veronica agrestis.* L. Sp. 18. Field Speedwell.
Jordan valley, Moab. Area, Europe, West Siberia, North Africa.

82. *Veronica cymbalaria.* Bod. Diss., p. 3.
Maritime and inland plains, Moab. Area, Mediterranean region.

83. *Veronica cymbalarioides.* Boiss. Fl. Or. iv., p. 468. Cultivated land on the northern coast.

84. *Veronica hederæfolia.* L. Sp. 19. Ivy-leaved Speedwell. General on cultivated land. Area, Europe, North Africa.

85. *Euphragia latifolia.* (L. Sp. 841.)
The Bukâà, plains of Moab. Area, Mediterranean region.

86. *Euphragia viscosa.* (L. Sp. 839. Sub *Bartsia.*) Yellow Bartsia.

On the coast and Upper Jordan valley. Area, Spain, South France, Switzerland, Canaries.

87. *Trixago apula.* Stev. Mém. Mosq. vi., p. 4.

Littoral districts. Area, Mediterranean region, the Cape.

88. *Odontitis aucheri.* Boiss. Diagn. Ser. i., iv., p. 74. Lebanon.

89. *Odontitis serotina.* (Lam. Fl. Fr. ii., p. 350.)

General. Area, Europe except extreme north, Siberia.

90. *Rhinanthus minor.* Ehrb. Beitr. vi., p. 144. Yellow Rattle.

In the north. Area, Europe, Siberia, North America.

91. *Rhinanthus major.* Ehrb. Beitr. vi., p. 144.

Plain of Huleh. Area, Europe, Siberia.

ORDER LXXXIX., OROBANCHACEÆ.

1. *Phelipæa cærulea.* (Vill. Dauph. ii., p. 406.) Blue Broomrape.

Mount Carmel, Lebanon, on *Anthemis.* Area, Central and Southern Europe, Siberia.

2. *Phelipæa lavandulacea.* (Rchb. Crit. vii., p. 48, pl. 935.)

Common on hill districts, on various plants. Area, Mediterranean region, Canaries.

3. *Phelipæa ramosa.* (L. Syll. 133.)

On various *Leguminosæ.* General; Lebanon, Gilead, etc. Area, Central and Southern Europe, North Africa, Abyssinia, Canaries, the Cape.

4. Ditto, var. *muteli* = *Orobanche muteli.* Schultz. General; on *Compositæ*, *Leguminosæ*, etc.

Ditto, var. *nana.* Sands near Sidon.

5. *Phelipæa ægyptiaca.* (Pers. Euch. ii., p. 181.)

On many plants, *Solanum,* etc. General; especially Jordan valley and Moab. Area, Sahara, India.

6. *Phelipæa gossypina.* Baker. M. S. Herb. Kew.
Near Heshbon. *Fide* Paine. P.

7. *Phelipæa lutea.* Desf. Atl. ii., p. 60, pl. 146.
South Judæa, Hebron, Jordan valley; on *Chenopodiaceæ.* Area, South Spain, North Africa.

8. *Phelipæa tubulosa.* Schenk. Pl. Spec., p. 23.
On Tamarisks. Round Dead Sea.

9. *Phelipæa salsa.* C. A. Mey. Fl. Alb. ii., p. 461.
On *Chenopodiaceæ.* East side of Dead Sea, Callirrhoe.

10. *Phelipæa incana.* Paine. Pal. Expl. Soc., pt. 3, p. 116.
Plains of Moab. P.

11. *Orobanche palæstina.* De Cand. Prodr. xi., p. 718.
Jerusalem, Gerizim, Banias, Moab, etc.; on *Anthriscus*, etc.

12. *Orobanche speciosa.* De Cand. Fl. Fr. Suppl., p. 393. = *O. pruinosa.* Lap.
On *Leguminosæ.* General. Area, Mediterranean region.

13. *Orobanche pubescens.* Urv. Enum., p. 76.
On *Compositæ.* General. Area, South France.

14. *Orobanche minor.* Sutton. Tr. Linn. iv., p. 178. Broom-rape.
On Mount Gilead. Area, Central and Southern Europe, North Africa.

15. *Orobanche cernua.* Loefl. It., p. 152. L. Sp. 882.
On *Achilleas* and *Artemisias*, etc. General. Area, Spain, South France, South Russia, Arabia, Sahara, Australia.

16. *Orobanche.* Sp.?
Very large, over three feet, east side of Dead Sea. On *Atriplex halimus*, with brilliant deep chrome-yellow blossoms. A most gorgeous plant, which I cannot identify.

17. *Orobanche.* Sp.?
Like the last, on *Atriplex halimus*, almost as large, with deep purple flowers, and not less gorgeous than the other. From two to three feet

high. Like the last, only found near the mouth of the Callirrhoe and on the Seisaban (the plain of Shittim), at the north-east end of the Dead Sea.

ORDER XC., ACANTHACEÆ.

1. *Blepharis edulis.* (Forsk. Eg. Arab., p. 114.) = *Acanthodium spicatum.* Del. Arab. شوك الداب, *Schok ed Dabb.*

Southern Desert, Moab, Ghor es Safieh, Seisaban. Area, Nubia Tropical Arabia, North-west India.

2. *Acanthus syriacus.* Boiss. Diagn. Ser. i., ii., p. 135.

Universal in dry places.

ORDER XCI., SELAGINACEÆ.

Absent.

ORDER XCII., GLOBULARIEÆ.

1. *Globularia arabica.* Jaub. et Sp. Ill. Or. iii., pl. 260.

Occurs in the Wâdy Arabah, south of Dead Sea.

2. *Globularia alypum.* L. Sp. 139. Area, Mediterranean region.

3. *Globularia orientalis.* L. Sp. 140. The Bukâà, near Hamath.

ORDER XCIII., VERBENACEÆ.

1. *Lippia nodiflora.* (L. Sp. 28.)

Esdraelon and maritime plains. Area, Mediterranean region, South Asia, the Cape, America.

2. *Verbena officinalis.* L. Sp. 29. Vervain.

Universal. Area, Europe, North Africa, Abyssinia, North India, Canaries, the Cape, America.

3. *Verbena supina.* L. Sp. 29.

Mount Carmel, Tabor, etc. Area, Mediterranean region, Nubia, Canaries.

4. *Vitex agnus-castus.* L. Sp. 890.

Plains and Jordan valley, Gilead. Area, Mediterranean region.

ORDER XCIV., LABIATÆ.

1. *Ocymum basilicum.* L. Sp. 833. Herb. Kew. *Non vidi.*

2. *Lavandula stæchas.* L. Sp. 800.

Universal, except in mountain region. Area, Mediterranean region, Portugal, Canaries, Madeira.

3. *Lavandula multifida.* L. Sp. 800.

Wâdy Zuweirah, south-west end of Dead Sea. Area, Portugal, Spain, South Italy, Sicily, North Africa.

4. *Lavandula coronopifolia.* Poir. Dict. ii., p. 398.

Southern Judæa, near Hebron. Area, Nubia, Tropical Arabia.

5. *Mentha sylvestris.* L. Sp. 804, and vars. Horse Mint.

Lebanon. Area, Europe, Siberia, North Africa, Abyssinia, North India, the Cape, Canaries.

Ditto, var. *Stenostachya.* Southern Judæa.

6. *Mentha aquatica.* L. Sp. 805. Water Mint.

Esdraelon. Area, Europe, North Africa, Madeira, the Cape, introduced in America.

7. *Mentha pulegium.* L. Sp. 307. Pennyroyal.

Wet places, littoral district. Area, Europe, Siberia, North Africa, Abyssinia, Canaries.

8. *Lycopus europæus.* L. Sp. 30. Gipsywort.

In the north by ditches. Area, Europe, Siberia, North Africa, North America, Australia.

9. *Origanum libanoticum.* Boiss. Diagn. Ser. i., v., p. 14.

Sub-alpine and lower Lebanon. P.

10. *Origanum ehrenbergii.* Boiss. Fl. Or. iv., p. 551.

Lebanon. P.

11. *Origanum dictamnus.* L. Sp. 823. Herb. Kew.

12. *Origanum vulgare.* L. Sp. 824. Marjoram.

Lebanon. Area, Europe, Siberia, Himalayas.

Ditto, var. *Viride* = *O. hirtum.* Auch.

13. *Origanum maru.* L. Sp. 826.

Through all the hill and mountain districts of Palestine.

14. Ditto, var. *Sinaicum* = *O. nervosum.* Vogel.

Southern Desert.

15. *Origanum onites.* L. Sp. 824.

Northern plains. Area, Sicily.

16. *Thymus serpyllum.* L. Sp. 482. Wild Thyme. Var. *Angustifolius* = *T. angustifolius.* Pers.

Desert near Gaza. Area, Europe, Siberia, Himalayas, Abyssinia, Greenland.

17. *Thymus syriacus.* Boiss. Diagn. Ser. i., xii., p. 47.

Base of Hermon, the Bukâà.

18. *Thymus capitatus.* L. Sp. 795.

Hill districts. Area, Portugal, Mediterranean region.

19. *Thymus billardieri.* Boiss. Diagn. Ser. ii., iv., p. 8.

Lebanon.

20. *Thymbra spicata.* L. Sp. 795. Galilee, Central Palestine.

21. *Satureia thymbra.* L. Sp. 794.

Mount Carmel and up to Hebron, coast and hills. Area, Sardinia.

22. *Satureia cuneifolia.* Ten. Fl. Nap. v., p. 3.

Sub-alpine Lebanon. Area, South Italy, Dalmatia.

23. *Micromeria nervosa.* (Desf. Atl. ii., p. 9, pl. 121.)

Littoral and hill districts, north and south, on both sides of Jordan. Area, South Italy, Balearics, North Africa.

24. *Micromeria juliana.* (L. Sp. 793.)

Hill country and Lebanon. Area, South France, Italy, Sicily, Dalmatia.

25. *Micromeria græca.* (L. Sp. 794.)

Littoral and Lebanon districts, Gaza. Area, Portugal, Mediterranean region.

26. *Micromeria nummularifolia.* Boiss. Diagn. Ser. i., xii., p. 49.
Fissures of rocks, Lebanon and Hermon. P.

27. *Micromeria libanotica.* Boiss. Diagn. Ser. i., xii., p. 50.
Alpine Lebanon, in fissures, 8,000 feet. P.

28. *Micromeria serpyllifolia.* (M. B. Taur. Cauc. ii., p. 40.)
Northern districts, not high up.

29. Ditto, var. *Barbata.* = *M. barbata.* Boiss.
Sub-alpine Lebanon, Mount Tabor.

30. *Calamintha florida.* Boiss. Diagn. Ser. i., xii., p. 51. Herb. Kew.

31. *Calamintha graveolens.* (M. B. Taur. Cauc. ii., p. 60.)
Northern hills.

32. *Calamintha incana.* (Sibth. Fl. Gr. vi., p. 62, pl. 577.)
Littoral and lower hill-districts.

33. *Calamintha organifolia.* (Lab. Dec. iv., p. 14, pl. 9.)
Lebanon, 7,000—9,000 feet.

34. *Calamintha clinopodium.* De Cand. Prodr. xii., p. 233. Wild Basil.
Lebanon, Galilee. Area, Europe, Siberia, Japan, North Africa, Canada.

35. *Melissa officinalis.* L. Sp. 827. Common Balm.
Littoral and lower mountain districts. Area, Central and Southern Europe, Dalmatian region.

36. *Hyssopus officinalis.* L. Sp. 796.
Area, South Europe, Danubian region, West Siberia.

37. *Zizyphora clinopodioides.* M. B. Taur. Cauc. i., p. 17. Var. *canescens.* = *Z. canescens.* Bth.
Lebanon and Hermon, 5,000—6,000 feet. Area, South Siberia.

38. *Zizyphora capitata.* L. Sp. 31.
Littoral and Southern Desert districts, Moab. Area, Dalmatia, South Russia.

39. *Zizyphora acutifolia.* Montbr. et Auch. Ann. Sc. N., 1836, i., p. 43. Herb. Kew.

40. *Zizyphora tenuior.* L. Sp. 31.
Northern plains. Area, South Siberia.

41. *Salvia grandiflora.* Ettling. Salv., No. 2. Lebanon.

42. *Salvia libanotica.* Boiss. Diagn. Ser. ii., iv., p. 16.
Lower slopes of Lebanon.

43. *Salvia triloba.* L. Fil. Suppl., p. 88.
Carmel, Galilæan and central hills. Area, Sicily, South Italy.

44. *Salvia pinnata.* L. Sp. 39.
Mount Carmel, central districts, Gilead.

45. *Salvia rubifolia.* Boiss. Diagn. Ser. i., xii., p. 96.
Lebanon and Anti-Lebanon. P.

46. *Salvia pomifera.* L. Sp. 34.
Wooded districts. *Non vidi.*

47. *Salvia bracteata.* Russ. Alep. ii., p. 242.
Waste places, northern plains.

48. *Salvia rascheyana.* Boiss. Diagn. Ser. i., xii., p. 58.
Anti-Lebanon, about Rascheya. P.

49. *Salvia pinardi.* Boiss. Diagn. Ser. i., xii., p. 59.
About the base of Anti-Lebanon. P.

50. *Salvia glutinosa.* L. Sp. 37.
Mount Hermon, 4,000 feet. Area, Central Europe, Danubian region.

51. *Salvia syriaca.* L. Sp. 36.
Plain of Gennesaret, Anti-Lebanon, Bukââ, Gilead.

52. *Salvia spinosa.* L. Mant., p. 54. Southern Desert.

53. *Salvia montbretii.* Bth. Ann. Sc. Nat., 1836, ii., p. 42.
Northern hills.

54. *Salvia spireæfolia.* Boiss. Diagn. Ser. i., v., p. 5.
Eastern plains.

55. *Salvia palæstina.* Bth. Lab., p. 718.
Northern Palestine, valley of the Leontes.

56. *Salvia graveolens.* Vahl. Enum. i., p. 273. = *S. commutata.* Bth.

Southern hills and plains, Jerusalem, Carmel, Moab.

57. *Salvia sclarea.* L. Sp. 38.
Lebanon and Hermon, 4,000 feet.

58. *Salvia verbascifolia.* M. B. Taur. Cauc. iii., p. 24. Var. *cana.* = *S. atomaria.* Boiss.
Alpine Lebanon and Hermon.

59. *Salvia brachycalyx.* Boiss. Fl. Or., iv., p. 625.
Jerusalem, hill-districts up to Hermon, Moab.

60. *Salvia pratensis.* L. Sp. 35. Meadow Sage.
Galilee. Area, Europe, except Mediterranean coasts.

61. *Salvia æthiopis.* L. Sp. 39.
Plains. Area, South Europe.

62. *Salvia ceratophylla.* L. Sp. 39. Bashan and Gilead.

63. *Salvia virgata.* Ait. Kew. i. 39.
General. Area, South Europe.

64. *Salvia hierosolymitana.* Boiss. Diagn. Ser. i., xii., p. 61.
Rocky places, Jerusalem and south; under Lebanon and Anti-Lebanon, Gilead. P.

65. *Salvia viscosa.* Jacq. Misc. ii., p. 328, Ic. Rar., pl. 5.
Lower and Middle Lebanon.

66. *Salvia verbenaca.* L. Sp. 35. Wild Sage. = *S. clandestina.* L. Sp. 36.
Central hills, littoral regions, Mount Carmel, Moab. Area, Central and South-western Europe, North Africa.

67. *Salvia controversa.* Ten. Syll., p. 18.

Desert places, south and north. Area, Spain, North Africa.

68. *Salvia horminum.* L. Sp. 34.

Littoral and Central Palestine, Jordan valleys, Ghor, Moab. Area, Italy, Dalmatia, North Africa.

69. *Salvia ægyptiaca.* L. Sp. 33.

Wâdy Zuweirah, south-west end of Dead Sea.

70. *Salvia verticillata.* L. Sp. 37.

Gennesaret. Area, Central Europe, Danubian region.

71. *Salvia jùdaica.* Boiss. Diagn. Ser. i., xii., p. 61.

Littoral districts, central hills, Jerusalem, Hebron.

72. *Salvia acetabulosa.* Vahl. En. i., p. 227. Var. *simplicifolia.* = *S. mollucellæ.* Benth.

Round Lebanon and Anti-Lebanon, up to 4,000 feet.

73. *Salvia viridis.* L. Sp. 34.

Mount Carmel. Area, Mediterranean region.

74. *Salvia peratica.* Paine. Pal. Expl. Soc., pt. 3, p. 118.

Heights of Mount Gilead. P.

75. *Rosmarinus officinalis.* L. Sp. 33. Rosemary.

Area, Mediterranean.

76. *Nepeta sibthorpii.* Bth. Lab., p. 474.

77. *Nepeta cataria.* L. Sp. 796. Catmint.

Lebanon. Area, Europe, Siberia, North-west India.

78. *Nepeta leucostegia.* Boiss. Diagn. Ser. i., xii., p. 63.

Anti-Lebanon.

79. *Nepeta orientalis.* Mill. Dict., No. 9. Lebanon.

80. *Nepeta cilicica.* Boiss. De Cand. Prodr., xii., p. 388.

Sub-alpine and alpine Lebanon and Anti-Lebanon, to 9,800 feet (on top of Hermon).

81. *Nepeta glomerata.* Montb. and Auch. Ann. Sc. Nat. 1836, p. 46.
Lebanon and Hermon, sub-alpine.

82. *Nepeta curviflora.* Boiss. Diagn. Ser. i., v., p. 22.
General in the hill-country, from Hebron to Lebanon. P.

83. *Nepeta heliotropifolia.* Lam. Dict. i., p. 711.

84. *Nepeta nuda.* L. Sp. 797. Var. *albiflora.* = *N. alba.* Desf.
Lebanon, 5,000—6,000 feet. Area, Central Europe, Danubian region, Siberia.

85. *Lallemantia iberica.* (M. B. Taur. Cauc. ii., p. 54.)
Wilderness of Judæa, the Bukâà, Moab.

86. *Lallemantia canescens.* (L. Sp. 831.) *Non vidi.*

87. *Scutellaria orientalis.* L. Sp. 834. Var. *alpina.*
Alpine Lebanon. Area, South Spain, Dalmatia, West Siberia, Turkomania.

88. *Scutellaria fruticosa.* Desf. Cat. Par., p. 63.
Hill-districts, Jerusalem, Gilead, Moab, etc., up to 6,000 feet on Hermon.

89. *Scutellaria peregrina.* L. Sp. 836. Var. *sibthorpii.*
Maritime plains, Jerusalem, Lebanon, Hermon, 4,000 feet. Area, Sicily.

90. *Scutellaria utriculata.* Lab. Ic. Syr. Dec. iv., p. 11, pl. 6.
Lebanon and Anti-Lebanon.

91. *Scutellaria heterophylla.* Montb. et Auch. Ann. Sc. N. 1836, p. 45.

92. *Scutellaria albida.* L. Mant., p. 248. *Non vidi.*

93. *Prunella vulgaris.* L. Sp. 837. Self-heal.
Littoral and mountain-districts. Area, Europe, Siberia, Japan, China, North India, North Africa, North America, Australia.

94. *Marrubium libanoticum.* Boiss. Diagn. Ser. i., xii., p. 73.
Lebanon, above the Cedars. P.

95. Ditto, var. *hermonis.* = *M. hermonis.* Boiss.
Hermon, by Rascheya.

96. *Marrubium alysson.* L. Sp. 815.

Southern Desert. Area, Mediterranean region.

97. *Marrubium vulgare.* L. Sp. 816. Horehound.

Waste places, general. Area, Europe, Northern and Tropical Africa.

98. *Marrubium cuneatum.* Russ. Alepp. ii., p. 255.

Lebanon and Anti-Lebanon districts.

99. *Marrubium micranthum.* Boiss. Diagn. Ser. i., xii., p. 73.

100. *Sideritis romana.* L. Sp. 802.

Littoral districts. Area, Mediterranean region.

101. *Sideritis montana.* L. Sp. 802.

General, from lowest to alpine districts. Area, Mediterranean and Danubian regions.

102. *Sideritis libanotica.* Labill. Ic. Syr. iv., p. 13, pl. 8.

Alpine Lebanon and Hermon.

103. *Sideritis pullulans.* Vent. Hort. Cels., pl. 98.

Galilee, Phœnicia, Lebanon.

104. *Sideritis perfoliata.* L. Sp. 892. Var. *condensata.*

Near Beyrout.

105. *Sideritis purpurea.* Bth. Lab., p. 742. Area, Dalmatia.

106. *Sideritis syriaca.* L. Sp. 801. *Non vidi.*

107. *Stachys libanotica.* De Cand. Prodr. xii., p. 462. = *S. ciliaris.* Boiss.

Northern Palestine, Lebanon, Hermon, to 4,000 feet; uplands of Moab. P.

108. *Stachys cretica.* Sibth. Fl. Gr. vi., p. 47, pl. 558. And var. *paniculata.* = *S. mersinæa.* Boiss.

All the hill-districts from Lower Lebanon to Hebron.

109. *Stachys ehrenbergii.* Boiss. Fl. Or. iv., p. 721.

Alpine Lebanon. P.

110. *Stachys cassia.* Boiss. Diagn. Ser. i., xii., p. 76.
Safed, Galilee.

111. *Stachys viticina.* Boiss. Diagn. Ser. i., xii., p. 77.
Phœnicia and Lebanon.

112. *Stachys longespicata.* Boiss. et Ky. Exs., 1859.
The Bukâà.

113. *Stachys diversifolia.* Boiss. Diagn. Ser. i. xii., p. 80.
Hill-country, in scrub.

114. *Stachys hydrophilæ.* Boiss. Diagn. Ser. i., xii., p. 81.
By streams and shaded places, Phœnicia and Lebanon. P.

115. *Stachys distans.* Bth. in De Cand., Prodr. xii., p. 472.
Lower Lebanon, Phœnicia.
Ditto, var. *oxyodonta.* Upper Jordan valley, below Banias.
Ditto, var. *teucriifolia.* = *S. teucriifolia.* Boiss. Sub-alpine Lebanon.

116. *Stachys nivea.* Labill. Ic. Syr. iii., p. 5, pl. 3.
Hills of Northern Palestine.

117. *Stachys palæstina.* L. Sp. 1674.
Universal among rocks. P.

118. *Stachys annua.* L. Sp. 813. Var. *ammophila.* Boiss.
Sands on the coast. Area, Central Europe, Danubian region.

119. *Stachys bombycina.* Boiss. Diagn. Ser. i., xii., p. 79.

120. *Stachys spectabilis.* Choisy. Pl. Rar. in Hort. Gen. i., p. 27. Herb. Kew.

121. *Stachys arabica.* Horn. Hafn., p. 554.
In rich alluvial soil, plains of Esdraelon and Sharon.

122. *Stachys neurocalycina.* Boiss. Diagn. Ser. i., xii., p. 85.
Dry places, central hills, Carmel, Gerizim, Lower Lebanon, etc. P.

123. *Stachys burgsdorffioides.* Boiss. Diagn. Ser. i., xii., p. 85.

124. *Stachys satureioides.* Montb. et Auch. A. Sc. N. Ser. ii., vi., p. 51.

125. *Betonica officinalis.* L. Sp. 810. Wood Betony.
Area, Europe, North Africa, West Siberia.

126. *Lamium striatum.* Sibth. Flor. Gr. vi., p. 46, pl. 557. Var. *minus.* = *L. nivale.* Boiss. *L. rectum.* Schenk.
Alpine and sub-alpine Lebanon and Anti-Lebanon, Gilead.

127. *Lamium amplexicaule.* L. Sp. 809. Henbit.
Gilead and Moab, Southern and Central Palestine. Area, Europe. Siberia, North Africa.

128. *Lamium ehrenbergii.* Boiss. Fl. Or. iv., p. 761.
Snow-line of Lebanon.

129. *Lamium veronicæfolium.* Benth. Lab., p. 510.

130. *Lamium moschatum.* Mill. Dict., No. 4. Var. *micranthum*, Boiss. Common and general in cultivated land.

131. *Lamium purpureum.* L. Sp. 809. Red Dead-nettle.
On cultivated land everywhere. Area, Europe, Canaries, Siberia.

132. *Lamium truncatum.* Boiss. Diagn. Ser. i., xii., p. 86.
Rocky shady places, Galilee, Lebanon, Anti-Lebanon. P.

133. *Molucella lævis.* L. Sp. 821.
Northern plains, hills, and mountains, Jordan valley.

134. *Molucella spinosa.* L. Sp. 821.
Coast and southern hills, base of Hermon. Area, South Spain, South Italy, Sicily.

135. *Ballota damascena.* Boiss. Diagn. Ser. i., xii., p. 87.
Sunny hills under Hermon, Damascus.

136. *Ballota undulata.* (Fres. Mus. Senck., p. 92.)
Universal in sunny stony places.

137. *Ballota saxatilis.* Sieb. in Bth. Lab., p. 596. = *B. obliqua.* Bth.
Universal in rocky places, not mountains.

138. *Ballota nigra.* L. Sp. 814. = *B. fœtida.* Lam. Black Horehound.

Waste land everywhere. Area, Europe, North Africa.

139. *Otostegia schimperi.* (De Cand. Prodr., xii., p. 519.)

Rocks, Wâdy Zuweirah, Dead Sea.

140. *Phlomis nissolii.* L. Sp. 819. Var. *leptorrhacos.* = *P. syriaca.* Boiss. Extremely abundant in dry stony places everywhere.

141. *Phlomis orientalis.* Mill. Dict. No. 9. Var. *brachyodon.* = *P. armeniaca.* Boiss. Judæan wilderness.

142. *Phlomis brevilabris.* Boiss. Fl. Or. iv., p. 782.

Lower, sub-alpine, and alpine Lebanon and Anti-Lebanon. P.

143. *Phlomis aurea.* Decaisn. Pl. Sinaic., p. 13. Arab. عورو, *Aoroa.* In Wâdys south of Dead Sea.

144. *Phlomis chrysophylla.* Boiss. Diagn. Ser. i., xii., p. 89.

Sub-alpine Lebanon and Anti-Lebanon. P.

145. *Phlomis viscosa.* Poir. Dict. v., p. 271.

Dry places; universal; North Judæa, Gilead, Moab, Jordan valley.

146. Ditto, var. *angustifolia.* = *P. longifolia.* Boiss. Under Lebanon.

147. *Phlomis rigida.* Labill. Dec. Syr. iii., p. 15, pl. 10.

Sandy dry places, Hermon, etc.

148. *Phlomis herba-venti.* L. Sp. 819. Var. *tomentosa* = *P. pungens.* Willd.

Highlands of Moab. Mediterranean region, South Siberia.

149. *Phlomis fruticosa.* L. Sp. 818. Herb. Kew.

Area, Sardinia, Sicily, South Italy, Dalmatia.

150. *Eremostachys laciniata.* (L. Sp. 819.)

Universal in warmer parts; Philistia, Jordan valley, Gilead, Moab, etc.

151. *Prasium majus.* L. Sp. 838.

Littoral district, Esdraelon, Galilee, Gilead. Area, Portugal, Mediterranean region, Madeira.

152. *Ajuga orientalis.* L. Sp. 785.

Lebanon and Anti-Lebanon, Bashan and Gilead, to 4,000 feet. Area, Calabria, Sicily.

153. *Ajuga iva.* (L. Sp. 787.)

Near Askalon. Area, Mediterranean region, Canaries, Madeira.

154. *Ajuga chia.* (Poir. Suppl. ii., p. 772.)

Littoral region and Jordan valley. Area, Mediterranean region.

155. Ditto, var. *suffrutescens.* = *A. palæstina.* Boiss.

Hill districts, Jerusalem, etc.

156. Ditto, var. *tridactylites.* = *A. tridactylites.* Ging.

Mountain region.

157. *Ajuga chamæpitys.* (L. Sp. 787.) Yellow Bugle, or Ground Pine.

Northern inland plains. Area, Central and Southern Europe, North Africa.

158. *Ajuga lævigata.* (Russ. Alep. ii., p. 255.) On the plains.

159. *Teucrium rosmarinifolium.* Lam. Dict. ii., p. 693.

Central hill district and Carmel, Galilee. Area, Island of Lampedusa, *auct.* Nyman.

160. *Teucrium orientale.* L. Sp. 786. Var. *nivale* = *T. nivale.* Boiss. Alpine Lebanon and Anti-Lebanon, 5,000—9,500 feet.

161. *Teucrium pruinosum.* Boiss. Fl. Or. iv., p. 808.

Anti-Lebanon.

162. *Teucrium procerum.* Boiss. Diagn. Ser. ii., iv., p. 56.

Northern district, behind Sidon ; Lebanon, above 6,000 feet.

163. *Teucrium parviflorum.* Schreb. Unilab., p. 31.

Waste places round Anti-Lebanon.

164. *Teucrium lamiifolium.* Urv. Enum., p. 64. Lebanon.

165. *Teucrium chamædrys.* L. Sp. 790. Wall Germander.

Area, Europe, West Siberia, North Africa.

166. *Teucrium flavum.* L. Sp. 791. Herb. Kew. Area, Mediterranean region.

167. *Teucrium iva.* Schreb. De Cand. Prodr. xii., p. 600.
Highlands of Moab. (*Fide* Paine.)

168. *Teucrium montbreti.* Benth. Ann. Sc. N. ii., vi., p. 56.

169. *Teucrium scordioides.* Schreb. Unilab., p. 37.
Wet places in Galilee and lower Lebanon. Area, Mediterranean region.

170. *Teucrium spinosum.* L. Sp. 793.
Galilee. Area, Mediterranean region, Canaries.

171. *Teucrium divaricatum.* Sieb. in Boiss. Fl. Or. iv., p. 816.
The hill and mountain districts, Jerusalem, Carmel, etc.

172. *Teucrium socinianum.* Boiss. Fl. Or. iv., p. 818.
Eastern slopes of Anti-Lebanon. P.

173. *Teucrium polium.* L. Sp. 792.
Dry rocky places, hill districts, Lebanon, Anti-Lebanon, Jordan valley. Area, Portugal, Mediterranean region, South Russia.

174. Ditto, var. *angustifolium* = *T. capitatum.* L. Sp. 792.
Mount Carmel.

175. *Teucrium multicaule.* Montb. et Auch. Ann. Sc. N. ii., vi., p. 54.

ORDER XCV., PLUMBAGINEÆ.

1. *Acantholimon libanoticum.* Boiss. in De Cand. Prodr. xii., p. 630.
Alpine and sub-alpine Lebanon and Anti-Lebanon. P.

2. Ditto, var. *ulicinum* = *Statice ulicina.* Willd.
Top of Lebanon and Hermon, 5,000—9,500 feet.

3. *Statice sinuata.* L. Sp. 397.
Littoral districts and east of Dead Sea. Area, Portugal, Mediterranean region.

4. *Statice thouini.* Viv. Cat. Hort. Negr., p. 34. = *S. ægytiaca.* Pers.
All round the Dead Sea. Area, South Spain, North African deserts.

5. *Statice limonium.* L. Sp. 394. Sea Lavender.
Round the Dead Sea, littoral sands, Sidon, etc. Area, coasts of Europe and North Africa.

6. *Statice sieberi.* Boiss. Voy. Esp., p. 350.
On the coast, northern Palestine. Area, Spain.

7. *Statice rorida.* Sibth. Fl. Gr. iii., p. 91.
Coast of Phœnicia.

8. *Statice virgata.* Willd. En. Berol. i., p. 336. = *S. oleifolia.* Sm.
Along the coast. Area, Mediterranean region.

9. *Statice pruinosa.* L. Mant., p. 59.
Common by the Dead Sea, Askalon.

10. *Statice spicata.* Willd. Sp. i., p. 1533.
North-east Ghor, by Dead Sea. Area, West Siberia.

11. *Armeria majellensis.* Boiss. De Cand. Prodr. xii., p. 685. Var. *leucantha.* = *S. undulata.* Bory et Chaub.
Alpine and sub-alpine Lebanon and Anti-Lebanon.

12. *Plumbago europæa.* L. Sp. 215.
General. Area, Southern Europe, North Africa.

13. *Limoniastrum monopetalum.* (L. Sp. i., 296.)
On the coast. Area, Mediterranean region.

ORDER XCVI., PLANTAGINEÆ.

1. *Plantago lanceolata.* L. Sp. 164. Ribwort Plantain.
General. Jordan valley, etc. Area, Europe, North Africa.
Ditto, var. *capitata.* Lebanon and Anti-Lebanon.

2. *Plantago major.* L. Sp. 163. Plaintain.
Plain of Sharon. Area, Europe, North Africa, Siberia.

3. *Plantago albicans.* L. Sp. 165.
Southern Judæa. Area, Mediterranean region.

4. *Plantago cretica.* L. Sp. 165.
All the plain and hill districts.

5. *Plantago ovata.* Forsk. Eg. Arab., p. 31.
Eastern desert of Bashan, and Moab. Area, South Spain.

6. *Plantago haussknechtii.* Vatke. Verh. Bot. Brand., 1874, p. 53.
Gilead and Moab.

7. *Plantago notata.* Lag. Gen. et Sp., No. 102.
Plain of Philistia, in sand. Area, South-east Spain, North Africa.

8. *Plantago lagopus.* L. Sp. 165.
General; especially Gilead, Moab, and Jordan valley. Area, Mediterranean region, South Russia, Canaries, Madeira.

9. Ditto, var. *major* = *P. lusitanica.* Willd. Littoral districts.

10. *Plantago ciliata.* Desf. Atl. i., p. 137, pl. 39.
Southern Desert. (?)

11. *Plantago maritima.* L. Sp. 165. Sea Plantain.
Littoral districts, on salt land. Area, Europe and Mediterranean, Siberia, North America.

12. *Plantago phæopus.* Paine. Pal. Expl. Soc., pt. 3, p. 120.
Eastern desert of Moab. P.

13. *Plantago arabica.* Boiss. Fl. Or. iv., p. 890.
South of Dead Sea.

14. *Plantago carinata.* Schrad. Cat. Gött. = *P. recurvata.* L. Mant. 198.
Sub-alpine Lebanon. Area, Southern Europe.

15. *Plantago psyllium.* L. Sp. 167.
General in lower ground. Area, Portugal, Mediterranean region.

16. *Plantago squarrosa.* Murr. Comm. Gött., 1781, p. 38, pl. 3.

In sands on the coast.

Ditto, var. *brachystachys.* All along the coast.

17. *Plantago arenaria.* W. K. Pl. Rar. Hung. i., p. 51, pl. 51.

The Ghor and eastern hills facing it. Area, Southern Europe, Danubian region.

18. *Plantago coronopus.* L. Sp. 166. Buckshorn Plantain.

Round the Dead Sea, maritime districts. Area, Central and Southern Europe, North Africa.

19. *Plantago phæostoma.* Boiss. Diagn. Ser. ii., iv., p. 71.

South end of Dead Sea, Wâdy Zuweirah.

PLANTÆ VASCULARES.

CLASS I., DICOTYLEDONEÆ, SUB-CLASS IV., MONOCHLAMYDEÆ.

ORDER XCVII., PHYTOLACCACEÆ.

1. *Phytolacca decandra.* L. Sp. 631. Pokeweed.

Near Beyrout. Introduced from North America.

ORDER XCVIII., CYNOCRAMBEÆ.

1. *Cynocrambe prostrata.* Gærtn. Fr. i., p. 362, pl. 75. = *Theligonum cynocrambe.* L.

On walls, general, Moab. Area, Mediterranean region, Canaries.

ORDER XCIX., SALSOLACEÆ.

1. *Beta maritima.* L. Sp. 322. Common Beet.

Littoral districts round Dead Sea. Area, coasts of all Europe, North Africa, Madeira, Canaries.

2. *Chenopodium vulvaria.* L. Sp. 321. Stinking Goose-foot.

Waste places, general. Area, Europe, North Africa.

3. *Chenopodium ficifolium.* Smith. Brit. i., p. 276. Fig-leaved Goose-foot.

Fields, plains. Area, Europe, West Siberia.

4. *Chenopodium album.* L. Sp. 219. White Goose-foot.

Universal. Area, world-wide.

5. *Chenopodium opulifolium.* De Cand. Prodr. v., p. 372.

On the coast. Area, Central and Southern Europe, North Africa, Abyssinia.

6. *Chenopodium murale.* L. Sp. 318. Nettle-leaved Goose-foot.

General. Area, almost world-wide. In North America introduced.

7. *Chenopodium botrys.* L. Sp. 320.

Lebanon, South Judæa. Central and Southern Europe, North and South Africa, West Siberia, North America.

8. *Chenopodium ambrosioides.* L. Sp. 302.

Near Beyrout. Area, Mediterranean. Introduced from America.

9. *Blitum virgatum.* L. Sp. 7.

Sub-alpine Lebanon. Area, Central and Southern Europe, North Africa, Siberia, India.

10. *Atriplex nitens.* Rebent. Prodr. Neom., p. 126.

In fields. Area, Germany, Danubian region, Siberia.

11. *Atriplex hastatum.* L. Sp. 1494. Var. *Salinum.* Walln. Common Orache. In waste and dry places. Area, Europe, Siberia.

12. *Atriplex patulum.* L. Sp. 1494.

Waste places, general. Area, Europe, Siberia, North Africa.

13. *Atriplex parvifolium.* Lowe. Fl. Mad., p. 16.

By the Dead Sea. Area, Madeira, Egypt.

14. *Atriplex dimorphostegium.* Kar. et Kir. Enum. Song., No. 714.

Barren salt plains, near Gaza and round Dead Sea. Area, Sahara.

15. *Atriplex tataricum.* L. Sp. 1053. Var. *virgatum.* = *A. lasianthum.* Boiss.

Barren and salt places, Higher Lebanon, and Southern Judæa. Area, Central and Southern Europe, North Africa, Siberia.

16. *Atriplex roseum.* L. Sp. 1493.

Littoral saline districts, saline districts east of Dead Sea. Area, Maritime Europe, Siberia.

17. *Atriplex portulacoides.* L. Sp. 1493. Sea Purslane.

On the coast and in the Ghor. Area, coasts of Europe and North Africa.

18. *Atriplex palæstinum.* Boiss. Diagn. Ser. i., xii., p. 96.

All the Judæan wilderness and Southern Desert.

19. *Atriplex halimus.* L. Sp. 1492.

On the coast. Especially plentiful round the Dead Sea. Area, salt districts, Mediterranean region.

20. *Atriplex farinosum.* Forsk. Fl. Eg. Arab., p. 123, No. 302.

Wâdy Akabah, south of Dead Sea. Area, Tropical Arabia.

21. *Camphorosma monspeliacum.* L. Sp. 178.

Salt marshes on the coast. Area, Mediterranean region.

22. *Chenolea arabica.* Boiss. Diagn. Ser. i., xii., p. 97.

Judæan wilderness, slopes west of Dead Sea.

23. *Kochia monticola.* Boiss. in Ky. Pl. Pers. Austr., 1845.

Barren places, alpine Lebanon and Anti-Lebanon.

24. *Kochia latifolia.* Fresen. Beitr., p. 179.

Judæan wilderness and Southern Desert.

25. *Kochia muricata.* (L. Mant., p. 54.) = *Echinopsilon muricata.* Mocq.

Wâdy Zuweirah, south-west end of Dead Sea, Moab.

26. *Arthrocnemum glaucum.* (Del. Fl. Eg. Ill., p. 69.)

By the Dead Sea. Area, Portugal, Mediterranean region.

27. *Salicornia fruticosa.* L. Sp. 5.

Maritime sands, round the Dead Sea. Area, coasts of Central and Southern Europe, Arabia, the Cape, America.

28. *Salicornia herbacea.* L. Sp. 5. Glasswort.

In salt places, common. Area, Europe, North Africa, the Cape, North America.

29. *Halopeplis amplexicaulis.* (Vahl. Symb. ii., p. 1.)

In salt places near Gaza.

30. *Suæda asphaltica.* Boiss. Diagn. Ser. i., xii., p. 98.

Marl hills, the Ghor. P.

31. *Suæda fruticosa.* (L. Sp. 324.) Shrubby Sea-blite.

Round the Dead Sea, on the coast. Area, coasts of Europe, North Africa, the Cape, North India, America, the Canaries, Madeira.

32. *Suæda monoica.* Forsk. Eg. Arab., p. 70.

By the Dead Sea. Area, Nubia, Abyssinia, Tropical Arabia, India.

33. *Suæda vermiculata.* Forsk. Eg. Arab., p. 70.

South of Dead Sea, Wâdy Ghurundel. Area, North Africa, Nubia, India, Canaries.

34. *Suæda maritima.* L.. Sp. 321. Sea-blite.

Maritime districts. Area, Mediterranean region, Siberia, North America, Canaries.

35. *Traganum nudatum.* Del. Fl. Eg. Ill., p. 230, pl. 22, f. 1.

Southern Desert. Area, Sahara.

36. *Haloxylon articulatum.* Cav. Ic. iii., p. 43, pl. 284.

Southern Desert and Northern Desert, Hauran. Area, South-east Spain, Sahara, Tropical Arabia.

37. *Haloxylon salicornioides.* Bge. Buhse. Aufz. 189.

Judæan wilderness.

38. *Seidlitzia florida.* M. B. Taur. Cauc. i., p. 190. Eastern Desert.

39. *Salsola kali.* L. Sp. 322. Arab., الكالي, *El Kali.* Salt Wort.

All round the Dead Sea. Area, on sea-shores, world-wide.

40. *Salsola tetragona.* Del. Fl. Eg., p. 228, pl. 21, f. 4.

Round the Dead Sea ; very common. Area, the Sahara.

41. *Salsola lancifolia.* (Boiss. Diagn. Ser. i., xii., p. 98.)

The northern Ghor. P.

42. *Salsola soda.* L. Sp. 323. Herb. Kew.

Area, Portugal, Mediterranean region.

43. *Salsola oppositifolia.* Desf. Fl. Atl. i., p. 219.

South-west of Dead Sea. Area, South and South-east Spain, Sicily, Tunisian Sahara.

44. *Salsola rigida.* Pall. Ill., p. 20, pl. 12. Var. *tenuifolia.* = *S. villosa.* Del.

In salt flats and marl hills, Jordan valley. Area, West Siberia.

45. *Noea spinosissima.* L. Fil. Suppl., p. 173.

Dry rocky places, Judæa, Lebanon, Anti-Lebanon. Area, Turcomania, Affghanistan.

46. *Anabasis articulata.* Forsk. Eg. Arab., p. 55, pl. 8, f. A.

Deserts, South Judæa, and Hauran. Area, South Spain, Sahara.

47. *Anabasis aphylla.* L. Amœn. Ac. ii., p. 347.

Salt marshes. Area, South-west Siberia.

48. *Anabasis.* Sp. ? Marsaba.

49. *Halostachys caspica.* Pall. Voy. App., p. 480. Herb. Kew.

Area, South-west Asia, south end of Caspian Sea.

ORDER C., AMARANTACEÆ.

1. *Amarantus hypochondriacus.* L. Sp. 1407.

Cultivated ground, Beyrout. Probably from cultivation. Area, North America.

2. *Amarantus chlorostachys.* Willd. Amar., p. 34, pl. 10, f. 19.

Phœnician plain. Probably from cultivation. Area, America.

3. *Amarantus retroflexus.* L. Sp. 1407.

General. Area, America, Central and Southern Europe and North Africa.

4. *Amarantus paniculatus.* L. Sp. 1406.

Littoral region. Introduced. Area, America.

5. *Amarantus sylvestris.* Desf. Cat. Hort. Par. 41. =*A. viridis.* L.

Waste places; general. Area, Central and Southern Europe, North Africa, India.

6. *Albersia blitum.* Kunth. Fl. Ber. ii., p. 144.

Littoral districts, Jordan valley. Area, Central and Southern Europe, North Africa, Abyssinia, India.

7. *Ærva javanica.* Juss. Ann. Mus. ii., p. 131.

Wâdy Zuweirah, Callirrhoe, Dead Sea.

8. *Achyranthes aspera.* L. Sp. 295.

The Ghor, near Jericho. Area, Arabia, Nubia, Abyssinia, India, the Cape, Madeira, Canaries, Azores.

9. *Gomphrena globosa.* L. Sp. 326.

Probably introduced. Area, Tropical East Asia.

ORDER CI., POLYGONEÆ.

1. *Calligonum comosum.* L'Her. Tr. Lin. Soc. Lond. i., p. 180.

Southern Desert. Area, the Sahara.

2. *Calligonum polygonoides.* L. Sp. 478.

Northern and eastern deserts.

3. *Rheum ribes.* Gron. Fl. Or., p. 130. Moab, Lebanon.

4. *Oxyria digyna.* (L. Sp. 480.) Mountain Sorrel.

Lebanon. Area, mountains of Central Europe, Siberia, and North America.

5. *Emex spinosus.* (L. Sp. 481.)

Littoral districts, inland plains, and Jordan valley. Area, Portugal, Mediterranean region.

6. *Rumex orientalis.* Benh. in Schult. Fil. Syst. vii., p. 1433.

Lebanon.

7. *Rumex aquaticus.* L. Sp. 479. Herb. Kew.

Area, Northern and Central Europe.

8. *Rumex crispus.* L. Sp., p. 476. Curled Dock.

Lebanon and the north. Area, Europe, Siberia, Japan, North Africa, introduced in America, Madeira.

9. *Rumex conglomeratus.* Murr. Prodr. Gött., p. 52. Sharp Dock.

Moist places on hills, Lebanon, Anti-Lebanon. Area, Central and Southern Europe, North Africa, Canaries, introduced in North America.

10. *Rumex obtusifolius.* L. Sp. 578. Broad Dock.

General in waste places. Area, Europe, Siberia, North Africa, Canaries, Madeira.

11. *Rumex nepalensis.* Spreng. Syst. ii., p. 159.

Higher and middle Lebanon. Area, Indian mountains, Java, the Cape.

12. *Rumex pulcher.* L. Sp. 477. Fiddle Dock.

General. Area, Central and Southern Europe, North Africa, Madeira, Canaries.

13. *Rumex bucephalophorus.* L. Sp. 479.

Maritime districts. Area, Mediterranean region, Canaries, Azores.

14. Ditto, var. *uncinatus.* = *R. aculeatus.* L. Sp. 481. General.

15. *Rumex acetosa.* L. Sp. i., 481. Sorrel.

Area, Europe, Siberia, North America.

16. *Rumex lacerus.* Balb. Misc., p. 19.

Philistia, Southern Desert.

Ditto, var. *macrocarpus.* Sands near Beyrout.

17. *Rumex vesicarius.* L. Sp. 479.

Callirrhoe, Moab. Area, North Africa, Madeira, Canaries.

18. *Rumex roseus.* L. Sp. 480. Rocky places, Galilee.

19. *Rumex acetoselloides.* Bal. Bull. Soc. Bot. Fr. i., p. 282. = *R. multifidus.* L.

Lower Lebanon. Area, South Italy, Sicily.

20. *Rumex cassius.* Boiss. Fl. Or. iv., p. 1013. (Doubtful.)

21. *Rumex dentatus.* L. Mant., p. 226.
Wet places in the Ghor. Area, North India.

22. *Atraphaxis billardieri.* Jaub. et Sp. Ill. Or. ii., p. 14, pl. 111.
Lower Lebanon and Anti-Lebanon.

23. Ditto, var. *heterantha.* = *A. variabilis.* Jaub. et Sp.
Middle Anti-Lebanon.

24. *Polygonum amphibium.* L. Sp. 517. Water Bistort.
Marshes, Lebanon. Area, Northern and Central Europe, North Africa, Siberia, China, the Cape, North America.

25. *Polygonum serrulatum.* Laq. Gen. et Sp. 14.
In water, Phœnicia, and plain of Sharon. Area, Southern Europe, North Africa, Abyssinia, North India, the Cape, Canaries, Madeira.

26. *Polygonum hydropiper.* L. Sp. 517. Biting Persicaria.
In ditches under Carmel. Area, Europe, Siberia, North Africa, North America.

27. *Polygonum persicaria.* L. Sp. 518. Spotted Persicaria.
Ditches, etc.; general. Area, Europe, Siberia, India, North America.

28. *Polygonum bistorta.* L. Sp. 516. Bistort.
Alpine Lebanon. Area, temperate regions of Europe, Asia, and North America.

29. *Polygonum lapathifolium.* L. Sp. 517. Pale-flowered Persicaria.
Ditches, north, and Lebanon. Area, Europe, Siberia, Japan, India, introduced in America.

30. *Polygonum convolvulus.* L. Sp. 522. Climbing Bindweed.
General in cultivated land. Area, Europe, North Africa, North Asia, Japan, introduced in North America.

31. *Polygonum polycnemoides.* Jaub. et Sp. Ill. Or., pl. 120.
Sub-alpine and alpine Lebanon.

32. *Polygonum bellardi.* All. Ped. ii., p. 205, pl. 90, f. 2.
Lebanon up to the Cedars. Area, Southern Europe, North Africa, Siberia, North India.

33. *Polygonum aviculare.* L. Sp. 519. Knotweed.

Cultivated land; general. Area, world-wide.

Ditto, var. *littorale.* Maritime plains.

34. *Polygonum equisetiforme.* Sibth. Fl. Gr. i., p. 266, pl. 364.

Plains and lower mountain districts, Jordan valley, Moab. Area, Spain, Portugal, Sicily, North Africa.

35. *Polygonum maritimum.* L. Sp. 519. Sea Knotweed.

Littoral districts. Area, Coasts of Europe, North Africa, Canaries, Madeira, Azores, America.

36. *Polygonum alpestre.* C. A. Mey. Enum., p. 157.

Lebanon, Anti-Lebanon, Hauran. Area, South Siberia, Himalayas.

37. *Polygonum herniarioides.* Del. Fl. Eg. Ill., No. 412.

Littoral districts. Area, South Italy, North-east Africa, Senegal, India.

38. *Polygonum libani.* Boiss. Diagn. Ser. i., xii., p. 99.

Lebanon and Hermon, 8,000—9,500 feet. P.

39. Ditto, var. *cedrorum* = *P. cedrorum.* Boiss.

Sub-alpine Lebanon and the Bukâà. P.

ORDER CII., NYCTAGINACEÆ.

1. *Boerhaavia plumbaginea.* Cav. Ic. ii., p. 7, pl. 112.

Maritime plains, Esdraelon, Jordan valley. Area, South-east Spain, Nubia, Abyssinia, Tropical Arabia, India, Australia, the Cape.

ORDER CIII., THYMELÆACEÆ.

1. *Daphne sericea.* Vahl. Symb. Bot. i., p. 28.

Lower wooded regions. Area, South Italy.

2. *Daphne oleoides.* Schreb. Dec. i., p. 13, pl. 7. Area, Spain.

3. *Daphne jasminea.* Sibth. Fl. Gr. iv., p. 50. *Non vidi.*

4. *Daphne glomerata.* Lam. Dict. iii. 438.

5. *Daphne gnidioides.* Jaub. et. Sp. Ill. Or. iv., pl. 304. Herb. Kew. Lebanon and Anti-Lebanon, alpine and sub-alpine. Area, Himalayas.

6. *Lygia aucheri.* (De Cand. Prodr. xiv., p. 552.)
Lower Lebanon, Galilee.

7. *Thymelæa hirsuta.* (L. Sp. 559. = Sub *Passerina.*)
Littoral and southern districts, abundant. Area, Mediterranean region.

8. *Thymelæa tartonraira.* (L. Sp. 356.) Herb. Kew.
Area, Portugal, Mediterranean region.

ORDER CIV., ELEAGNACEÆ.

1. *Eleagnus angustifolius.* L. Sp. 176. Oleaster. Heb., עֵץ שֶׁמֶן, *Oil-tree.* (Is. xli. 19.) Arab, شنلسد, *Sindshid.*

Abundant in every part of Palestine above the Jordan valley, especially about Hebron, Samaria, and Tabor. It has a fine, hard wood, and yields an inferior oil.

It is probably not a native of South Europe, where it is cultivated; but in Asia its range extends as far as China.

ORDER CV., LAURINEÆ.

1. *Laurus nobilis.* L. Sp. 529. The Bay-tree. Heb., אֶזְרָח. (Ps. xxxvii. 35.)

Carmel, Tabor, Gilead. Area, Mediterranean region.

ORDER CVI., SANTALACEÆ.

1. *Osyris alba.* L. Sp. 1450.
Gilead, Moab, Galilee. Area, South Europe, North Africa.

2. *Thesium divaricatum.* Jan. M. and K. Deutschl. Fl. ii., p. 286.
Sub-alpine Lebanon. Area, South Europe, North Africa.

3. *Thesium libanoticum.* De Cand. Prodr. xiv., p. 648.
Alpine Lebanon, on the snow-line. P.

4. *Thesium heterophyllum.* Boiss. Ann. Sc. Nat. Ser. iv., ii., p. 354. Var. *billardieri* = *T. billardieri.* Boiss. Anti-Lebanon.

5. *Thesium humile.* Vahl. Symb. iii., p. 43.

Littoral and lower hill-districts. Area, Mediterranean region.

6. *Thesium ramosum.* Hayne. Schrad. Jour., p. 30. Herb. Kew.

7. *Thesium bergeri.* Zucc. Pl. Nov. Fasc. ii., p. 16. = *T. græcum.* Boiss.

Littoral districts, Galilee and Lebanon.

ORDER CVII., LORANTHACEÆ.

1. *Viscum album.* L. Sp. 1451. Mistletoe.

Hermon and Lebanon, from 4,000 feet upwards. Area, Europe, Siberia, Japan, North Africa.

2. *Viscum cruciatum.* Boiss. Voy. Esp., p. 274.

On the Olive. Extremely abundant, especially at Gaza, Jerusalem, Nablus. Area, South Spain, but occurs nowhere else.

3. *Arceuthobium oxycedri.* (De Cand. Fl. Fr. iv., p. 274.)

On Junipers, sub-alpine Lebanon. Area, South Europe, North Africa.

4. *Loranthus acaciæ.* Zucc. Abh. Münch. Ac. iii., p. 249, pl. 2, f. 3.

On Zizyphus and Acacia, Jordan valley, Moab, Hebron. Area, Nubia.

ORDER CVIII., CYTINACEÆ.

1. *Cytinus hypocistis.* L. Sp. 826.

Littoral districts on Cistus. Area, Mediterranean region, Canaries.

ORDER CIX., BALANOPHORACEÆ.

1. *Cynomorium coccineum.* L. Sp. 1875.

On sands on the coast; parasitic on roots. Area, Mediterranean coasts, Canaries.

ORDER CX., ARISTOLOCHIACEÆ.

1. *Aristolochia altissima.* Desf. Fl. Atl. ii., p. 324, pl. 249. = *A. sempervirens.*

Lower Lebanon and Anti-Lebanon. Area, Sicily, North Africa.

2. *Aristolochia hirta.* L. Sp. 961.

Mount Hermon, 4,000 feet.

3. *Aristolochia pœcilantha.* Boiss. Diagn. Ser. i., xii., p. 104.

Common in every part of the country, except the mountains. P.

4. Ditto, var. *scabridula.* = *A. scabridula.* Boiss.

Anti-Lebanon. P.

5. *Aristolochia maurorum.* L. Sp. 1303.

Gilead and Bashan.

6. *Aristolochia parvifolia.* Sibth. Fl. Gr. x., p. 27.

General, coast and hills.

ORDER CXI., EUPHORBIACEÆ.

1. *Euphorbia peplis.* L. Amœn. iii., p. 113. Purple Spurge.

On the coast, common, Moab. Area, Mediterranean coast, West France and England, Western Siberia, Canaries, Azores.

2. *Euphorbia granulata.* Forsk. Fl. Eg. Arab., p. 94.

Wâdy Akabah, south of Dead Sea.

3. *Euphorbia chamæsyce.* L. Amœn. 115.

Waysides, common. Area, Mediterranean region.

4. *Euphorbia lanata.* Sieb. in Spreng. Syst. iii., p. 792.

Galilee, waste places.

5. *Euphorbia cornuta.* Pers. Syr. ii., p. 17.

Southern Desert. Area, Sahara.

6. *Euphorbia altissima.* Boiss. Diagn. Ser. i., p. 52.

Moist places under Anti-Lebanon.

7. *Euphorbia arguta.* Soland. Russ. Alep. ii., p. 252.
In the plains.

8. *Euphorbia gaillardoti.* Boiss. Diagn. Ser. ii., iv., p. 84.
Cultivated land, Bukâà, and Anti-Lebanon.

9. *Euphorbia cybirensis.* Boiss. Diagn. Ser. i., vii., p. 89.
General in cultivated land, east and west of Jordan.

10. *Euphorbia apios.* L. Syst., p. 375. Var. *lamprocarpa.* Boiss.
Under Anti-Lebanon.

11. *Euphorbia thamnoides.* Boiss. Cent. Euphorb., p. 33. = *E. dumosa.* Boiss.

Littoral and plain districts, Moab.

12. *Euphorbia hierosolymitana.* Boiss. Diagn. Ser. i., xii., p. 110.
Jerusalem, Anti-Lebanon. P.

13. *Euphorbia erinacea.* Boiss. Diagn. Ser. ii., iv., p. 87.
Alpine Anti-Lebanon, 5,000—9,000 feet. P.

14. *Euphorbia pubescens.* Vahl. Symb. ii., p. 55.
General. Area, Portugal, Mediterranean region, Canaries.

15. *Euphorbia helioscopia.* L. Sp. 658. Sun Spurge.
Waste and cultivated plains; universal. Area, Europe, North Africa, Japan.

16. *Euphorbia berythæa.* Boiss. Diagn. Ser. ii., iv., p. 82.

Rocks on Phœnician coast. P.

17. *Euphorbia cassia.* Boiss. Diagn. Ser. i., xii., p. 108.
Lower and middle Lebanon.

18. *Euphorbia aleppica.* L. Amœn. iii., p. 122.
Littoral and interior plains, Moab, etc. Area, South Italy, Dalmatia.

19. *Euphorbia exigua.* L. Amœn. iii., p. 118. Dwarf Spurge.
Cultivated and waste places; universal. Area, Europe, North Africa, Canaries.

20. *Euphorbia falcata.* L. Sp. 654.

Cultivated land; general. Area, Central and Southern Europe, North Africa.

21. Ditto, var. *galilæa.* Boiss. Phœnicia and Galilee.

22. *Euphorbia aulacosperma.* Boiss. Diagn. Ser. i., xii., p. 117.

Phœnicia, Jerusalem, etc.

23. *Euphorbia peplus.* L. Sp. 658. Petty Spurge.

Coast and south. Area, Central and Southern Europe, North Africa, Canaries.

24. *Euphorbia peploides.* Gou. Fl. Monsp., p. 174.

The coast, deserts of Moab. Area, Mediterranean region.

25. *Euphorbia chamæpeplus.* Boiss. Diagn. Ser. ii., iv., p. 98.

Near Jerusalem, etc. P.

26. *Euphorbia szovitsii.* F. and M. Ind. Petrop. i., p. 27.

The Bukâà.

27. *Euphorbia reuteriana.* Boiss. Diagn. Ser. i., xii., p. 115.

Cultivated land; general. P.

28. *Euphorbia chesneyi.* (Kl. and Geke. Tric., p. 99.)

Anti-Lebanon.

29. *Euphorbia caudiculosa.* Boiss. De Cand. Prodr. xv., p. 154.

Top of Lebanon and Hermon on snow-line. P.

30. *Euphorbia segetalis.* L. Sp. 657.

Area, Central and Southern Europe, North Africa, Madeira, Canaries.

31. *Euphorbia terracina.* L. Sp. 654.

On the coast and plains. Area, Portugal, Mediterranean region, Canaries, Madeira, Azores.

32. *Euphorbia tinctoria.* Boiss. De Cand. Prodr. xv., p. 106. Var. *schizocerus.* Boiss.

Lebanon and Anti-Lebanon, highlands of Moab.

33. *Euphorbia paralias.* L. Sp. 657. Sea Spurge.
On maritime sands. Area, Western and Southern Europe, Dalmatia.

34. *Euphorbia myrsinites.* L. Sp. 661. Herb. Kew.
Area, Italy, Sicily, Dalmatia.

35. *Euphorbia phymatosperma.* Boiss. Diagn. Ser. ii., iv., p. 83.
Anti-Lebanon.

36. *Euphorbia platyphylla.* L. Sp. 660. Broad-leaved Spurge.
Area, Central and Southern Europe, North Africa.

37. *Euphorbia acanthothamnos.* Held. Sart. Suppl. 54.=*E. spinosa.* Sibth.

38. *Andrachne telephioides.* L. Sp. 1014.
Waste ground everywhere. Area, Mediterranean region, Cape de Verde.

39. *Andrachne aspera.* Spreng. Syst. iii., p. 884.
Southern Desert. Area, Nubia, Abyssinia, Tropical Arabia.

40. *Crozophora tinctoria.* (L. Sp. 1425.)
Plains and cultivated land. Area, Mediterranean region.

41. *Crozophora verbascifolia.* (Willd. Spec. iv., p. 539.)
Waste and cultivated lands; general. Area, South-east Spain, North Africa.

42. *Crozophora obliqua.* (Vahl. Symb. i., p. 78.)
South of Dead Sea. Area, Nubia, Abyssinia, Tropical Arabia, North India.

43. *Mercurialis annua.* L. Sp. 1465. Mercury.
In cultivated land; common. Area, Europe, North Africa.

44. *Ricinus communis.* L. Sp. 1430. Castor Oil Plant.
Jordan valley and hot plains and valleys; common. Area, Tropical and Sub-tropical Asia, Africa, America.

ORDER CXII., BUXACEÆ.

1. *Buxus longifolia.* Boiss. Diagn. Ser. i., xii., p. 107. Oriental Box Tree. Heb. תְּאַשּׁוּר. Is. xli., 19.

On Lebanon (sub-alpine).

ORDER CXIII., EMPETRACEÆ.

Absent.

ORDER CXIV., URTICACEÆ.

1. *Urtica urens.* L. Sp. 1396. Small Nettle.

Waste places. Area, Europe, North Africa, Abyssinia, East Asia.

2. *Urtica dioica.* L. Sp. 1396. Common Nettle.

Waste places, Lebanon and Anti-Lebanon. Area, Europe, North and South Africa, North Asia, North America, Andes, Australia.

3. *Urtica pilulifera.* L. Sp. 1395. Roman Nettle.

Lower districts among ruins; reaches a height of six feet in the Jordan valley. Area, Central and Southern Europe, North Africa.

4. *Urtica membranacea.* Poir. Dict. iv., p. 638.

On the coast. Area, Portugal, Mediterranean region.

5. *Parietaria officinalis.* L. Sp. 1492. Wall Pellitory.

Waste places and walls, Sidon, etc. Area, Europe, West Siberia, North Africa.

6. *Parietaria judaica.* L. Sp. 1492. = *P. diffusa.* Koch.

Walls, rocks, and ruins; common everywhere except Jordan valley.

7. *Parietaria lusitanica.* L. Sp. 1492.

Walls and rocks; general. Area, Portugal, Mediterranean and Danubian regions.

8. *Parietaria alsinefolia.* Del. Fl. Eg., p. 137, pl. 5, f. 2.

Dry walls and rocks, Hebron, Jerusalem, Korak, Moab.

9. *Forskahlea tenacissima.* L. Mant., p. 11.

The Ghor, east and west sides of Dead Sea. Area, Sahara, North-west India.

10. *Humulus lupulus.* L. Sp. 1457. Hop.

Anti-Lebanon, Galilee. Area, Temperate Europe, Asia, North America.

11. *Cannabis sativa.* L. Sp. 1457. Hemp.

Cultivated, but becomes wild in places.

12. *Morus nigra.* L. Sp. 1398. Black Mulberry.

Cultivated everywhere. Introduced from Persia.

13. *Morus alba.* L. Sp. 1398. White Mulberry.

Cultivated everywhere, especially on Lebanon and Anti-Lebanon. Introduced from China.

14. *Ficus carica.* L. Sp. 1513. Fig. Hebr. תְּאֵנָה. Arab. تين, *Tîn.*

Wild in fissures of rocks, walls, etc., from Lebanon to the south of the Dead Sea; cultivated everywhere. Area, Southern Europe, North Africa, North India.

15. *Ficus pseudosycomorus.* Decaisn. Fl. Sin., p. 4.

The Ghor, east of Dead Sea.

16. *Ficus sycomorus.* L. Hort. Cliff., p. 471. Sycomore Fig. Hebr. שִׁקְמִים.

On the coast and in the Jordan valley. Area, Nubia, Abyssinia.

17. *Celtis australis.* L. Sp. 1478.

Gilead, upper Jordan valley, Banias. Area, Southern Europe, North Africa.

18. *Ulmus campestris.* L. Sp. 327. Common Elm.

In the north, rare. Area, Europe, North Africa, Siberia, Himalayas.

ORDER CXV., JUGLANDACEÆ.

1. *Juglans regia.* L. Sp. 1415. Walnut. Heb. אֱגוֹז, *Nuts.* A.V. Cant. vi. 11. Arab. جوزة, *Jawzah.*

All the glens and lower slopes of Lebanon and Hermon, Galilee; cultivated. Well established and spontaneous in Southern Europe.

ORDER CXVI., PLATANACEÆ.

1. *Platanus orientalis.* L. Sp. 417. Oriental Plane. Heb. עַרְמוֹן. Trans. A.V. *Chestnut*, Gen. xxx. 37.

By the banks of mountain streams and in woods, Lebanon, etc.; but cultivated everywhere near water. Area, Italy, Sicily.

ORDER CXVII. CUPULIFERÆ.

1. *Quercus sessiliflora.* Sm. Brit. iii., 1026. Oak. Heb. אֵלָה. Area, Europe.

2. Var. *cedrorum.* De Cand. Higher parts of Lebanon, the Hauran, Gilead.

3. Var. *pinnatifida.* Lebanon, 5,000 feet. Area, Central and Southern Europe.

4. *Quercus infectoria.* Oliv. Voy. i., p. 252, pl. 14 and 15. = *Q. lusitanica.* Lam.

Northern and Eastern Palestine. Area, Portugal, Spain, Oran.

5. Ditto, var. *boissieri.* De Cand. Lebanon.

6. Ditto, var. *latifolia.* = *Q. syriaca.* Ky. Lebanon and Anti-Lebanon.

7. *Quercus ilex.* L. Sp. 1412. Holm Oak.

Near the coast; rare. ? Indigenous. Area, Mediterranean region.

8. *Quercus coccifera.* L. Sp. 1413. Prickly-leaved Evergreen Oak. Heb. אלון. Area, Mediterranean region.

9. Var. *Calliprinos.* Webb. It. 15. Lebanon, Mount Tabor, Gilead.

10. Var. *Pseudococcifera.* Desf. Fl. Atl. ii., p. 349. Carmel, Galilee, Gilead, Bashan. Area, Sardinia, Southern Italy, Sicily.

This variety is the most common, and sometimes attains a magnificent growth, as the oak of Libbeya in Galilee.

11. Var. *palæstina.* Ky. Southern Palestine, Western hills of Judæa, Hebron.

12. *Quercus cerris.* L. Sp. 1415.

Lower and sub-alpine Lebanon. Area, Southern Europe and Mediterranean.

13. *Quercus ehrenbergii.* Ky. Eich., pl. 15.

Lower Lebanon, base of Anti-Lebanon.

14. *Quercus ægilops.* L. Sp. 1414. Var. *Ithaburensis.* Decaisn.

Carmel, Tabor, all the hills of Galilee and Samaria.

15. *Quercus balout.* Griff. Itin. Not., p. 328. *Non vidi.*

16. *Quercus look.* Ky. Eich., pl. 31.

On the east side of Hermon, 5,000 feet, the Hauran, Bashan. Probably a variety of *Q. ægilops.*

17. *Quercus libani.* Oliv. Voy., pl. 32. Northern Lebanon.

18. *Castanea vulgaris.* Lam. Dict. i., p. 708. Spanish Chestnut.

Lebanon, Anti-Lebanon, etc.; probably introduced. Area, Southern Europe and Mediterranean.

ORDER CXVIII., BETULACEÆ.

1. *Alnus orientalis.* Decaisn. Fl. Sin. Ann. Sc. Nat. Ser. ii., iv., p. 348. Oriental Alder. Phœnician plain, on the Litany.

2. *Alnus glutinosa.* Willd. Sp. 334. Common Alder.

Area, Europe, North Africa, Northern Asia to Japan.

ORDER CXIX., SALICINEÆ.

1. *Salix safsaf.* Forsk. Eg. Arab., p. 76. Heb. צַפְצָפָה. Willow. Arab. صفصاف, *Safsaf.*

Banks of the Jordan, Arnon, and Callirrhoe. Area, Nubia, Abyssinia, Senegal.

2. *Salix fragilis.* L. Sp. 1443. Crack Willow or Withy.

By streams, Lebanon. Area, Europe, West Siberia.

3. *Salix alba.* L. Sp. 1449. White Willow.

By streams, coast, interior Lebanon. Area, Europe, North Africa, Siberia.

4. *Salix caprœa.* L. Sp. 1448. Sallow.

Lebanon. Area, Europe, Siberia.

5. *Salix pedicillata.* Desf. Atl. ii., p. 362.

Lebanon. Area, Mediterranean region.

N.B.—Several other willows occur, which cannot be identified, from the imperfect specimens.

6. *Populus alba.* L. Sp. 1463. White Poplar.

General in the north. Area, Europe, North Africa, West Siberia.

7. *Populus euphratica.* Oliv. Voy. iii., p. 449, pl. 456. Heb. בְּכָאִים. A.V. *Mulberry.* 2 Sam. v. 23, etc., but more probably this poplar.

Banks of the Jordan and all other rivers. Area, North Africa, South Siberia, Scinde, Himalayas.

8. *Populus nigra.* L. Sp. 1464. Black Poplar.

By water in the north. Area, Central and Southern Europe, North Africa, West Siberia.

9. *Populus pyramidalis.* Rozier in Lam. Dict. v., p. 235.

Most abundant in the valleys of Anti-Lebanon and the Bukâà.

Area, cultivated everywhere in Central and Southern Europe.

PLANTÆ VASCULARES.

CLASS II., MONOCOTYLEDONEÆ.

ORDER CXX., HYDROCHARITACEÆ.

Absent.

ORDER CXXI., ALISMACEÆ.

1. *Alisma plantago.* L. Sp. 486. Water Plantain.

In pools and ditches, Lake Huleh, etc. Area, Europe, Siberia, North India, Abyssinia, North America, Australia.

2. *Damasonium bourgæi.* Coss. Nat. Pl. Crit. Esp., p. 47.

In dried ditches. Area, Mediterranean region.

3. *Sagittaria sagittifolia.* L. Sp. 1410. Arrowhead.

In pools. Area, Europe, North Asia, North-west India.

ORDER CXXII., BUTOMACEÆ.

1. *Butomus umbellatus.* L. Sp. 532. Flowering Rush.

In sluggish streams, Lake Huleh, Esdraelon, the Bukâà. Area, Europe, Siberia, North-west India, China.

ORDER CXXIII., JUNCAGINEÆ.

1. *Triglochin palustre.* L. Sp. 482. Marsh Arrow Grass.

Said to occur in the marshes of the Zerka river south of Carmel. Area, Europe, North Africa, Siberia, North-west India, China, Thibet, North and South America.

ORDER CXXIV., POTAMEÆ.

1. *Zannichellia palustris.* L. Sp. 1475. Horned Pondweed.

Lake Phiala, Huleh marshes. Area, almost world-wide.

2. *Potamogeton natans.* L. Sp. 182. Broad-leaved Pondweed.

Stagnant waters ; general. Area, temperate and sub-tropical regions.

3. *Potamogeton lucens.* L. Sp. 183. Shining Pondweed.

Lake Yamouni, Lebanon. Area, Europe, Siberia, India, North Africa, North America.

4. *Potamogeton crispus.* L. Sp. 183. Curly Pondweed.

In waters in the north. Area, Europe, Siberia, Japan, North India, the Cape, North America, Australia.

5. *Potamogeton densus.* L. Sp. 182. Opposite-leaved Pondweed.

In water, Lebanon district, etc. Area, Central and Southern Europe, North Africa, North America.

6. *Ruppia spiralis.* L'Herb. Dum. Fl. Belg., p. 164. Tassel Pondweed.

Salt pools on the coast. Area, all shores, temperate and tropical.

ORDER CXXV., NAIADACEÆ.

1. *Naias graminea.* Del. Fl. Eg., p. 377, pl. 50, f. 3.

Ditches, plain of Sharon. Area, North Italy, Africa, India and islands, Japan, generally with rice.

ORDER CXXVI., LEMNACEÆ.

1. *Lemna minor.* L. Sp. 1376. Duckweed.

In ditches ; general. Area, temperate and tropical regions.

2. *Wolffia hyalina.* (Del. Ill., No. 877.)

In stagnant waters.

ORDER CXXVII., ARACEÆ.

1. *Biarum pyrami.* (Schott. Prodr. 66.) By the Lake of Galilee.

2. *Biarum angustatum.* (Hooker. Bot. Mag., pl. 6355.) Collected by Sir J. D. Hooker; locality not marked. P.

3. *Biarum bovei.* Decaisn. Ann. Sc. Nat. 1835, p. 4. Lebanon.

4. *Biarum alexandrinum.* Boiss. Diagn. Ser. i., xiii., p. 6. Gennesaret.

5. *Arum detruncatum.* C. A. Mey in Schott. Prodr., p. 80. Cedars of Lebanon.

6. *Arum colocasia.* L. Sp. 1368. Upper Jordan valley. Area, India.

7. *Arum dioscoridis.* Sibth. Prodr. ii., p. 245. Var. *syriacum.* In the neighbourhood of Beyrout.

8. Ditto, var. *philisteum.* = *A. philisteum.* Ky. Philistian plain.

9. Ditto, var. *spectabile.* Boiss. Phœnician plain.

10. *Arum melanopus.* Boiss. Fl. Or. v., 40. Area, Central and Southern Europe, North Africa.

11. *Arum italicum.* Mill. Dict. No. 2. Herb. Kew.

12. *Arum palæstinum.* Boiss. Diagn. Ser. i., xiii., p. 6. Central Palestine, general; Jerusalem, Nablus, Carmel, Moab. P.

13. *Arum hygrophilum.* Boiss. Diagn. Ser. i., xiii., p. 7. Moist places in littoral districts.

14. Ditto, var. *rupicola.* = *A. rupicola.* Boiss. Shady rocky places; Anti-Lebanon.

15. *Arum orientale.* M. B. Taur. Cauc. ii., p. 407. Var. *gratum.* Schott. Lebanon, 5,000 feet. Area, South Russia.

16. *Helicophyllum crassipes.* (Boiss. Diagn. Ser. i., xiii., p. 9.) In cultivated ground; universal; mountains, plains, Jordan valley.

17. *Arisarum vulgare.* Targ. Tozz. Ann. Mus. Flor. ii., p. 617. =*Arum arisarum.* L.

General; littoral and central districts. Area, Portugal, Mediterranean region, Canaries.

18. *Dracunculus vulgaris.* Schott. Melet. i., 17. Herb. Kew.

Area, Mediterranean region.

19. *Dracunculus crinitus.* (Ait. Kew. iii., 314.)

Area, Mediterranean Islands.

20. *Acorus calamus.* L. Sp. 462. Sweet Flag.

Margin of streams. Area, Europe, Siberia, Himalayas, Japan, North America.

ORDER CXXVIII., PALMÆ.

1. *Phœnix dactylifera.* L. Hort. Cliff. 482. The Date Palm. Heb. תּוֹמֶר. Arab. نخلة, *Nakhleh.* The date, تمر, *Tamar.*

Indigenous in the ravines east of the Dead Sea, especially Callirrhoe. Formerly abundant at Engedi, where it is sub-fossil, and all along the Jordan valley. Cultivated in the plains of Palestine. Area, interior of North Africa.

2. *Chamærops humilis.* L. Hort. Cliff. 482.

Said to be found in the Jordan valley; but has not been met with by me or other collectors. Area, Mediterranean region.

ORDER CXXIX., TYPHACEÆ.

1. *Sparganium ramosum.* Hudson. Angl., 401. Burweed.

On the edges of lakes, ponds, and streams; general. Area, Europe, Siberia, North Africa, North America.

2. *Typha latifolia.* L. Sp. 1377. Bulrush.

By sluggish and stagnant water; general. Area, Europe, Siberia, Abyssinia, North America.

3. *Typha angustifolia.* L. Sp. 1377. Lesser Bulrush.

By streams, Lebanon. Area, Europe, Siberia, Africa, North America.

ORDER CXXX., ORCHIDEÆ.

1. *Serapias pseudocordigera.* Mon. Fl. Ven. 374. = *S. longipetala.* Poll.

Coast, Central and Southern Palestine, Esdraelon to Hebron. Area, South Europe.

2. *Anacamptis pyramidalis.* (L. Sp. 1332.) Pyramidal Orchis.

Coast and hill-districts. Area, Central and Southern Europe, North Africa.

3. *Orchis papilionacea.* L. Sp. 1331.

Moist places, Central Palestine. Area, South Europe, North Africa.

4. *Orchis morio.* L. Sp. 1333. Green-winged Orchis. Var. *albiflora.* = *O. syriaca.* Boiss.

Lebanon and Phœnician Plain. Area, Europe, Siberia.

Ditto, var. *picta.* Lois.

5. *Orchis coriophora.* L. Sp. 1332. Var. *fragrans* = *O. fragrans.* Poll.

Littoral districts. Area, Central and Southern Europe.

6. *Orchis punctulata.* Stev. in Lind. Ord. 273. Herb. Kew.

7. *Orchis longicruris.* Link. in Schrad. Journ. ii., p. 323.

Maritime plains. Area, Mediterranean region.

8. *Orchis militaris.* L. Cod. 6816, Excl. Var. β. Military Orchis.

Sides of ravines, Mount Gilead. Area, Central Europe, Danubian region, Siberia.

9. *Orchis incarnata.* L. Fl. Suec., p. 312. Marsh Orchis.

The Bukâà. Area, Europe.

10. *Orchis sancta.* L. Sp. 1330. Galilee, Tabor, Lebanon.

11. *Orchis tridentata.* Scop. Carn. 190.

General, east and west. Area, East Europe, Italy, Danubian region.

12. *Orchis lactea.* Poir. Dict. iv., 594.

Phœnicia. Area, Mediterranean region.

13. *Orchis simia.* Lam. Fl. Fr. iii., p. 507. Monkey Orchis.

Hill and mountain-districts, north to south, and Moab and Gilead. Area, Central and Southern Europe.

14. *Orchis angustifolia.* M. B. Taur. Cauc. ii., p. 368.

Lebanon and Anti-Lebanon. Area, Central Europe.

15. *Orchis saccata.* Ten. Nap. Prodr. 53.

Littoral and central districts. Area, Mediterranean region.

16. *Orchis mascula.* L. Sp. 1333. Early Purple Orchis.

Lebanon, in open places. Area, Central and Southern Europe, North Africa, West Siberia.

17. *Orchis anatolica.* Boiss. Diagn. Ser. i., v., p. 56.

General, Lower Lebanon; littoral, Jerusalem.

18. *Orchis laxiflora.* Lam. Fl. Fr. iii., p. 504.

General in low, moist places; hills of Gilead. Area, South Europe.

19. *Orchis palustris.* Jacq. Call. i., p. 75.

Esdraelon, by the Kishon. Area, Central and Southern Europe, North Africa.

20. *Orchis pseudosambucina.* Ten. Syn. Nap., p. 72.

Lebanon and Anti-Lebanon. Area, Mediterranean region.

21. *Orchis maculata.* L. Sp. 1335. Var. *saccigera.* Rchb. Spotted Orchis.

Sub-alpine Lebanon. Area, Europe, Siberia.

22. *Ophrys fusca.* Link. in Schrad. Journ. 1799, ii., p. 325.

Common on dry hills, east and west. Area, Portugal, Mediterranean region.

23. *Ophrys lutea.* Cav. Ic. ii., p. 46, pl. 160.

Very common in dry places. Area, Portugal, Mediterranean region.

24. *Ophrys speculum.* Link. in Schrad. Journ. 1799, ii., p. 324.

Gilead. Area, Portugal, Mediterranean region.

25. *Ophrys tenthredinifera.* Willd. Sp. iv., p. 67.

Chalky hills, common. Area, Portugal, Mediterranean region.

26. *Ophrys arachnites.* (Scop. Carn. ii., p. 194, Var. 2.) Spider Orchis.

Lebanon, Phœnicia, etc. Area, Central and Southern Europe.

27. *Ophrys atrata.* Lindl. Bot. Reg., pl. 1807.

Mount Tabor, etc. Area, South Europe.

28. *Ophrys ferrum-equinum.* Desf. Cor. Tourn., p. 9, pl. 5.

29. *Ophrys hiulca.* Rahb. Germ., p. 93, pl. 101, f. 2.

Littoral district.

30. *Ophrys æstrifera.* M. B. Taur. Cauc. ii., p. 369.

Littoral district, Lebanon. Area, Dalmatia, Servia.

31. *Cephalanthera pallens.* (Willd. Sp. iv., p. 85.) = *C. grandiflora.* Bab. Large White Helleborine.

Wâdys of Gilead. Area, Central and Southern Europe.

32. *Cephalanthera ensifolia.* (Murr. Linn. Syst. ed. xv., p. 670.) Narrow-leaved Helleborine.

Sub-alpine Lebanon, Gilead. Area, Central and Southern Europe, Morocco, West Siberia.

33. *Epipactis veratrifolia.* Boiss. Diagn. Ser. i., xiii., p. 11.

Lebanon, Phœnicia, Galilee.

34. *Epipactis latifolia.* All. Ped. ii., p. 151. Broad-leaved Helleborine.

Lebanon. Area, Europe, Siberia, Japan, Himalayas, North Africa.

35. *Spiranthes autumnalis.* Rich. Orch. Eur. 28. Lady's Tresses.

Area, Central and Southern Europe, North Africa.

ORDER CXXXI., IRIDACEÆ.

CROCUS. Heb. (*generic*) כַּרְכֹּום. Arab. كركم, *Karkum.*

1. *Crocus ochroleucus.* Boiss. Diagn. Ser. ii., iv., p. 93.

Rocky places; Lower Lebanon and Anti-Lebanon, Galilee. P.

2. *Crocus zonatus.* J. Gay in Bal. Exs. 1855. Lebanon.

3. *Crocus cancellatus.* Herb. Bot. Mag., pl. 3864. Var. *damascenus.* Lebanon and Anti-Lebanon.

4. *Crocus gaillardoti.* Boiss. Fl. Or. v., p. 105.
General in grassy places northwards.

5. *Crocus hyemalis.* Boiss. Diagn. Ser. ii., iv., p. 93.
Coast and hill districts; general; southwards. P.
Ditto, var. *foxii.* G. Maw. Wilderness of Judæa.

6. *Crocus sativus.* L. Sp. 50. Cultivated in Europe.

7. *Crocus vitellinus.* Wahl. Isis xvi., p. 106. = *C. syriacus.* Boiss. Littoral districts.

8. *Crocus hermoneus.* Ky. Sched. It. Syr. 1855.
Hermon, snow-line 9,000 feet.

9. *Crocus aureus.* Sibth. Fl. Græc. i., p. 25, pl. 35.
Near Tyre. Area, Servia, Transylvania.

10. *Romulea bulbocodium.* (L. Sp. 51.)
Littoral and central districts, Lower Lebanon. Area, Mediterranean region. Ditto, var. *grandiflora.*

11. *Romulea nivalis.* (Boiss. Diagn. Ser. ii., iv., p. 92.)
Lebanon and Anti-Lebanon, 6,000—9,500 feet. P.

12. *Romulea columnæ.* Seb. et Maur. Fl. Rom. Prodr., p. 18.
Phœnician coast. Area, Western and Southern Europe.

13. *Iris sisyrinchium.* L. Sp. 59. Arab. زمبق, *Zạmbac.*
Universally abundant in dry places. Area, Portugal, Mediterranean region.

14. *Iris histrio.* Rchb. fil. Bot. Zeit. 1872, p. 388.
All hill and mountain districts.

15. *Iris palæstina.* (Baker. Seem. Journ. 1871, p. 108.) Universal.

16. *Iris pseudacorus.* L. Sp. 56. Yellow Flag.
Wet places, coast, Huleh. Area, Europe, North Africa.

17. *Iris lortetii.* Boiss. Fl. Or. v., p. 131.
Dry thickets, Lebanon, 2,000 feet. P.

18. *Iris helena.* Boiss. Fl. Or. v., p. 132. Southern Desert.

19. *Iris persica.* L. Sp. 59. *Non vidi.*

20. *Iris florentina.* L. Sp. 55.

21. *Iris tuberosa.* L. Sp. 58.
Hilly districts. Area, Mediterranean region.

22. *Iris germanica.* L. Sp. 55.
Plains; common. Area, Central and Southern Europe.

23. *Iris pallida.* Lam. Dict. iii., p. 294. Galilee.

24. *Iris haynei.* Baker. Gard. Chron. vi. (1876), p. 710.
Mount Gilboa. P.

Named by Mr. Baker after my fellow-traveller, its discoverer, the late W. Amherst Hayne, Trin. Coll., Cambr., to whom I am indebted for most of the Moab and Gilead localities.

25. *Iris reticulata.* M. B. Taur. Cauc. i., p. 34.
Bashan, Gilead, Galilee.

26. *Iris caucasica.* Hoffm. Com. Soc. Ph. Mosq. i., p. 40.
Hebron, Judæan wilderness.

27. *Iris fœtidissima.* L. Sp. 57. Fetid Iris, or Gladdon.
Area, Central and Southern Europe, North Africa, Canaries.

28. *Iris heylandiana.* Boiss. Fl. Or. v., p. 130.
Plain of the Orontes.

29. *Iris sari.* Baker. Gardn. Chron. v., 1876, p. 780.
Lake Huleh. P.

There are at least three other species of Iris in Gilead, one of them the largest and finest of the genus I ever saw, the flower a deep uniform maroon purple; but none of which have I been able to identify. There are also two other unidentified species from the woods of Galilee.

30. *Gladiolus segetum.* Gawl. Bot. Mag., pl. 719.
In corn-fields. Area, Mediterranean region, Madeira, Canaries.

31. *Gladiolus byzantinus.* Mill. Dict. iii., Bot. Mag., pl. 874.
Galilæan hills. Area, Corsica, Sardinia, Sicily, North Africa.

32. *Gladiolus illyricus.* Koch. Sturn. 83, Ic.
Lower Lebanon. Area, Southern Europe.

33. *Gladiolus imbricatus.* L. Sp. 52. Var. *libanoticus.* Boiss.
Moist places, Anti-Lebanon.

34. *Gladiolus atroviolaceus.* Boiss. Diagn. Ser. i., xiii., p. 14.
Dry stony places and corn-fields ; universal.

35. *Gladiolus communis.* L. Sp. 52. Perhaps introduced.
Area, South France, Corsica, Dalmatia.

ORDER CXXXII., AMARYLLIDACEÆ.

1. *Sternbergia clusiana.* Gawl. Schult. Syst. vii., 794.
Universal, except Jordan valley and desert.

2. *Sternbergia pulchella.* Boiss. Diagn. Ser. ii., iv., p. 97.
Lower Lebanon.

3. *Sternbergia lutea.* (L. Sp. 420.) Herb. Kew.
Area, Mediterranean region.

4. *Narcissus tazetta.* L. Sp. 416.
Cultivated land and lower hills from Gaza to Lebanon.
Ditto, var. *syriacus.* Boiss. Phœnician plain, Anti-Lebanon.
Area, Portugal, Mediterranean region.

5. *Pancratium maritimum.* L. Sp. 418.
Sands by Gaza. Area, Portugal, Mediterranean region.

6. *Pancratium parviflorum.* Decaisn. Pl. Palest. Ann. Sc. Nat. Déc., 1835. Rocks and walls, Beyrout and Sidon.

7. *Pancratium sickenbergeri.* Asch. et Schw. Gartenzeit. cum Icone.
Desert near Sebbeh.

8. *Ixiolirion montanum.* (Labill. Déc. ii., p. 5.)
Hills and fields ; local from Gaza to Lebanon, 6,000 feet.

ORDER CXXXIII., COLCHICACEÆ.

1. *Colchicum decaisnei.* Boiss. Fl. Or. v., p. 157.
Lebanon and Anti-Lebanon, lower to sub-alpine regions.

2. *Colchicum fasciculare.* (L. Sp. 439.) The Bukâ̂, in dry fields.

3. *Colchicum ritchii.* R. Br. App. ad Denh. et Clapp. 241.
Wâdys by Dead Sea, Anti-Lebanon, near Damascus. Area, North-east Africa.

4. *Colchicum lætum.* Stev. Mem. Mosq., p. 66, pl. 13. Herb. Kew.
Area, South Astrachan.

5. *Colchicum variegatum.* L. Sp. 485.

6. *Colchicum steveni.* Kunth. Enum. iv., p. 144.
Maritime plains. Area, Atlas mountains.

7. *Colchicum libanoticum.* Boiss. Fl. Or. v., p. 166.
Crest of Lebanon on the snow-line. P.

8. *Colchicum candidum.* Ky. Pl. Cilic., No. 91. Cat. Kew.

9. *Colchicum montanum.* L. Sp. 485.
Hill and mountain districts. Area, Spain, North Africa, South Russia.

10. *Merendera sobolifera.* C. A. Mey. Ind. Petrop. i., 1834, p. 24.
Northern plains.

11. *Merendera caucasica.* M. B. Taur. Cauc. i., p. 293.
Snow-line of Lebanon.

12. *Erythrosticus palæstinus.* Boiss. Baker. J. Lin. Soc. xvii., p. 45.
All round Dead Sea, Ghor, Engedi, Callirrhoe, etc. P.

ORDER CXXXIV., LILIACEÆ.

1. *Lilium candidum.* L. Sp. 433.
Lebanon ; rare. Area, South Europe.

2. *Lilium chalcedonicum.* L. Sp. 434. Cat. Kew. *Non vidi.*

54

3. *Fritillaria acmopetala.* Boiss. Diagn. Ser. i., vii., p. 104.
Sub-alpine Lebanon.

4.. *Fritillaria crassifolia.* Boiss. Diagn. Ser. ii., iv., p. 103.
Lebanon, eastern slopes, Ainât.

5. Ditto, var. *hermonis.* Fenzl.
Summits of Hermon, 9,000—9,500 feet.

6. *Fritillaria persica.* L. Sp. 436.
Esdraelon, in rich wet soil. (May be a variety of the next species.)

7. *Fritillaria libanotica.* (Boiss. Diagn. Ser. i., xiii.)
Dry, stony places, Lebanon, Anti-Lebanon, Samaria, near Jerusalem. P.

8. *Fritillaria messanensis.* Rafin. Prèc., p. 44. Cat. Kew. *Non vidi.*
Area, Sicily, Italy, North Africa.

9. *Fritillaria tulipifolia.* M. B. Taur. Cauc. i., p. 270.
Northern hills. Area, South-west Siberia.

10. *Tulipa gesneriana.* L. Sp. 438, pt.
Grassy and rocky hills, Gilead. Area, Italy, South Russia, West Siberia.

11. *Tulipa oculus-solis.* St. Amand. Rec. Soc. Agen. i. 75.
Fields and hills, Galilee, Lower Lebanon. Area, South France, Italy.

12. *Tulipa præcox.* Ten. Fl. Nap. i., p. 170, pl. 32.
Gilead and Moab. Area, South France, Italy.

13. *Tulipa montana.* Lindl. Bot. Reg., pl. 1106.
Universal, plains, hills, and valleys.
Ditto, var. *julia.* Koch. Lebanon.

14. *Tulipa lownei.* Baker. J. Lin. Soc. xiv., p. 294.
Snow-line, Lebanon and Hermon. P.

15. *Lloydia rubroviridis.* (Boiss. Diagn. Ser. ii., iv., p. 106.)
Lebanon and Hermon to 8,000 feet, Jerusalem, etc.

16. *Lloydia græca.* (L. Sp. 444.) Phœnicia.

17. *Gagea arvensis.* (Pers. in Ust. Ann. v., p. 8, pl. 1.)

Northern district, fields and brushwood. Area, Central and Southern Europe, North Africa.

18. *Gagea foliosa.* (Presl. Del. Prag., p. 149.)

Littoral districts, Middle Lebanon. Area, Sicily.

19. Ditto, var. *micrantha.* Boiss.

Alpine Lebanon and Hermon to 9,500 feet.

20. *Gagea bohemica.* (Zauschn. Deu. Fl. ii., p. 544.) = *G. billardieri.* Kth.

Near Jerusalem. Area, Central France, South Germany, Servia, South Persia.

21. *Gagea reticulata.* (Pall. It. iii., p. 553.)

Coast and hill-districts, Jerusalem, etc. Area, North Africa, Bulgaria to Turkestan.

22. Ditto, var. *tenuifolia.* = *Ornith. circinnatum.* L. Suppl. 199.

Lebanon, Southern Desert.

23. *Gagea damascena.* Boiss. Diagn. Ser. ii., iv., p. 105.

In limestone rocks everywhere, Beyrout, Jerusalem, Moab, etc.

24. *Gagea rigida.* Boiss. Diagn. Ser. i., vii., p. 108.

On bare hills, Judæan wilderness.

25. *Gagea chlorantha.* (M. B. Taur. Cauc. iii., p. 264.)

Dry, rocky places, Jerusalem. Area, Turkestan.

26. *Gagea bulbifera.* (L. Suppl. 149.)

Galilee. Area, South and East Russia, West Siberia, Turkestan.

27. *Gagea persica.* Boiss. Diagn. Ser. i., vii., p. 108.

Sub-alpine and alpine Lebanon.

28. *Ornithogalum libanoticum.* Boiss. Diagn. Ser. ii., iv., p. 106.

Sub-alpine Lebanon. P.

29. *Ornithogalum fuscescens.* Boiss. Diagn. Ser. ii., iv., p. 107.

Cultivated land under Rascheya. P.

30. *Ornithogalum narbonense.* L. Sp. 440.

General in cultivated and grassy land. Area, Mediterranean region, Portugal, Canaries.

31. *Ornithogalum arabicum.* L. Sp. 441.

The Ghor. Area, Mediterranean region, Madeira, Canaries.

32. *Ornithogalum densum.* Boiss. Diagn. Ser. ii., iv., p. 107.

Near Beyrout, oak groves on Lebanon.

33. *Ornithogalum lanceolatum.* Labill. Pl. Rar. Syr. Dec. v., p. 11, pl. 8.

Littoral and central hill-districts, Lebanon, Anti-Lebanon to 9,000 feet.

34. *Ornithogalum montanum.* Cyr. in Ten. Fl. Nap. i., p. 176.

Lebanon, Galilee. Area, South Italy, Sicily.

35. *Ornithogalum umbellatum.* L. Sp. 441. Star of Bethlehem.

All Palestine; universally abundant. Area, Central and Southern Europe, North Africa.

36. *Ornithogalum tenuifolium.* Guss. Prodr. i., 413.

Under Anti-Lebanon. Area, South France, Italy.

37. *Ornithogalum fimbriatum.* Willd. Nov. Ac. Berol. 3.

Lebanon.

38. *Ornithogalum trichophyllum.* Boiss. Diagn. Ser. ii., iv., p. 108.

Beersheba.

39. *Ornithogalum neurostegium.* Boiss. Fl. Or. v., p. 222.

Alpine Lebanon. P.

40. *Urginea maritima.* (L. Sp. 442.) = *U. scilla.* Steinh.

Maritime sands. Area, Mediterranean coasts, Canaries.

41. *Scilla autumnalis.* L. Sp. 443. Autumn Squill.

Littoral and hill-districts, rocky places; general. Area, Central and Southern Europe, North Africa.

42. *Scilla hanburyi.* Baker. J. Linn. Soc. xiii., p. 235.

Anti-Lebanon, 4,000 feet. P.

43. *Scilla hyacinthoides.* L. Syst. xiii., p. 272.
All the littoral districts, Jordan valley. Area, South Europe.

44. *Scilla cernua.* Red. Lil., pl. 298. =*S. sibirica.* Ait.
Anti-Lebanon, Tabor, Carmel. Area, South-east Europe.

45. *Scilla amœna.* L. Sp. 443.
Esdraelon. ? = *S. cernua.* Area, South France, Italy.

46. *Scilla bifolia.* L. Sp. 443.
Snow-line, summit of Jebel Makmel. Area, Mediterranean and Danubian regions.

47. *Allium scorodoprasum.* L. Sp. 425. Var. (?) *caudatum.* Boiss. Sand Garlic.
Near the cedars of Lebanon. Area, Central Europe, Danubian region.

48. *Allium ampeloprasum.* L. Sp. 423. Wild Leek.
In cultivated ground; general. Area, Mediterranean region.

49. *Allium rotundum.* L. Sp. 423.
Cultivated ground; general. Area, Central Europe, Danubian region.

50. *Allium phaneranthemum.* Boiss. Fl. Or. v., p. 235.
Lebanon, 2,500 feet.

51. *Allium sphærocephalum.* L. Sp. 426. Round-headed Garlic.
Lower Lebanon, Galilee. Area, Central and Southern Europe, North Africa.

52. *Allium descendens.* Sibth. Fl. Gr. iv., p. 15, pl. 316.
Littoral and central districts. Area, South France, South Italy, Sicily.

53. *Allium calyptratum.* Boiss. Diagn. Ser. i., xiii., p. 30.
Northern Lebanon.

54. *Allium cilicium.* Boiss. Diagn. Ser. i., vii., p. 115.
Alpine and sub-alpine Anti-Lebanon.

55. *Allium sinaiticum.* Boiss. Diagn. Ser. i., v., p. 117. =*A. deserti.* Boiss. Hill desert of Judah, south of Hebron, Engedi.

56. *Allium hierochuntinum.* Boiss. Fl. Or. v., p. 244.
Ravines above Jericho; ravine of Kedron.

57. *Allium curtum.* Boiss. Diagn. Ser. ii., iv., p. 116.
Judæan wilderness, Phœnician hills.

58. *Allium cepa.* L. Sp. 431. Onion, cultivated.

59. *Allium ascalonicum.* L. Sp. 429. Heb. שׁוּם, Shallot.
Sands at Askalon; probably from cultivation.

60. *Allium stamineum.* Boiss. Diagn. Ser. ii., iv., p. 119.
Hills of Palestine from north to south, Esdraelon, west base of Moab and Gilead hills.

61. *Allium chloranthemum.* Boiss. Diagn. Ser. i., xiii., p. 33.
Hill districts, Gilead.

62. *Allium paniculatum.* L. Sp. 428.
Cultivated land; universal. Area, Southern Europe, Madeira, Canaries.

63. *Allium modestum.* Boiss. Diagn. Ser. i., xiii., p. 33.
Southern Desert.

64. *Allium libani.* Boiss. Diagn. Ser. i., xiii., p. 26.
Dry hill districts, from Judæa northwards to Lebanon and top of Hermon, 9,000 feet. P.

65. *Allium erdelii.* Zucc. Abh. Bay. Ac. iii., p. 236, pl. 5. = *A. philistæum.* Boiss. South Judæa, Hebron, Gilead, Moab.

66. *Allium subhirsutum.* L. Sp. 424.
Littoral and hill districts, Lebanon to Jerusalem. Area, Mediterranean region.

67. *Allium trifoliatum.* Cyr. Pl. Rar., fasc. ii., p. 11, pl. 3.
Littoral and low hills, Beyrout to Hebron. Area, Spain, Italy and its islands.

68. *Allium hirsutum.* Zucc. Abh. Bay. Ac. iii., p. 232, pl. 2.
Rocky places, Benjamin, Jerusalem, Hebron. P.

69. *Allium papillare.* Boiss. Diagn. Ser. i., xiii., p. 27.
Philistian plain, Southern Desert, near Gaza. P.

70. *Allium cassium.* Boiss. Diagn. Ser. i., xiii., p. 28. Var. *hirtellum.* Cedars of Lebanon, the Bukâà.

71. *Allium zebdanense.* Boiss. Diagn. Ser. ii., iv., p. 113.
Anti-Lebanon and Lebanon.

72. *Allium carmeli.* Boiss. Diagn. Ser. i., xiii., p. 28.
Mount Carmel, among rocks. P.

73. *Allium roseum.* L. Sp. 432.
Plains of Moab. Area, Portugal, Mediterranean region.

74. *Allium neapolitanum.* Cyr. Pl. Rar. Nap. i., p. 13, pl. 4.
Littoral districts, about Jerusalem, Gilead. Area, Southern Europe.

75. *Allium schuberti.* Zucc. Abh. Münch. Ac. iii., p. 234, pl. 3.
Inland plains, lower hills.

76. *Allium nigrum.* L. Sp. 440.
Lebanon, Anti-Lebanon, Galilee, Tabor. Area, Southern Europe, North Africa, Canaries.

77. *Allium orientale.* Boiss. Diagn. Ser. i., xiii., p. 25.
Lower cultivated and grassy hills.

78. *Allium aschersonianum.* W. Barbey. Herb. Levant., p. 163, pl. 3.
Southern Palestine, from Samaria to Judæan wilderness.

79. *Allium rothii.* Zucc. Abh. Münch. Ac. iii., p. 235, pl. 4.
Judæan wilderness and Southern Desert. P.

80. *Allium lachnophyllum.* Paine. Pal. Expl. Soc., No. 3, p. 125.
Eastern plain of Moab, at Ziza. P.

81. *Allium scabriflorum.* Boiss. Diagn. Ser. i., v., p. 60.
North-west hills of Moab, in marl.

82. *Allium dyctioprasum.* C. A. Mey. Kth. Enum. iv., p. 390.
North-west plains of Moab.

83. *Muscari pinardi.* (Boiss. Diagn. Ser. i., v., p. 62.)
Lebanon, Anti-Lebanon, Phœnicia.

84. *Muscari longipes.* Boiss. Diagn. Ser. i., xiii., p. 37.
Fields, Judæa, plains of Philistia and Sharon.

85. *Muscari comosum.* Mill. Dict. 2.
In cultivated land; frequent. Area, Central and Southern Europe, North Africa.

86. *Muscari maritimum.* Desf. Atl. i., p. 308.
Sandy districts along the whole coast. Area, Sicily, North Africa.

87. *Muscari racemosum.* (L. Sp. 455.) Grape Hyacinth.
Among rocks in hill districts, from Lebanon south, Gilead, Moab. Area, Central and Southern Europe, North Africa.

88. *Muscari pulchellum.* Heldr. Boiss. Diagn. Ser. ii., iv., p. 109.
Anti-Lebanon, near Baalbek.

89. *Muscari commutatum.* Guss. Prodr. Sic. i., p. 426.
Esdraelon and Galilee. Area, South Italy, Sicily, Balearic Isles.

90. *Muscari parviflorum.* Desf. Atl., pl. 309.
Littoral plains, Ghor near Jericho. Lebanon, *teste* Kunth. Area, North Africa, Sicily, Balearic Isles, Malta.

91. *Muscari palleus.* M. B. Taur. Cauc. i., p. 283. Cat. Kew.
Non vidi. Area, Dalmatia.

92. *Bellevalia ciliata.* (Cyr. Nap. ii., p. 22, pl. 10.)
Southern Desert, near the Dead Sea. Area, South Italy, North Africa, South Russia.

93. *Bellevalia trifoliata.* (Ten. Nap. iii., p. 376, pl. 136.)
Littoral districts and Southern Desert. Area, France, South Italy.

94. *Bellevalia densiflora.* Boiss. Diagn. Ser. i., vii., p. 109.
The Hauran.

95. *Bellevalia macrobotrys.* Boiss. Diagn. Ser. i., xiii., p. 35.
Maritime and inland plains, Gennesaret, etc., southern hills; general.

96. *Bellevalia flexuosa.* Boiss. Diagn. Ser. i., xiii., p. 36.
Walls and dry places, Jerusalem, Gilead, etc. ; general.

97. *Bellevalia nivalis.* Boiss. Diagn. Ser. ii., iv., p. 110.
Alpine Anti-Lebanon.

98. *Bellevalia haynei.* Baker. Journ. Bot. xii., p. 7.
Gorge of the Calirrhoe, Moab. P.

99. *Bellevalia nervosa.* (Bertol. Misc. i., p. 21.)
Hills of the Hauran.

100. *Bellevalia romana.* (L. Mant., p. 224.)
Doubtful. Area, Southern Europe.

101. *Hyacinthus orientalis.* L. Sp. 454.
Lower slopes of Lebanon towards Sidon. Area, Sicily, Dalmatia.

102. *Puschkinia scilloides.* Ad. Nov. Act. Petrop. xiv., p. 164. Var. *libanotica.* Zucc. Alpine Lebanon.

103. *Asphodelus microcarpus.* Viv. Fl. Cors. Diagn., p. 5.
Common in all the plains. Area, Mediterranean region, Madeira, Canaries.

104. *Asphodelus fistulosus.* L. Sp. 444.
Common on plains and hills, west and east of Jordan. Area, Mediterranean region, Nubia, Mauritius.

105. *Asphodelus tenuifolius.* Cav. An. Cienc. iii., 46, pl. 27.
South of Beersheba. Area, South-east Spain, North Africa, Nubia, Arabia, India, Canaries, Madeira.

106. *Asphodelus visicidulus.* Boiss. Diagn. Ser. i., vii., p. 118.
Southern Desert.

107. *Asphodelus ramosus.* L. Sp. 444.
Abundant ; Moab, Beersheba, plains, littoral and central. Area, Mediterranean region, Canaries.

108. *Asphodeline lutea.* (L. Sp. 443.)
Esdraelon, central hills, etc. Area, Italy, Sicily, Dalmatia, Algeria.

109. *Asphodeline brevicaulis.* (Bert. Misc. i., p. 20.)
Lebanon, hill districts.

110. *Asphodeline taurica.* (Pall. Ind. Taur. Cauc. i., p. 279.)
Alpine and sub-alpine Lebanon and Anti-Lebanon.

111. *Asphodeline damascena.* Boiss. Diagn. Ser. i., xiii., p. 22.
Wâdy Barada, Anti-Lebanon, Gilead, Moab. P.

112. *Eremurus spectabilis.* M. B. Taur. Cauc. iii., p. 269. = *E. libanoticus.* Boiss.
Lebanon and Anti-Lebanon.

113. *Aloe vera.* L. Sp. 458 = *Aloe vulgaris.* Lam.
Said to be found on rocks by the sea coast; Jordan valley, near Beisan. Originally American.

ORDER CXXXV., ASPARAGACEÆ.

1. *Asparagus palæstinus.* Baker. Linn. Soc. Jour. xiv., p. 602.
All along the Jordan, from Lake Huleh to Jericho. P.

2. *Asparagus lownei.* Baker. Linn. Soc. Jour. xiv., p. 601.
Jericho. P.

3. *Asparagus acutifolius.* L. Sp. 449.
Lower Lebanon. Area, Mediterranean region, Canaries.

4. *Asparagus aphyllus.* L. Sp. 450.
Hills near Nazareth. Area, Portugal, Spain, Sardinia, Sicily, North Africa, Canaries.

5. *Asparagus stipularis.* Forsk. Fl. Eg. Arab., p. 72 = *A. horridus.* L.
Littoral districts, Southern Desert. Area, Portugal, South Spain, Sicily, North Africa, Canaries.

6. *Asparagus tenuifolius.* Lam. Dict. i., 294. Herb. Kew.
Area, South France, Italy, Danubian region.

7. *Ruscus aculeatus.* L. Sp. 1474. Butcher's Broom.

Central and Galilæan hills, lower Lebanon. Area, Central and Southern Europe, North Africa.

ORDER CXXXVI., SMILACEÆ.

1. *Smilax aspera.* L. Sp. 1458.

Littoral districts, lower hills, Samaria, etc., Gilead. Area, Mediterranean region, South-west France, Canaries, Madeira, Abyssinia, North India.

Ditto, var. *mauritanica.* Desf. Near Sidon and Beyrout.

ORDER CXXXVII., DIOSCOREACEÆ.

Tamus communis. L. Sp. 1458. Black Bryony.

Common in woods, glens, and brushwood. Area, Central and Southern Europe, North Africa.

Ditto, var. *cretica.* L. Lebanon and Phœnician coast.

ORDER CXXXVIII., COMMELYNACEÆ.

Absent.

ORDER CXXXIX., JUNCACEÆ.

1. *Juncus effusus.* L. Sp. 464. Common Rush.

Wet places; general; and in Jordan valley. Area, Europe, Siberia, Japan, India, North America, Australia.

2. *Juncus glaucus.* Ehr. Beit. vi., p. 83. Hard Rush.

Wet places, coast, to sub-alpine Lebanon. Area, Europe, West Siberia, North Africa, Madeira.

3. *Juncus acutus.* L. Sp. 463. Var. *L.* Great Sea Rush.

Common in wet places. Area, Central and Southern Europe, North Africa, Canaries, Madeira, Azores.

4. *Juncus subulatus.* Forsk. Eg. Arab., p. 75. = *J. multiflorus.* Desf.

Round Dead Sea Area, Mediterranean region.

5. *Juncus gerardi.* Loisel. Not., p. 60.

Wet places, top of Lebanon. Area, Central and North-eastern Europe, Coast of Spain and Italy, Danubian region, North Africa, North America.

6. *Juncus maritimus.* Lam. Dict. iii., p. 264. Lesser Sea Rush.

Near the coast. Area, world-wide in temperate regions.

7. *Juncus punctorius.* L. Fil. Suppl., p. 208.

Eastern Lebanon, in water. Area, North Africa, Abyssinia, the Cape.

8. *Juncus lamprocarpus.* Ehr. Calam., No. 126. Jointed Rush.

Wet places; general. Area, Europe, Siberia, North Africa, North America.

9. *Juncus pyramidatus.* Lah. Junc., p. 40.

Marshes, Phœnicia, the Bukâà.

10. *Juncus bufonius.* L. Sp. 466. Toad Rush.

Wet places; general; from Huleh upwards.

Area, Europe, North Africa, Siberia, North America.

11. *Juncus tenageia.* Ehr. Beitr. iv., p. 148.

Lebanon, the Bukâà. Area, Central and Southern Europe, West Siberia, North Africa.

12. *Juncus acutiflorus.* Ehr. Beitr. vi., p. 86. Sharp-flowered Jointed Rush.

Northern marshes. Area, Northern and Central Europe, Siberia, North America.

ORDER CXL., CYPERACEÆ.

1. *Cyperus flavescens.* L. Sp. 68.

General. Area, Central and Southern Europe, Siberia, North Africa, India, America, Australia.

2. *Cyperus globosus.* All. Auct. 49.

North Palestine. Area, Southern Europe, India, Java, Tropical Africa, Japan, Australia.

3. *Cyperus distachyus.* All. Auct. 48, pl. 2, f. 5. = *C. junciformis.* Cav. On the coast. Area, Mediterranean region, Canaries, South Africa, Central and Southern America.

4. *Cyperus pygmæus.* Rottb. Descr. 20, pl. 14, f. 4.

On the coast. Area, North and Tropical Africa, India.

5. *Cyperus schœnoides.* Gris. Spic. Rum. 421.

On the coast. Area, Mediterranean region, Canaries.

6. *Cyperus alopecuroides.* Rottb. Descr. 38, pl. 8, f. 2.

Lake Huleh. Area, Canaries, North-east Africa, India, Australia.

7. *Cyperus conglomeratus.* Rottb. Descr., p. 21.

Coast near Gaza. Area, Nubia, Abyssinia, Senegal, India, Madagascar.

8. *Cyperus fuscus.* L. Sp. 69.

General. Area, Central and Southern Europe, Siberia, North Africa.

9. *Cyperus glaber.* L. Mant. Alt., p. 179.

The coast, Bethlehem, etc. Area, Italy, Sicily, Danubian region, North Africa, Siberia.

10. *Cyperus longus.* L. Sp. 67. Galingale.

General. Area, Central and Southern Europe, North Africa, India.

11. Ditto, var. *heildrichianus.* Boiss. Diagn. Ser. i., xiii., p. 39.

Lake of Gennesaret.

12. *Cyperus rotundus.* L. Syst. 98.

Dry places; general. Area, Mediterranean region, Tropical Africa, Arabia, India, China, Australia, America.

13. Ditto, var. *tetrastachys.* Desf. On the coast.

14. *Cyperus esculentus.* L. Sp. 67.

Phœnician coast. Area, Abyssinia, Madeira, South Africa, India.

15. *Papyrus antiquorum.* Willd. Bruce. It., pl. 1. The Papyrus. Heb. גֹּמֶא, A.V. *Rush.* Arab. بابير, *Babeer.*

North-east corner of Gennesaret. Many acres north of Lake Huleh. Swamps in plain of Sharon. Formerly in Egypt, where now extinct. Introduced into Sicily. Found in Nubia, Abyssinia, and Tropical Africa.

16. *Scirpus setaceus.* L. Sp. 73. Mud Rush.

Lebanon. Area, Europe, Siberia, North Africa, North India, Australia, Madeira.

17. *Scirpus savii.* Seb. et Maur. Fl. Rom. 22.

West Lebanon. Area, Western and Southern Europe, North Africa, Madeira, Canaries, Azores, South America, Australia, New Zealand.

18. *Scirpus holoschœnus.* L. Sp. 72. Cluster-headed Mud Rush.

On the coast, Mount Tabor, etc. Area, Central and Southern Europe, Siberia, North Africa, Canaries.

19. *Scirpus lacustris.* L. Sp. 72. Lake Club Rush.

General. Area, all temperate, many tropical regions.

20. *Scirpus littoralis.* Schrad. Germ. i., 142, pl. 5, f. 7.

Litany river. Area, South France, Italy, North Africa, India.

21. *Scirpus maritimus.* L. Sp. 74. Sea Club Rush.

General. Area, world-wide.

22. *Scirpus australis.* L. Syst., p. 86.

In gravelly places; frequent. Area, Europe, North Africa, Siberia.

23. *Scirpus parvulus.* R. et Sch. i., p. 125.

By the Dead Sea. Area, Central Europe.

24. *Scirpus triqueter.* L. Mant. 105. Triangular Club Rush.

Marshes of Huleh. Area, Central and Southern Europe, North America.

25. *Scirpus mucronatus.* L. Sp. 73.

Near Lake Huleh. Area, Central and Southern Europe, India, Mauritius, Australia, North America.

26. *Blysmus compressus.* (L. Sp. 65.)

Moab highlands, Lebanon, in wet places. Area, Northern and Central Europe, Himalayas.

27. *Heleocharis palustris.* (L. Sp. 70.) Spike Rush.

Marshes and wet places; general. Area, almost world-wide.

28. *Heleocharis macrantha.* Böckel. Linn. xxxvi., p. 453.

Coast near Beyrout. P.

29. *Fimbristylis ferruginea.* (L. Sp. 74.)

Gravelly streams and marshes, coast. Area, India, China, Nubia, Abyssinia, Senegal, Tropical America, Australia.

30. *Fimbristylis dichotoma.* (Rottb. Gram, p. 57, pl. 13, f. 1.)

Marshes, littoral districts. Area, Southern Europe, North-east Africa, Senegal, India, Canaries.

31. *Schœnus nigricans.* L. Sp. 64. Bog Rush.

General in moist places. Area, Europe, North Africa, Siberia.

32. *Carex stenophylla.* Wahl. Act. Holm. 1803, No. 21.

Moist places, sub-alpine and alpine Lebanon. Area, Danube region, Siberia, Thibet, North America.

33. Ditto, var. *planifolia.* Boiss. = *C. pachystylis.* Gay.

Judæa, Philistia, Southern Desert.

34. *Carex divisa.* Huds. Fl. Angl. i., p. 348. Bracteate Marsh Sedge.

General in marshes. Area, Central and Southern Europe, West Siberia, North Africa, Madeira, Canaries, North-west India.

35. *Carex vulpina.* L. Sp. 1382. Great Sedge.

General in marshes, and by rivers. Area, Europe, West Siberia, North Africa, North America.

36. *Carex muricata.* L. Sp. 1382. Prickly Sedge.

General, up to 6,000 feet on Lebanon. Area, North temperate zone.

37. *Carex divulsa.* Gooden. Tr. Linn. Soc. ii., p. 160. Grey Sedge.

Moist wooded places, littoral. Area, Central and Southern Europe, Siberia, North Africa, Madeira, Canaries, North America.

38. *Carex remota.* L. Sp. 1383. Distant Spiked Sedge.

Northern districts, in wet, shady places. Area, Northern and Central Europe, Siberia, Atlas Mountains, Himalayas, North America.

39. *Carex glauca.* Scop. Fl. Carn. ii., p. 223. Heath Sedge.

Common in moist places. Area, Europe, North Africa, Siberia.

40. *Carex echinata.* Desf. Atl. 338.

Marshes on the coast. Area, Mediterranean region.

41. *Carex maxima.* Scop. Fl. Carn. ii., p. 219.=*C. pendula.* Huds. Great Pendulous Sedge.

Sub-alpine and alpine Lebanon. Area, Central and Southern Europe, North Africa, Madeira.

42. *Carex acuta.* Auct. L. Sp. 1388, *ex parte.*

General. Area, Europe, Siberia, Arctic America.

43. *Carex sylvatica.* Huds. Fl. Angl. 353. Wood Sedge.

Galilee, Lebanon. Area, Central Europe, Siberia.

44. *Carex extensa.* Gooden. Tr. Linn. Soc. ii., p. 17, pl. 21.

Plains near Beersheba. Area, European, North African, and American coasts.

45. *Carex distans.* L. Sp. 1387. Loose Sedge.

General from Lebanon to Philistia. Area, Europe, North Africa.

46. *Carex eremitica.* Paine. Pal. Expl. Soc., No. 3, p. 126.

Upland plain of Moab. P.

ORDER CXLI., GRAMINEÆ.

1. *Oryza sativa.* L. Sp. 465. Rice.

Cultivated. Asia, Africa, America.

2. *Phalaris arundinacea.* L. Sp. 80 Reed Grass.
Area, Europe, North Africa, Siberia, North America.

3. *Phalaris minor.* Retz. Obs. iii., 8 Moab highlands.
Area, Portugal, Mediterranean region, South Africa.

4. *Phalaris paradoxa.* L. fil. Dec. 35, pl. 18.
Moab, western slopes, Seisaban. Area, Central and Southern Europe.

5. *Phalaris canariensis.* L. Sp. 79. Canary Grass. Galilee.
Area, Mediterranean region.

6. *Phalaris aquatica.* Sibth. Fl. Gr. i., 42. =*Ph. nodosa.* L.
Area, South Europe.

7. *Phalaris bulbosa.* Cav. et Schenchz. Agrost. 53, pl. 2, f. 3.
Area, Mediterranean region.

8. *Phalaris ambigua.* Figari. Acta. Tor., 1854, p. 327.

9. *Alopecurus agrestis.* L. Sp. 89. Slender Fox-tail. General.
Area, Central and Southern Europe, Danube region, South Siberia.

10. *Alopecurus anthoxanthoides.* Boiss. Diagn. Ser. i., xiii., p. 42.
Alpine Lebanon, Moab, Callirrhoe.

11. *Alopecurus cassius.* Boiss. Diagn. Ser. i., xiii., p. 41.
Northern hills.

12. *Alopecurus utriculatus.* Pers. Syn. i., 80.
Area, Southern Europe.

13. *Phleum.* ? Mount Tabor.

14. *Phleum schœnoides.* L. Sp. 88. On the coast.
Area, Mediterranean region, Siberia.

15. *Phleum tenue.* Schrad. Germ. i., 191. Area, Southern Europe.

16. *Psilurus nardoides.* Trin. Fund. 93. Area, Southern Europe.

17. *Pennisetum cenchroides.* Rchb. in Pers. Syn. i., 72.
Moab, Wâdy Zuweirah. Area, North and South Africa, Canaries.

18. *Pennisetum tiberiadis.* Boiss. Diagn. Ser. i., xiii., p. 43. P.

19. *Pennisetum.* ? Engedi.

20. *Pennisetum rufescens.* Spreng. Syst. i., 302.
Area, North Africa.

21. *Pennisetum orientale.* Pers. Syn. i., 72.

22. *Crypsis alopecuroides.* Schrad. Germ. i., 167.
Area, Southern Europe, Siberia.

23. *Gastridium.* Sp. ?

24. *Stipa tortilis.* Desf. Atl. i., 99, t. 31, f. 1.
Area, Mediterranean region.

25. *Stipa pennata.* L. Sp. 115. Feather Grass. Under Heshbon.
Area, Europe, North Africa, Siberia.

26. *Stipa sibthorpii.* Boiss. Gilead.

27. *Stipa damascena.* Boiss. Diagn. Ser. i., xiii., p. 45.
Anti-Lebanon. P.

28. *Stipa lagascæ.* Rœm. et Schult. Syst. ii., 333. Area, Spain.

29. *Piptatherum cærulescens.* Beauv. Agrost. 18, t. 5, f. 10.
Mount Tabor. Area, Mediterranean region.

30. *Piptatherum blanchianum.* Boiss. Diagn. Ser. ii., iv., p. 127
Phœnician plain. P.

31. *Piptatherum multiflorum.* Beauv. Agrost. 18.
Area, South Europe, Siberia.

32. *Aristida cærulescens.* Desf. Atl. i., 109, pl. 21, f. 2.
The Ghor, east of Jordan. Area, Spain, North Africa, Brazil.

33. *Aristida ciliata.* Desf. Emend. in Schrad. N. Journ. iii., 255.
Seisaban. Area, North Africa.

34. *Aristida adscensionis.* L. Sp. 121.
Callirrhoe, Wâdy Zuweirah. Area, Ascension Island.

35. *Aristida.* ? Mount Tabor.

36. *Aristida plumosa.* L. Sp. 1666. Wâdy Zuweirah.
Area, North-east Africa.

37. *Hemarthria fasciculata.* Künth. Gram. i., 153.
Area, Mediterranean region.

38. *Arundo donax.* L. Sp. 120. Heb. אגמון, A.V. *Bulrush.* Arab. قصب, *Kassab.*
Area, Mediterranean region, South Siberia.

39. *Arundo mauritanica.* Desf. Atl. i., 106. = *A. plinii.* Turr.
Area, Spain, Sicily, North Africa.

40. *Phragmites communis.* Trin. Fund. 134. Common Reed.
Area, almost world-wide.

41. *Panicum turgidum.* Forsk. Descr. 18. Wâdy Zuweirah.

42. *Panicum patens.* L. Sp. 86. Beyrout. Area, India.

43. *Panicum teneriffa.* R. Br. Prodr. i., 39. Wâdy Zuweirah.
Area, Sicily, South Italy, Madeira.

44. *Panicum verticillatum.* L. Sp. 82. = *P. setaria*, var. Beauv.
The littoral districts. Area, all warm climates.

45. *Panicum miliaceum.* L. Sp. 86. Millet. Heb. דֹּחַן. Arab. دخان, *Duchan.*

Cultivated. Native oı India. Introduced 'ab antiquissimis temporibus.'

46. *Panicum italicum.* L. Sp. 83.
Area, India. Elsewhere introduced.

47. *Panicum repens.* L. Sp. 87.
Area, Southern Europe, Central America.

48. *Neurachne alopecuroides.* Brown. Prodr. i., 196.
Area, Australia, *teste* Kunth.

49. *Polypogon monspeliensis.* Desf. Atl. i., 66. Annual Bean Grass. Hills of Gilead, Mount Tabor.

Area, Central and Southern Europe, North Africa, North Asia, Japan, India.

50. *Polypogon.* ? Seisaban.

51. *Polypogon.* ? Callirrhoe.

52. *Cynodon dactylon.* Pers. Syn. i., 85. Dog's-tooth Grass, Doab, Bermuda Grass. Gennesaret.

Area, Central and Southern Europe, Asia, Africa. Introduced in America. The chief pasture in many dry climates.

53. *Avena strigosa.* Schreb. Spicil. 52. Bristle Oat. Moab.
Area, Europe, North Africa, Siberia, North-west India.

54. *Avena sterilis.* L. Sp. 118. Mount Gilead.
Area, Southern Europe.

55. *Avena elatior.* L. Sp. 117. False Oat Grass.
Area, Europe, North Africa.

56. *Avena.* ? Moab.

57. *Avena carmeli.* Boiss. Diagn. Ser. i., xiii., p. 50.
Mount Carmel. P.

58. *Avena fatua.* L. Sp. 118. Wild Oat, or Havers.
Area, Europe, North Africa, Siberia, North-west India.

59. *Avena hirsuta.* Roth. Cat. iii., 19. Area, Southern Europe.

60. *Chloris villosa.* Pers. Syn. i., 87. Wâdys of Moab.
Area, Interior of North Africa.

61. *Kœleria berythea.* Boiss. Diagn. Ser. ii., iv., p. 135.
Phœnicia. P.

62. *Kœleria phleoides.* Pers. Syn. i., 97.
Area, Mediterranean region.

63. *Kœleria sinaica.* Boiss. Diagn. Ser. i., xiii., p. 53.
South of Dead Sea.

64. *Digitaria sanguinalis.* Scop. Carn. ii., 72. Zara, Moab.
Area, Central and Southern Europe, Asia, America.

65. *Æleuropus lævis.* Trin. Fund. 143.
Area, Malabar, *fide* Kunth.

66. *Trisetum neglectum.* Rœm. et Schult. Syst. ii., 660.
Mount Carmel. Area, Mediterranean region.

67. *Trisetum glumaceum.* Boiss. Diagn. Ser. i., xiii., p. 49.
Southern Desert.

68. *Trisetum lineare.* Boiss. Diagn. Ser. i., xiii., p. 49.
Southern Desert.

69. *Trisetum macrochætum.* Boiss. Diagn. Ser. i., xiii., p. 48.
Near Jerusalem. P.

70. *Nardurus orientalis.* Boiss. Diagn. Ser. i., vii., p. 127.
The Bukâà.

71. *Poa bulbosa.* L. Sp. 102. Bulbous Meadow Grass.
Wâdy Heshban. Area, Europe, North Africa, Siberia.

72. *Poa pratensis.* L. Sp. 99. Smooth Meadow Grass.
Esdraelon. Area, North Temperate and Arctic regions.

73. *Poa pilosa.* L. Sp. 100. Willd. Sp. i., 391. Nazareth.
Area, Southern Europe, Siberia.

74. *Poa angustifolia.* L. Sp. 99. Area, Europe, Siberia.

75. *Poa annua.* L. Sp. 99. Annual Meadow Grass.
Area, North Temperate Europe, North Africa, Asia.

76. *Poa Alpina.* L. Sp. 99. Var. *vivipara.* Willd. Enum. 103. Alpine Meadow Grass.

Area, Alpine and Arctic Europe, Siberia, Himalayas, North America.

77. *Briza maxima.* L. Sp. 103. Moab.
Area, South Europe, Cape, India.

78. *Briza bipinnata.* (L. Syst. Nat. ii., 885.) Dead Sea.
Area, India.

79. *Briza minor.* L. Sp. 102. Least Quaking Grass.
Area, Central and Southern Europe, North Africa.

80. *Melica minuta.* L. Mant. 32. Area, Southern Europe.

81. *Melica trachyantha.* Boiss. in Kotschy. Pl. Pers. Austr. Exs. Feb. 1845. Banias.

82. *Melica angustifolia.* Boiss. et Bl. Diagn. Ser. ii., iv., p. 132.
Lower Lebanon.

83. *Melica boissieri.* Reut.

84. *Melica cretica.* Boiss. Diagn. Ser. i., xiii., p. 54.
Alpine Lebanon.

85. *Melica pannosa.* Boiss. Diagn. Ser. i., xiii., p. 55.
Rocks in lower Anti-Lebanon, under Wâdy Barrada. P.

86. *Ammochloa palæstina.* Boiss. Diagn. Ser. i., xiii., p. 52.
Desert near Gaza. P.

87. *Eragrostis cynosuroides.* Retz. Obs. iv., 20. The Ghor.
Area, North-east Africa, India.

88. *Eragrostis plumosa.* Retz. Obs. iv., 20.
South end of Dead Sea. Area, India.

89. *Cynosurus echinatus.* L. Sp. 105. Var. *elegans.* Rough Dog's-tail Grass. Mount Tabor.
Area, Central and Southern Europe, North Africa.

90. *Nephelochloa tripolitana.* Boiss. et Bl. Diagn. Ser. ii., iv., p. 133.
Coast near Beyrout.

91. *Lagurus ovatus.* L. Sp. 119. Hare's-tail Grass. Galilee.
Area, Central and Southern Europe, North Africa.

92. *Dactylis glomerata.* L. Sp. 105. Cock's-foot Grass. In waste land.

Area, Europe, North Africa, Siberia, North India.

93. *Dactylis.* ? Mount Tabor.

94. *Agrostis spica-venti.* L. Sp. 91. Bent Grass.
Area, Europe, North Africa, Siberia.

95. *Schismus marginatus.* Beauv. Agrost. 74, pl. 15, f. 4.
Desert east of Moab. Area, Southern Europe, the Cape.

96. *Schismus minutus.* Rœm. et Schult. Syst. ii., 584.
Wâdy Zuweirah.

97. *Lamarckia aurea.* Mœnch. Meth. 201.
Plains of Moab, Callirrhoe, Jericho. Area, Mediterranean region.

98. *Arrhenatherum palæstinum.* Boiss. Diagn. Ser. i., xiii., p. 51.
Barren plains. P.

99. *Setaria glauca.* Beauv. Agrost. 51. Beyrout.
Area, almost world-wide.

100. *Catabrosa aquatica.* Beauv. Agrost., 97, pl. 19, f. 8. Whorl Grass. Magdala.

Area, Europe, North Africa, Siberia, Himalayas, North America.

101. *Festuca maritima.* De Cand. Gall. iii., 47. Sea Fescue Grass. Caiffa. Area, Europe, North Africa, Arabia.

102. *Festuca rigida.* Kunth. Gram. i., 129. Hard Fescue Grass. Caiffa. Area, Europe, North Africa.

103. *Festuca fusca.* L. Sp. 109. Area, North-east Africa.

104. *Festuca myurus.* L. Sp. 109. Wall Fescue Grass.
Area, Europe, North Africa.

105. *Festuca bromoides.* L. Sp. 110.
Area, Europe, North Africa.

106. *Bromus scoparius.* L. Amœn. Acad. iv., 266.
Mount Gilead, Seisaban, Tabor, Gennesaret.
Area, Mediterranean region.

107. *Bromus erectus.* Huds. Angl. 49. Upright Brome Grass.
Callirrhoe. Area, Europe, North Africa.

108. *Bromus madritensis.* L. Sp. 114. Zara.
Area, Europe, North Africa.

109. *Bromus argyphæus.* Paine. Pal. Expl. Soc. No. 3, p. 128.
Pine forests of Gilead. P.

110. *Bromus.* ? Seisaban.

111. *Bromus danthoniæ.* Fr. Mount Hermon, 6,000 feet.

112. *Bromus syriacus.* Boiss. Diagn. Ser. ii., iv., p. 139.
Stony hills, Phœnicia, and lower Lebanon.

113. *Bromus divaricatus.* Rhode. Lois. Journ. de Bot. ii., 214.
Jericho. Area, Southern Europe.

114. *Bromus tectorum.* L. Sp. 114. Caiffa.
Area, Europe, Siberia.

115. *Bromus mollis.* L. Sp. 112. Soft Brome Grass.
Area, Europe, North Africa, West Siberia.

116. *Bromus distachyus.* L. Sp. 115.
Area, Mediterranean region.

117. *Bromus maximus.* Desf. Atl. i., 95, pl. 26. Great Brome Grass.
Area, Mediterranean region.

118. *Bromus sterilis.* L. Sp. 113. Barren Brome Grass.
Area, Europe, North Africa, West Siberia.

119. *Milium syriacum.* Boiss.

120. *Milium trichopodum.* Boiss. Diagn. Ser. i., xiii., p. 45.
Wet places, northern plains.

121. *Lolium perenne.* Perennial Rye Grass. L. Sp. 122. Gorge of Callirrhoe. Area, Mediterranean region.

122. *Lolium multiflorum.* Lam. Gall. iii., 621. Galilee. Area, Southern Europe.

123. *Lolium temulentum.* L. Sp. 122. Darnel. Area, Europe, North Africa, West Siberia, India, Japan.

124. *Lolium speciosum.* Stev. in Bieb. Fl. i., 80. Safed. Area, Southern Europe.

125. *Vulpia brevis.* Boiss. Diagn. Ser. ii., iv., p. 139. Lebanon, 5,000 feet. P.

126. *Vulpia patens.* Boiss. Diagn. Ser. i., xiii., p. 62. Southern Desert.

127. *Hordeum murinum.* L. Sp. 126. Barley Grass, Weybent. Moab, Seisaban. Area, Europe, North Africa.

128. *Hordeum bulbosum.* L. Sp. 125. Seisaban. Area, Mediterranean region.

129. *Hordeum hexasticum.* L. Sp. 125. Plains of Moab.

130. *Hordeum distichum.* L. Sp. 125. Var. Area, Tartary.

131. *Hordeum ithaburense.* Boiss. Diagn. Ser. i., xiii., p. 70. Mount Tabor. P.

132. *Triticum aucheri.* Parl. Fl. Ital. i., 508.

133. *Triticum spelta.* L. Sp. 127. Spelt.

134. *Triticum vulgare.* Vill. Delph. ii., 153.

135. *Scleropoa rigida.* (Kunth. Gram. i., 129.) Gennesaret.

136. *Scleropoa philistæa.* Boiss. Diagn. Ser. i., xiii., p. 60. Loose sands near Gaza. P.

137. *Scleropoa pumila.* Boiss. Diagn. Ser. i., xiii., p. 61. Southern Desert.

138. *Scleropoa memphitica.* Boiss. Diagn. Ser. i., xiii., p. 62.
Desert south of Dead Sea. P.

139. *Elycus crinitus.* Schreb. Gram. 15, pl. 24, f. 1.
Area, Danubian region, North Africa.

140. *Anthistiria glauca.* Desf. Atl. ii., 380, pl. 254.
Area, North and West Africa.

141. *Anthistiria syriaca.* Boiss. Diagn. Ser. i., xiii., p. 72.
Near the coast.

142. *Anthistiria brachyantha.* Boiss. Diagn. Ser. i., xiii., p. 71.
Dry places near the coast.

143. *Erianthus ravennæ.* Beauv. Agrost. 14.
Area, Shores of Mediterranean and Caspian.

144. *Saccharum ægyptiacum.* Willd. Enum. 82. Beyrout.
Area, Sicily.

145. *Saccharum officinarum.* L. Sp. 79. Sugar Cane. Cultivated.

146. *Imperata arundinacea.* Cyrill. Ic. ii., pl. 11. Gennesaret.
Area, Mediterranean region, Senegal, India, China.

147. *Rottboellia digitata.* Sibth. Græc. i., pl. 92.
Eastern slopes towards Dead Sea.

148. *Lasiurus hirsutus.* Boiss. Diagn. Ser. ii., iv., p. 146.
Southern Desert.

149. *Cornucopiæ cucullatum.* L. Sp. 79. Banias.

150. *Sorghum halepense.* Pers. (Forsk.) Syn. i., 101.
Lake Huleh. Area, Mediterranean region.

151. *Sorghum vulgare.* Pers. Syn. i., 101. Arab. دخن, *dakkn.*
Area, India, where it is called Durrha, Jordaree, or Jondla.

152. *Pollinia distachya.* Spr. Syst. i., 288.
Area, Mediterranean region.

153. *Andropogon hirtum.* L Sp. 1482. Callirrhoe.
Area, Mediterranean region, the Cape.

154. *Andropogon annulatum.* Forsk. Descr. 173.
Callirrhoe, Wâdy Zuweirah. Area, India.

155. *Andropogon schœnanthum.* L. Sp. 1481. Gennesaret.
Area, North India, Arabia, the Cape.

156. *Ægilops ovata.* L. Sp. 1489. Mount Tabor.
Area, Southern Europe.

157. *Ægilops triuncialis.* L. Sp. 1489. Gilead.
Area, Southern Europe.

158. *Ægilops aucheri.* Boiss. Diagn. Ser. i., v., p. 74.
Northern plains.

ORDER CXLII., CONIFERÆ.

1. *Juniperus excelsa.* M. B. Fl. Cauc. ii., p. 245.=*J. fœtidissima.*
Lebanon, Gilead.

2. *Juniperus communis.* L. Sp. 1470. Juniper. North Lebanon.
Area, most mountainous countries, as Atlas, Himalayas, Japan, Rocky Mountains.

3. *Juniperus drupaceæ.* Labill. Fl. Syr. Dec. ii., p. 14, pl. 8.
Lebanon and Anti-Lebanon ; common.

4. *Juniperus oxycedrus.* L. Sp. 1470. Lebanon, Anti-Lebanon, Galilee, Bashan, Gilead. Area, Southern Europe.

5. *Juniperus phœnicea.* L. Sp. 1471. Lebanon, *non vidi.*
Area, Southern Europe.

6. *Juniperus sabina.* L. Sp. 1472. Savin. Lebanon, 6,000 feet.
Area, Southern Europe.

7. *Juniperus thurifera.* L. Sp. 1471. On the coast.
Area, Portugal, Spain.

8. *Cupressus sempervirens.* L. Sp. 1422. Cypress.
General on the mountains.

9. *Pinus pinea.* L. Sp. pl., p. 1419. Stone Pine.
Northern slopes of Lebanon. Area, Portugal, Spain.

10. *Pinus halepensis.* Mill. Dict. N. 8., Ic. t. 216. Aleppo Pine.
Near Hebron, heights of Judæan wilderness, coast, etc.
Area, Southern Europe.

11. Ditto, var. *carica.* Don. Ex. Carr. Mount Gilead.

12. *Pinus pyrenaica.* Lapeyr. Abr. Pl. Pyren., p. 146. Pyrenean Pine. On Lebanon, 6,000 feet. Area, the Pyrenees.

13. *Pinus pinaster.* Sclard. in Ait. H. Kew. ed. i., vol. iii., p. 367. The Pinaster. Near Beyrout ; probably introduced.
Area, Mediterranean region, Himalayas, China.

14. *Cedrus libani.* Parr. Icon. 499. Cedar of Lebanon. Heb. אֶרֶז. Arab. ارز, *Arz.*
Lebanon only, but scattered over many portions of the range.
Area, Mount Atlas.

ORDER CXLIII., GNETACEÆ.

1. *Ephedra fragilis.* Desf. Fl. Atl. ii., p. 372.
Rocky ledges in Moab. Area, Southern Portugal, Spain, Sicily.

2. *Ephedra alte.* C. A. Mey. Eph., p. 75, pl. 3, f. 4.
Rocky ravines, wilderness of Judæa, edge of the plain of Sharon.

3. *Ephedra alata.* Déc. Fl. Sin. Ann. Sc. Nat. ii., Ser. i., p. 236.
The Southern Desert.

CRYPTOGAMIA.

ORDER, LYCOPODIACEÆ.

1. *Selaginella denticulata.* (L. Sp. 1569.)
Lebanon. Area, Mediterranean region.

ORDER, FILICES.

1. *Cystopteris fragilis.* (L. Sp. 1553.) Brittle Bladder Fern.

Lebanon, 6,000 feet. Area, world-wide, except lowlands in the Tropics, and Australia.

2. *Adiantum capillus-veneris.* L. Sp. 1558. Maiden Hair Fern.

In wells, caves, fissures of moist rocks, under cliffs by streams, and wherever there is dropping moisture ; in every part of the country except the mountain region, and very abundant. Area, Central and Southern Europe, Asia south of Siberia, Africa, Polynesia, and America, from Florida to the Amazon.

3. *Cheilanthes fragrans.* (Desf. Fl. Atl. ii., p. 408, pl. 257.)

Among rocks in the hill country and Judæan wilderness, Gerizim. Area, Mediterranean region, Canaries, Madeira, Himalayas at 5,000 feet.

4. *Pteris longifolia.* L. Sp. 1531.

Lower slopes and base of Lebanon. Area, tropical and warm temperate regions all round the world.

5. *Pteris cretica.* L. Mantiss. 130.

Lebanon. Area, Italy, Corsica, South Africa, South Siberia, Japan, Himalayas and Neilgherris, Philippines, Fiji, Sandwich Islands, Florida, Mexico, and Guatemala.

6. *Pteris aquilina.* L. Sp. 1533. Bracken.

Galilee, by the Leontes river, Lebanon. Area, world-wide, from just within the Arctic Circle, through the whole of both Temperate Zones and the Tropics, reaching up to 8,000 feet in tropical mountain regions, attaining on the Andes a height of 14 feet.

7. *Lomaria spicant.* (L. Sp. 1522.) Hard Fern.

Lebanon. Area, all Europe, Northern and Central Asia, Japan, North-west America.

8. *Asplenium trichomanes.* L. Sp. 1540. Common Spleenwort.

Lebanon. Area, Europe, Siberia, America, Australia.

9. *Asplenium adiantum-nigrum.* L. Sp. 1541. Black Spleenwort.

Lebanon. Area, Europe, North Africa, Azores, Canaries, Abyssinia, the Cape, Siberia, Himalayas, Sandwich Islands.

10. *Asplenium fontanum.* (L. Sp. 1550.) Var. *bourgæi.* Boiss. Rock Spleenwort.

Gorge of the Litany. Area, Central and Southern Europe, Himalayas.

11. *Asplenium lanceolatum.* Huds. 454. Lanceolate Spleenwort.

West slopes of Lebanon. Area, Mediterranean region, Madeira, Azores, St. Helena.

12. *Asplenium filix-fœmina.* (L. Sp. 1551.) Lady Fern.

Lebanon. Area, Europe, North Asia, Japan, Himalayas, Africa, North America, Cuba, Venezuela.

13. *Asplenium ceterach.* L. Sp. 1538. Scaly Ceterach, or Scale Fern. On rocks, Southern Judæa, Gilead, Moab.

Area, Central and Southern Europe, Himalayas, Canaries, Madeira.

14. *Scolopendrium vulgare.* Sm. E. B. iv., p. 301. Hart's Tongue.

On damp cliffs, Lebanon; caves, Central Palestine. Area, Central and Southern Europe, West Central Asia, Japan, Madeira, Azores.

15. *Scolopendrium hemionitis.* (L. Sp. 1536.)

Cisterns and caves, Mount Gerizim. Area, Southern Europe.

16. *Aspidium aculeatum.* L. Sp. 1552. Prickly Shield Fern.

Lebanon. Area, world-wide in all zones and hemispheres.

17. *Lastrea filix-mas.* (L. Sp. 1551.) Male Fern.

Lebanon, Galilee. Area, Europe, Asia, Rocky Mountains, Andes.

18. *Lastrea rigida.* (Swartz. Syn. Fil. 53.) Var. *pallida.* Link. Rigid Buckler Fern.

Lebanon. Area, Central and Southern Europe.

19. *Polypodium vulgare.* L. Sp. 1544. Common Polypody.

Woods on Lebanon. Area, all Europe, North Africa, Azores, Madeira, Japan, North America, as far south as Mexico.

20. *Nothochlæna lanuginosa.* (Desf. Fl. Atl. ii., p. 400, pl. 256.)

Moab, in deep glens. Area, Mediterranean region, Madeira, Canaries, Cape Verde, Australia.

21. *Gymnogramme leptophylla.* (L. Sp. 1553.)

Mount Carmel and central hills. Area, Mediterranean region, Madeira, Canaries, Azores, Abyssinia, the Cape, Neilgherris, New South Wales, Tasmania, New Zealand, Andes.

22. *Osmunda regalis.* L. Sp. 1521. Royal Flowering Fern.

Banks of the Leontes, Upper Jordan. Area, Europe, Asia, North and South Africa, North and South America.

THE END.

BILLING AND SONS, PRINTERS, GUILDFORD AND LONDON.

Printed in the United States
By Bookmasters